The How and the Why

The How and the Why

AN ESSAY ON THE ORIGINS

AND DEVELOPMENT

OF PHYSICAL

THEORY

David Park

With drawings by ROBIN BRICKMAN

PRINCETON UNIVERSITY PRESS

PRINCETON, NEW JERSEY

COPYRIGHT © 1988 BY PRINCETON UNIVERSITY PRESS
PUBLISHED BY PRINCETON UNIVERSITY PRESS,
41 WILLIAM STREET, PRINCETON, NEW JERSEY 08540
IN THE UNITED KINGDOM:
PRINCETON UNIVERSITY PRESS, CHICHESTER, WEST SUSSEX

LIBRARY OF CONGRESS CATALOGING-IN-PUBLICATION DATA

PARK, DAVID ALLEN, 1919-
THE HOW AND THE WHY.

BIBLIOGRAPHY: P.
INCLUDES INDEX.
I. SCIENCE—HISTORY. 2. PHYSICS—HISTORY.
3. ASTRONOMY—HISTORY. I. TITLE.
Q125.P324 1988   509   87-29135
ISBN 0-691-08492-0
ISBN 0-691-02508-8 (PBK.)

SECOND PRINTING, 1989

THIRD PRINTING, WITH CORRECTIONS, 1990

THIS BOOK HAS BEEN COMPOSED IN LINOTRON SABON

PRINCETON UNIVERSITY PRESS BOOKS ARE PRINTED ON ACID-FREE PAPER
AND MEET THE GUIDELINES FOR PERMANENCE AND DURABILITY OF THE
COMMITTEE ON PRODUCTION GUIDELINES FOR BOOK LONGEVITY OF THE
COUNCIL ON LIBRARY RESOURCES

THE QUOTATION FROM RICHARD WILBUR'S "EPISTEMOLOGY"
THAT BEGINS CHAPTER 15 IS FROM
Ceremony and Other Poems,
© 1950, 1978 BY RICHARD WILBUR.
REPRINTED BY PERMISSION OF HARCOURT BRACE JOVANOVICH, INC.

PRINTED IN THE UNITED STATES OF AMERICA BY PRINCETON ACADEMIC PRESS

10   9   8   7   6

TO

Clara Claiborne Park

WHO, IN ADDITION TO PROVIDING FACTS,

INTERPRETATION, AND CRITICISM,

HAS OCCASIONALLY BEEN ABLE

TO MAKE ME THINK CLEARLY

Contents

List of Illustrations

List of Tables

Acknowledgments

FOR their generous help I am grateful to Beth Ebel, Meredith Hoppin, Robert Mills, Clara Park, and William Wootters, who kindly read certain sections, to Katharine Park for many suggestions, and to many helpful people in the Williams College Library, especially Robert Voltz and Wayne Hammond in the Chapin Library of Rare Books.

Note on References

IN the bibliography a reference such as "Cicero 1928, *Republic* VI.9" invites you to look at the volume of Cicero that the bibliography lists as having been published in 1928; find a work in it called *The Republic*; and look at Chapter 9 of Book 6.

A word on classical texts. For the Presocratics I have used G. S. Kirk, J. E. Raven, and M. Schofield's *The Presocratic Philosophers* (1983), whence I have taken translations, and often their thoughtful and learned arrangements and interpretations. For other texts I have used Diels's *Fragmente der Vorsokratiker* (1959). When a quotation from an ancient text is labelled "fr. N," it refers to the Nth entry in Diels's list under the author's name.

Freeman, in *Ancilla to the Presocratic Philosophers* (1978), translates some of Diels's fragments.

I have taken most of my translations of Plato from Hamilton and Cairns's *Plato: The Collected Dialogues* (1961), and of Aristotle from Barnes's *Complete Works of Aristotle* (1983).

The texts are footnoted as they occur. Translations not otherwise attributed are my own, and I have without comment retranslated some passages from Loeb and other bilingual editions.

Introduction

IT turns into a computer and makes a mistake in the gas bill; in the hands of a doctor it saves the life of a child. It turns into a bomb. What is science anyhow? Most of us perceive it as a series of accomplished facts that may or may not affect our lives; in school it was a barrier to be climbed over. How can we talk about what it is? For one thing it is too vast in scope to be contemplated as a whole. All right, let us talk about the part of it that deals with the nature of the universe and of matter. Where shall we begin?

Perhaps the other night you missed the Bach but got there in time for the Brahms, but even so the program prepared you for Persichetti by reminding you once again, for the thousandth time, of the tradition in which a modern composer writes and a modern listener listens. In school everyone reads Shakespeare and perhaps even Chaucer; our teachers try to make the past of our artistic culture as real to us as the present. There is hardly a college that has not a columned portico somewhere. Every culture is a history, but you would hardly guess from the way we learn about science that science is a culture. Somebody has discovered a hole in the atmosphere; someone has noticed a new supernova, somebody has found a new treatment for cancer. Mostly, science impinges on us not as a coherent picture of the universe but as an incoherent collection of results.

Of course, one might say that art is also a collection of results, but if we think about these things we know better. It is also a collection of relationships. We cannot walk through a picture gallery without sensing the interplay in line and color, in surface texture and pictorial depth, in form and content, of one century with another. Art is a culture with a history of effort, taste, style, opinion—but what do most of us know of the culture of science? The physics book has a picture of Galileo with his rough beard, looking angry, and later of Isaac Newton looking fat. The biology book shows Mendel with buttons down his front. Isn't there more to it than beards and buttons? What, to Galileo, was a good scientific question? What did he think he was doing? What intellectual tools did he command? What did his contemporaries think? What ideas did he inherit from the past? How did he change things in his own day? How much of what he thought do we still think?

Scientists make telescopes, weapons, life-saving drugs; each lasts for a while. Scientists also carry on a traffic of ideas, theories, explanations of the world. These are usually regarded as less suitable for the public view, but it seems to me that it is when seen out of context that they seem dull and complicated and obscure. We would not expect someone from a far-distant culture (even though living next door to us) to profit from one hearing of a sophisticated piece of music; it is hardly surprising that most literate people find articles in the *Scientific American* heavy going.

Can a book open the scientific culture to minds inexperienced in scientific thought? I don't know, but let me explain what I was trying to do; it may save you the trouble of reading further. My subject is the effort to understand the physical world. Of course, "understand" has meant different things to different people, and we will have to judge what they meant from the words they spoke and the contexts in which they spoke them. Science aims at universal statements: to find out what is true of all thunderstorms or all cases of cholera. During most of the period we study, a scientific proposition was considered true if it could be deduced from unquestionable postulates by rigorous logic. Today we reach toward scientific statements by studying individual cases, but then, science was considered to deal only with universals. One might declare the truth about thunderstorms in general, but what is happening now, the experience of *this* thunderstorm, was outside the scope of such a science, and it was only long after Aristotle that new perspectives brought theory closer to experience. They were at their closest in the eighteenth century, but from the seventeenth, at an increasing rate, traditionalists were affronted by the introduction of new entities: first atoms, then energy, then the idea that heat is a form of energy, then the electromagnetic field. These were not and could not be experienced by sight and touch, but their existence and properties were deduced from rigorous arguments based on experience.

Computers and the rest of the things science produces can mostly be seen and handled, but to invent them and understand them one has to think about electrons and holes and Brillouin zones and other things belonging to more abstract levels of ideas. None of the entities that appear in fundamental physical theory today are accessible to the senses. Even more, as we shall see in the chapters on quantum theory, there are phenomena that apparently are not *in any way* amenable to explanation in terms of things, even invisible things, that move in the space and time defined by the laboratory. We can explain these phenomena in terms of a reasonable and coherent physical theory, but the things the theory talks about are mathematical things that do not refer directly to any visible object, and only at the end are there recipes for saying how the results of the theory are related to experience. When we have learned to calculate these effects, are they understood? A theory is judged by seeing whether it fits the facts, but is that the only criterion? These are not scientific questions and do not have scientific answers, but if the purpose of science is to observe the world and then explain what has been seen in a way that satisfies the mind, there is some reason to ask them. I think that if we study the development of science from this point of view we find that questions of understanding are very old questions and that modern experience, instead of answering them, only provides clearer examples.

To design a computer chip it is necessary to understand how electrons move through a semiconductor, and there is a theory that explains this. It assumes that there are electrons. One can also ask why there are electrons. There is a cliché to the effect that science tells how but not why, a cliché so powerful that it is sometimes used to define the boundaries of science: " 'Why' does not begin a scientific question." But most physicists are very interested in questions of why. Why *do* electrons exist? Why does anything exist? Why does nature obey the laws we know, rather than some other laws? A question like this last one cannot be answered by pointing to the laws themselves; there must be some other considerations. What could they be? But let's not pretend these are not scientific questions. Let's think about them.

In the narrative that follows I have tried to address a wide circle of readers: a friendly layman who would like to know about scientific thinking; someone who studies the history of ideas (especially scientific ideas); or a scientific colleague not satisfied with the historical background given in most books. My account is historical, but it is not a history. Some great events and the scientists responsible for them do not appear here at all, since I found that I could tell my story without troubling them to get their wigs on. I have also avoided unnecessary technicalities. To those readers who think there are still too many I apologize; for colleagues who would like some more details there are notes at the back of the book that, at gradually increasing levels of sophistication, work out some calculations. Please believe, though, that you miss very little if you skip them. The same applies to the few mathematical arguments in the text. Their purpose is to illustrate, not to expound, and there will be sympathy between those who do not read mathematics and those who have to miss the implications of certain words and phrases in Latin.

Science, like any other intellectual product, reflects the social and intellectual climate in which it was produced. It has not been possible to say much about social climate, and I hope interested readers will supply this knowledge from their own store. I have tried to suggest the intellectual climate by quoting words that reveal ways of thought and by referring to a few famous controversies. I have also tried to indicate the kinds of knowledge available to educated people at various times by citing encyclopedias and other collections of knowledge. But looking at what I have written I see that it is only a sketch of what might have been done.

The story is told from a physicist's point of view. It is organized so as to emphasize questions that have some relevance to questions of current interest discussed in later chapters. For readers who may want to read further I have made a fairly extensive bibliography. Lack of space, and the wish not to lose nonmathematical readers, have led me to shorten discussions that should have been longer and leave out some topics that ought to have been included. The bibliography will tell where more detail can be found.

One unifying idea in the diversity that follows is the idea of form. An artist—a poet or painter, a musician, or a teller of funny stories—creates within a world of formal principles. A work in words or sound has at the very least a beginning, a middle, and an end; a painting or sculpture is anchored by its relations with the edges that enclose it or the ground on which it rests (this is less universally true now than it once was, but even if the object violates these principles they are there in the background of the work). The form of ballet is movement as well as configuration, and this is true of many of the forms of nature. A creator of scientific theories also works within formal constraints. Art and science are very different enterprises but in both, constraint imposes loss as well as gain. For direct, spontaneous experience of muscle and sense is substituted an artifice, but this artifice, constructed on some formal principles, be they obvious or deeply hidden, aims at a different level of reality.

Just as in the old days, modern scientific explanations represent experience but do not ordinarily describe it. They aim at generality and also at formal clarity. The remarkable fact, quite unexpected to anyone without the experience, is that the two are related—there is something about the world and our representations of it such that the more carefully we study it and the more we understand, the more surely we discover formal principles that add light and meaning to our understanding.

Is form real? Certainly snowflakes had six points and planets moved in elliptical orbits, at speeds regulated by a simple formal principle, before Kepler was born. A task of science is to invent abstract forms related to the forms we see, to find out whether perhaps a very few abstract forms suffice to explain all the rest. The general propositions of science have been developed out of uncountable contacts with the specific, but they are more than mere generalizations. They are conditioned also by our sense of how we would like things to be, how we think they ought to be, and I think that part of that sense is an appreciation of form. We enjoy a flower because we think it is beautiful; we pick up a stub of hexagonal quartz crystal on the beach perhaps because it reminds us that the veil that covers nature's mathematical structure has a few thin places. One of the strangest aspects of the story I shall tell is how long it took, after the existence of the mathematical structure had been conjectured, before anyone could begin to show what it is actually like. Partly, this is because it was hard to understand that forms exist in time as well as in space. This discovery was Galileo's as much as anybody's and that is why the modern age in physical science is often said to have begun with him. Still, I have tried not to overemphasize questions of form because of course much more than form is involved in the subtle intellectual process that leads someone to say "I understand."

The ideas of early science may sometimes have been expressed in terms that were poetic or mystical or derived from idiosyncratic philosophical conceptions, but they were never very far removed from their origin in contemplation of things we see as we look around us. It was really not until the rise of experimental science in the seventeenth century that scientists began to write about matters of which we have no direct experience at all. In order to write about the new kinds of thoughts that began then, I must start, in about Chapter 13, to describe the world of fact that is reached only through the door of a laboratory. This will slow down the narrative of ideas but it is necessary if I am to communicate anything of the great intellectual structures of modern physics, for these are based just as firmly on experience as, let us say, Plato's theory of ideas. But, you may protest, what a terrible comparison! The theory of ideas is an intellectual construction, and not very successful at that. Yes, but it is a construction based on experience of life, of beauty, of living in society, of learning what others have thought, of trying to think for oneself. Of course it was not forced on Plato by experience; in that sense it was a free creation, but experience provided the material. The analogy with the process that has led to modern scientific theory is almost perfect. I shall try to show that even the outcome has similarities and that the physical theory of our own time has a decidedly Platonic structure—not that it can very well be expressed in Plato's language but that its content of truth is necessarily related to the nature of its form.

Finally, having started with the cosmological theories of the ancients, I give a brief outline of the generally accepted modern theory of the nature and origin of the universe and its history till now, showing how theories dealing with the two extreme limits of size—the particle and the cosmos—are being welded together so that each leans on the other for support. And in the middle, on a scale midway between the two extremes, sits our new and fragile race: how did we get here, what special properties did the world have to have in order for us to be here, and how long will the cosmos allow us to remain if we have the qualities that permit us to remain?

These signposts may help guide the way through the variety that follows, but most of all I hope to encourage some people who would not otherwise have done so to start reading some of the original texts for themselves.

The How and the Why

CHAPTER I

What Is the World?

Now it is well known that in the heavens nothing happens by chance or at random, and that all things above proceed in orderly fashion according to divine law. Therefore it is unquestionably right to assume that harmonious sounds come forth from the rotation of the heavenly spheres, for sound has to come from motion, and Reason, which is present in the divine, is responsible for the sounds being melodious.

—MACROBIUS

Prelude: Pastoral

Scene: A hillside. Sun, a shady tree. Under it a youth and a maiden. Phyllis wears a dress of homespun linen with an embroidered blouse and has plaited colored ribbons into her hair; Corydon wears buckskin knee-breeches and a simple light shirt. His ponytail is held with a silver buckle. Sheep graze in the distance.

PHYLLIS. The water in my little pail, Corydon, what is it made of?

CORYDON. Molecules, as you know very well.

PHYL. What are the molecules made of?

COR. Two hydrogen atoms and one oxygen atom. What's all this about?

PHYL. (*still ignoring his questions*) One of these molecules—is it wet?

COR. No. You need to have a lot of molecules before you can say they are wet. It's like popularity. You can't say somebody's popular just because two people happen to like them.

PHYL. I see. Those atoms—what are they made of?

3

COR. Nuclei and electrons. The electrons go around the nucleus. Think of the horses in a merry-go-round. And the nucleus is made of neutrons and protons and they are made of quarks and there are gluons in there . . .

PHYL. Gluons?

COR. Wait till Chapter 17.

PHYL. I want to ask a question about something simple.

COR. Electrons are simple.

PHYL. Not made of quarks or anything?

COR. No.

PHYL. All right, electrons, then. Is an electron a thing?

COR. You've been reading.

PHYL. Yes, I read. These sheep, you know . . .

COR. I know. What's a thing?

PHYL. My pail of water is a thing.

COR. Is that all you can say?

PHYL. Yes. It's here, we can look at it, I carried it here, it can't be confused with any other thing. You know what a thing is.

COR. That's not much of a definition. Can't you clean it up? Electrons are a borderline case.

PHYL. I remember that when that friend of yours was here he said "People's reason for wanting a definition is to take care of the borderline case and this is what a definition, as if by definition, will not do" (Nemerov 1975).

COR. Is an electron a thing?

PHYL. I don't think so. People have found that it can be in two places, lots of places, at the same time. If that is so it can't possibly be my kind of a thing.

COR. Why do you care whether it is a thing or not? Can't you just decide to include or exclude electrons in your definition?

PHYL. Playing with words doesn't help me to think.

COR. What's the problem?

PHYL. By the same test, neutrons and protons aren't things either. Is a water molecule a thing? If it's not made of things? Is that why it isn't wet?

COR. No. . . . I don't like this. Now you're going to start talking about your pail of water.

PHYL. Well that's the example of a thing I started with. Let me ask you this, and excuse the stupid language: Can you make the world out of things that aren't things?

COR. I don't know. What is the world anyway?

PHYL. What kind of a question is that? Do you want me to show it to you?

COR. Thank you, that wouldn't answer the question. Maybe it's a collection of objects.

PHYL. That's no good. It's also a collection of events.

COR. Suppose I show you a movie film of the world, frame by frame. Each picture shows an arrangement of objects and your events are just gradual changes in the arrangement. The world is a collection of objects with the time dimension added.

PHYL. I know there's more to it than that. It isn't just that things move around; I should have said a collection of processes. Processes are just as real as things.

COR. Or a collection of reasons?

PHYL. Whose reasons?

(*There is a pause. A breeze rustles the leaves overhead.*)

COR. Here is another dumb answer: It's a collection of my sensations. That's all I can ever know about it. I'm playing safe.

PHYL. Not very safe. In the first place it's not clear. Your sensations or my sensations, or both, and why stop there? If it's not just your sensations why not all the sensations of all the funny little people in the universe? And if you haven't met them how can you say anything about their sensations?

COR. In the second place?

PHYL. In the second place, what if you get very sick or something? You can't even be sure about your sensations.

COR. What can I be sure of?

PHYL. What you think. The world is a collection of ideas.

COR. Is that just a polite way of saying that the world is a mental construction?

PHYL. I don't know. That depends on other assumptions. We'll just have to see. (*Another pause.*) What is the world trying to tell us?

COR. Phyllis, a big pile of rocks is not trying to tell us anything.

PHYL. But I just told you it's more than that.

COR. So are you going to listen to it?

PHYL. (*ignoring him*) So far, we have objects, sensations, processes, reasons, ideas, and some kind of a message. As far as I can see, they all miss the point.

COR. What point?

PHYL. (*ignoring him*) What happens if they all get rolled together?

COR. We get the world, I suppose. But do we get any real explanation of anything?

PHYL. You don't know what the point is. I don't know what an explanation is. Excuse me, Corydon, but I want to get back to my book.

(*She reads.*)

Questions

Look out at the world for a moment: the scene darkens as a cloud moves
overhead; a red car stops across the street; someone is playing a radio
with the window open. The world is sending out signals. Most people
attend mainly to the ones obviously destined for them—this is called
taking life as it comes. Some few, and perhaps all of us at certain
moments, try to read more of the signals and see whether perhaps they
bear some coherent message as to the world's nature, its origin, its intentions, its purpose. Underlying this ambition is the assumption that the
world has these things, or some of them, that they do not change all the
time, that true understanding might reveal that behind the flow of appearance and event there stands a fixed and comprehensible reality.

We speculate; we ask questions. The world does not supply them; we
must make up our own. Since the world is not open with its secrets we
must make a guess, and to find out whether the guess is true requires a
method. As we read the earliest recorded thinkers we see questions,
guesses, and methods proliferating almost incoherently. Later, people
began to agree on a group of central questions as well as the terms in
which they might be answered. The earliest speculative thinkers we know
about were mostly "Greeks," with quotes to remind us that what united
them was not so much a place as a language (in several dialects) and some
commonly held ideas about man, nature, and divine power. Most of the
first people to be mentioned lived in Asia Minor or in Italy, not in what
is now Greece; later there was also a great North African intellectual
center in Alexandria. Theirs were not the only intellects in that part of the
world. They inherited mathematical methods and astronomical data from
the Babylonians. The life stories of many Greek philosophers include a
year or so in Egypt where, as everybody knew, the culture was older and
deeper and wiser than their own. (According to some early Christian
texts, when Plato went to Egypt he studied with the prophet Jeremiah.)
But perhaps exactly because they were late starters, the Greeks developed
very quickly in art, in literature, and in philosophical speculation. Later,
most of the Greek world was absorbed into the Roman Empire, and as
we trace the adventures of some Greek ideas we have to follow them
across the map.

In the pages to follow I will summarize some of the ideas by which
Greek thinkers sought to understand the world around them. The ideas
are arranged by theme more than by chronology. They really cannot be
understood chronologically—the order in which they appear in the texts
that have come down to us is clearly not always the order in which they

became current. Also, a chronological account would tend to ignore that between Asia Minor and western Italy stretched nine hundred miles of slow sailing that few people traversed often and most did not traverse at all. In the early days it is best to think in terms of isolated schools of thought.

The aim of this chapter and the next is to show how the oldest recorded speculations about the world originated certain main themes which, as we shall see later, are present in the scientific culture of today. We begin in the sixth century B.C. and survey about two hundred years, until the first phase of development is complete (See Figure 1.1). In the following chapters I will show how some of these speculations entered the culture of medieval Europe, and what happened to them when they arrived.

The World as a Noun

Of course there was no first philosopher, but if history is to be neatly packaged there ought to be one, and the honor goes by acclamation to Thales, who lived in the great city of Miletus in Asia Minor early in the sixth century B.C. He seems to have been of Phoenician descent, and the stories that accumulated around him feature him as a military commander and engineer. He learned enough about astronomy, possibly during a visit to Egypt, to predict the solar eclipse of 585 B.C., which serves to date him; he stated, though he did not necessarily prove, some simple facts of geometry and used them for surveying; he looked at the stars as he walked and so fell into a well, but he got out again.

Thales was principally remembered by posterity for his saying, "the world is alive and full of gods," which was quoted again and again in antiquity as an example of the highest wisdom. No famous oracular words written down later should be taken very literally, and certainly not these, just as nobody takes us literally if we say a room is full of people. Thales probably said, or meant, that divinity pervades the world. In saying so he is searching for a principle that will give a unified explanation of the world's diversity: human nature, what happens in history, the weather, the action of a magnet. But what about these actual things that are full of gods? Is there a unifying principle to explain them? Thales says it is water. I do not think we need accuse him of claiming that everything is ultimately made of what comes out of the tap. We shall see that a century later the four elements were named after common substances but were not supposed to be anything we know with our senses. Aristotle, reasonably, suggests that Thales is arguing from the fact that wherever there is life there is moisture, and that the extension to the rest of nature is merely by way of analogy (*Meteorology*, 983b).

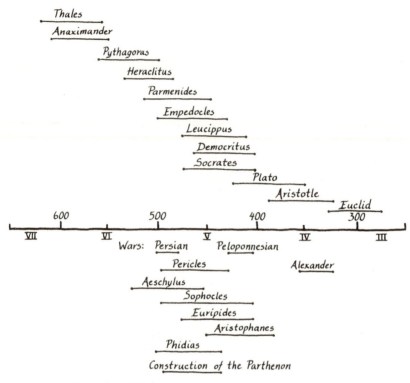

FIGURE I.I. Time chart: The ancient philosophers (many dates are approximate).

Thales is probably not an invention of later writers, but he is clearly a reconstruction based on little evidence. A generation afterwards Anaximines, also of Miletus, speculated that the fundamental substance might be air in different states of condensation and rarefaction, and after him Xenophanes proposed earth and water. But arguments about a fundamental substance do not bring us much closer to an understanding of nature because nature does not just sit there; it happens.

The World as a Verb

For minds seeking to understand what happens in the world and perhaps to read in nature some rules for the conduct of life, it is not very helpful to be told that everything is made of water. The world can also be discussed in terms of process governed by law, and the first one to do this was named Anaximander; he too lived in Miletus, probably a generation after Thales, around 560 B.C. He is said to have set up a sundial in Sparta that told the time of day and showed the sun's passage along the ecliptic; he may have invented the idea of geographical scale and drawn the first map that was not merely a schematic diagram; he speculated on the nature of earth, sun, and moon, and he held ideas on evolution that would shock a creationist today. His cosmology filled the universe with what he called *to apeiron*, the Boundless, a word hard for us to define, but which seems to imply that it fills all space. Although not directly perceived, the Boundless gives rise to all the properties of different substances; in this role it is a more general and abstract substitute for Thales's water. But it manifests itself not just as matter but also as motion, energy, law, perhaps purpose. It has no beginning but is the beginning and cause of everything else. Aristotle says simply that it is the Divine, and it anticipates what was later expressed by Heraclitus, and later still by Saint John the Evangelist, in the word *logos*.

Anaximander also wrote a sentence that impressed even writers who came long after him, and it comes down to us in several versions. In one of them he writes of the Boundless:

> The source from which existing things derive their existence is also that to which they return at their destruction, for they pay penalty and retribution to each other for their injustice according to the assessment of Time. (fr. 1)

The terms sound poetic but they come from the law court. Aristotle, in the *Nicomachean Ethics*, defines injustice as that which is unlawful but also as that which in unequal, or unfair. Anaximander is telling us that the world is just. The seasons alternate, the generations of plants and animals and men rise and die away. Whatever disturbs the evenness of nature—flood or pestilence, a life or a death, or even something much smaller like the ripples in a pond after a stone drops in—endures for a time and is gone; for nature, like man, lives under the commandment "Nothing too much."

Anaximander's works are lost, and the words of the famous quotation are the only ones we have that are claimed to be his. For Anaximander law is, I think, inherent in the nature of the Boundless, and Time is the judge.

Heraclitus

Heraclitus of Ephesus, a solitary and arrogant man, lived a little before 500 B.C. Even the ancient commentators do not claim to have seen the book he is supposed to have written, but fragments and epigrams attached to his name preserve his legend. He argues nothing; the flat,

uncompromising statements have many interpretations or else none, and
we can see why he was known as The Obscure. Anaximander unifies substance and principle in the Boundless; Heraclitus is rather casual about
substance, and he explains principle in a series of utterances concerning
process and logos. The word is Greek, but the Greeks, especially those of
Asia Minor, were not an encapsulated race. The Indians to the east and
the Jews to the south expressed similar ideas in terms of *dharma* and
Wisdom. Let us visit with Wisdom for a moment. The Apocrypha has a
whole book devoted to her who "knoweth the subtilities of speeches and
can expound dark sentences: she forseeth signs and wonders, and the
events of seasons and times" (Wisdom 8:8). In Proverbs she claims this
for herself and much more:

> She standeth in the top of high places. . . . She crieth at the gates. . . . The
> Lord possessed me in the beginning of his way, before his works of old. I was
> set up from everlasting, from the beginning, or ever the earth was. When
> there were no depths I was brought forth; when there were no fountains
> abounding with water. . . . When he prepared the heavens, I was there: when
> he set a compass upon the face of the depth. . . . When he gave to the sea his
> decree, that the water should not pass his commandment: when he appointed
> the foundations of the earth: then I was by him as one brought up with him,
> and I was daily in his sight. . . . (Proverbs 8)

As the Old Testament goes, this is strong language, for no one else ever
speaks of the Lord so nearly as an equal. I think that Solomon's Wisdom
is an order that flowed from God into the world he created, immanent,
not imposed, and that Heraclitus means something similar by logos.

Heraclitus is sometimes confusing when he speaks of logos because he
uses the word also to refer to his own writings. It seems to me, however,
that the more religious meaning is primary. Here, given as far as I can in
his own words, is a summary of his teaching. The logos is a hidden,
changeless principle that manifests itself in process more than in substance.

> Nature likes to hide. (fr. 123)

But still,

> When you have listened not to me but to the logos, it is wise to agree that all
> things are one. (fr. 50)

What appear to be opposites are joined in the logos:

> Immortals are mortal, mortals are immortal: each lives the death of the other,
> and dies their life. (fr. 61)

One should know that war is universal and jurisdiction is strife, and everything comes about by way of strife and necessity. (fr. 80)

That which is in opposition is in concert, and from that and from things that differ comes the most beautiful harmony. (fr. 8)

Again, extremes strive against each other, and logos, like Anaximander's Time, decides the case.

The world is in continual transformation:

Fire lives the death of earth, and air lives the death of fire; water lives the death of air; earth that of water. (fr. 76)

but strife never results in victory; there is a measure in the universe that controls the outcome.

The sun will not transgress his measures; otherwise the Furies, ministers of Justice, will find him out. (fr. 94)

The active principle of the world is fire, but just as Thales' water is not that of rivers, this is a transcendent fire.

This ordered universe [the Greek is *cosmos*] which is the same for all, was not created by any one of the gods or of mankind, but was ever and is and shall be ever-living Fire, kindled in measure and quenched in measure. (fr. 30)

The world governed by the unchanging One is a world of change:

We step and do not step into the same river; we are and are not. (fr. 49a)

The sun is new each day. (fr. 6)

A flame, the visible manifestation of fire, is an example of nature's continuing flux. It is a process. Fuel is consumed at the bottom of it and smoke issues from the top; yet the flame holds its identity and we speak of it as if it were a thing.

Finally, the logos is divine, but is it God? Whatever the case, it is not the divinity of those operatic brawlers of legend and epic:

Homer deserves to be flung out of the contests and given a beating; and also Archilochus. (fr. 42)

That which alone is wise is one; it is willing and unwilling to be called by the name of Zeus. (fr. 32)

Socrates is said to have remarked to Euripides, "What I understood is noble, and also, I think, what I did not understand" (Diogenes Laertius 1925, II).

The words are filtered through our understanding, and are limited by the accidents that have led to the preservation of one sentence and the loss of another. There is enough confusion and mystery in our knowledge of the Presocratics so that for almost any of them, the discovery of one indubitable new text could force a reassessment of all the others. Still, these fragments were kept because for followers of Heraclitus they epitomized his meaning. As well as we can know it, they outline the Heraclitean solution of the problem: The world is changing and diverse and yet we see in it order and lawfulness. Logos imposes a unity of process. Is there also a unity of structure?

The World as a Number: Pythagoras

According to tradition, Pythagoras was born on the island of Samos in about 570 B.C., so that he flourished about midway between Thales and Heraclitus. Diogenes Laertius (1925, VIII) tells us that after Pythagoras had traveled and learned he and some three hundred followers settled among the Greek colonists of Croton in southern Italy, wrote a constitution for the colony, and established a commune in which women participated as full members. This happened, perhaps, late in the sixth century B.C. As believers in the transmigration of souls the communicants ate no meat. They regarded beans as sacred and did not eat them, they wore no wool, and they expressed their rules of conduct in a series of curious maxims that in the Middle Ages were known as the "symbols" (in the sense of passwords) of the Pythagoreans: Don't stir the fire with a knife; don't step over the beam of a balance; when you go abroad don't turn around at the frontier, don't piss toward the sun, and many others. Interpreted, these are: Don't involve yourself in the quarrels of the powerful; don't overstep the bounds of equity and justice; as your life nears its end do not try to hold onto the world; and the last (see also Hesiod, *Works and Days*, l. 727) is left as an exercise for the reader.

Pythagoras left us legends but no texts. Once when he went into the river Nessus it rose up between its banks and a number of people heard it welcome him. Another time, when he laid his clothes aside, it was seen that his thigh was of solid gold. We cannot tell now how much of the school's doctrine was his work; in fact, much of what we know of these doctrines comes from lectures in which Aristotle points out that they do not make much sense. It is probably significant that he directs his criticisms not against Pythagoras but against "the Pythagoreans," who may have been a much more recent group.

Several lines of investigation pursued by the Pythagoreans concern us. One is that of mathematical proof, an idea they seem to have invented.

Perhaps the best known example of their proofs is the theorem named after Pythagoras: that the square on the hypotenuse of a right triangle equals the sum of the squares on the other two sides. We do not know the original form of the proof. The one Euclid gives is so neatly fitted into his system of axioms and postulates that it cannot be the original. Of all the proofs I have seen, and there is a book of them (Loomis 1927), the one that seems plainest to me was given by Jacob Bronowski (1974) and is shown in Figure 1.2.

Start with the square figure and cut out the five figures it contains. Rearrange them as shown and look carefully at the lengths of the lines as they have been labeled. The original square, with side c equal to the hypotenuse of one of the triangles, is now reassembled to make two smaller squares with sides a and b. But moving a figure does not change its area, and so (in modern notation)

$$a^2 + b^2 = c^2.$$

I reproduce this proof because it is authentic in spirit even if not authenticated. It is an example of one of two sharply different styles of proof that we shall encounter. This one proceeds by geometrical construction and could be explained without saying a word. The next proof, in a couple of pages, will proceed like an argument out of one of Plato's dialogues.

The theorem just proved is named after Pythagoras but it goes back much further. In the collection of Columbia University is an Old Babylonian cuneiform tablet called Plympton 322, dated at least a thousand years before Pythagoras, that lists fifteen solutions to the equation in terms of integers (Neugebauer 1962). Many people know the smallest solution,

$$3^2 + 4^2 = 5^2,$$

but the Babylonian calculator was not interested in trivia. In our notation the first two examples in the tablet are

$$119^2 + 120^2 = 169^2$$
$$\text{and}$$
$$3367^2 + 3456^2 = 4825^2.$$

What they illustrate is not just an interesting numerical relation, for though the clay tablet is broken, the word "diagonal" can be read at the top. The calculator knew the numbers' geometrical meaning, but neither here nor anywhere else among the Babylonian mathematical tablets is any

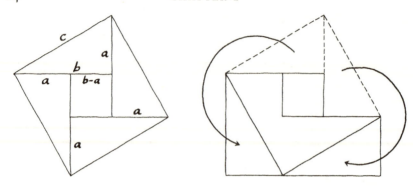

The whole area is c^2 Move two triangles

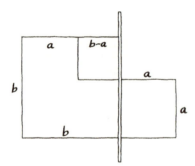

The whole area is $a^2 + b^2$

FIGURE 1.2. Simple constructive proof of the Pythagorean theorem.

proof exhibited. The theorem was known in China at about the time of
Pythagoras, and a proof resembling the one given here appears in the
Chinese classic *Chou Pei Suan Ching* (The Arithmetical Classic of the
Gnomon and the Circular Paths of Heaven; see Needham 1954, vol. 3, p.
22). Like the Babylonians, the Chinese did not ordinarily pay much attention to proof, and neither developed a body of mathematics in the systematic way, setting down definitions and axioms and requiring that one
proof stand on the shoulders of another. It seems that interest in the
process of proof, in logic as in mathematics, began in the Greek world
and for a long time flourished nowhere else.

The Pythagorean relation is not restricted to integers, of course. Consider the right triangle that is half of a unit square. Evidently the hypotenuse c satisfies

$$c^2 = 1^2 + 1^2 = 2, \quad \text{or} \quad c = \sqrt{2}.$$

How do we evaluate $\sqrt{2}$? Let us look at an ancient approximation, $17/12$. This gives

$$(17/12)^2 = 1 + 1 \quad \text{or} \quad 17^2 = 12^2 + 12^2.$$

Work it out:

$$289 = 144 + 144.$$

It is good but not perfect; $577/408$ is better, and $1393/985$ is better still, but can $\sqrt{2}$ be represented exactly by the ratio of two integers? The Pythagoreans proved that it cannot. We do not know their proof, but it was probably the same as the one that appears as the last proposition in an appendix to Book 10 of Euclid's *Elements*. For some reason it was probably not part of the original book, even though tradition says that the result had long been known. The proof is a classic of mathematics and proceeds by a method Euclid often used: assume the contrary and see that you are led into a contradiction.

Assume, therefore, that there exist two integers p and q such that

$$\sqrt{2} = p/q \quad \text{or} \quad p^2 = 2q^2 \, ,$$

and that the fraction p/q has been reduced to its lowest terms; that is, p and q have no common factors. Thus p and q cannot both be even numbers. Evidently p is even, since its square is even (the square of an odd number is always odd). This means that q must be odd. If p is even it can be represented as $2r$, where r is some other integer, even or odd. Put this into the equation above:

$$(2r)^2 = 2q^2 \quad \text{or} \quad 2r^2 = q^2 \, .$$

As before, since q^2 is even, q itself must be even. But we have already required that it be odd. Since no number can be both even and odd, we have arrived at a contradiction that invalidates the original assumption that $\sqrt{2}$ can be represented as a ratio of integers. QED. The seventeenth-century translator Isaac Barrow (Euclid 1660) adds, "this Theorem was of great note with the ancient Philosophers; so that he that understood it not was esteemed by Plato undeserving the name of a man, but rather to be reckoned among brutes" (he was referring to Plato's *Laws*, 819d). Euclid said nothing along those lines.

In Euclid's time all numbers were either integers or fractions, the ratios of integers. There were no other kinds; $\sqrt{2}$ was not a number and for a while a quantity of this kind was called *arrheton*, not to be spoken of. Later the term *alogon* was used, meaning that it could not be expressed as a logos, or ratio. Hence our term "irrational," which means the same thing and not that the number behaves unreasonably. But it can be represented as a geometrical quantity: draw the triangle and it is the length of the hypotenuse. Thus geometrical quantities were more general than the quantities representable by numbers, for they included also magnitudes like $\sqrt{2}$.[1]

While Italian shopkeepers were making change and calculating interest with an abacus or in their heads, Pythagoras was forming patterns with pebbles in the sand, a primitive arithmetic of integers but with something visual about it. He and his followers were the first to designate numbers as lines, triangles, squares, rectangles, and cubes. Examples of the first four are shown in Figure 1.3. The triangular 10, known as the *tetraktys*, was especially esteemed. They swore by it and used it as a sign by which to recognize other members of the community. The number 10 is sacred and universal; not only is it evident in the fingers by which we count, but there are ten celestial bodies as well: the seven planets (sun and moon were always included), the earth, the sphere of fixed stars, plus another, the anti-earth, invisible to us and invented, Aristotle says, to round out the number 10 (*Metaphysics*, 986a).

To understand how the Pythagoreans thought of number let us look at the uses to which these diagrams can be put. In Figure 1.3 a large square is represented by rows of dots, and I have drawn some lines to guide the eye. Obviously,

$$1 + 3 = 2^2, \qquad 1 + 3 + 5 = 3^2, \qquad 1 + 3 + 5 + 7 = 4^2,$$

and in general, we see that the sum of the first N odd numbers is N^2. The result is not quite so obvious without the diagram. Look again at the rectangular numbers of Figure 1.2. Obviously,

$$2 + 4 = 2 \times 3, \qquad 2 + 4 + 6 = 3 \times 4, \qquad 2 + 4 + 6 + 8 = 4 \times 5,$$

and again, the result can at once be generalized to

$$2 + 4 + 6 + \ldots \text{ (the first } N \text{ even integers)} = N(N + 1).$$

---

[1] There is of course nothing special about 2 or the square root. The same reasoning shows that if the *n*th root of any integer is not an integer it is irrational.

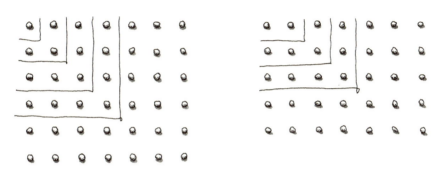

FIGURE 1.3. Illustrating Pythagorean number theory.

If every term is divided by 2 the result is

$$1 + 2 + 3 + \ldots \text{(the first } N \text{ integers)} = N(N + 1)/2,$$

and this relation is beautifully embodied in the *tetraktys*. The transparency of the results obtained in this way, and the ease with which specific examples are generalized to N, make it easier to understand the next development of Pythagorean mathematics.

Aristotle writes (*Metaphysics*, 1080b) that "they construct the whole universe out of numbers—not only numbers consisting of abstract units; they suppose the units to have spatial magnitude. But how the first 1 was constructed so as to have magnitude, they seem unable to say." He strikes a second blow when he adds that "natural bodies are manifestly endowed with weight and lightness, but an assemblage of units can neither be composed to form a body nor possess weight." Aristotle's remark about "the first 1" points to one of the Pythagoreans' preoccupations: the One. By adding ones the successive integers are generated, and for them the successive integers—alternately even and odd, hot and cold, male and female—generated the world.

One

What is this One from which all other quantities are constructed? In the Pythagorean representation it is a pebble on the sand. The One represents unity: perfect, indivisible, productive of multiplicity. But considered as a quantity, in what sense is it indivisible? Every carpenter, every shopkeeper divides it; yet if it is divided, in what sense is it the basic unit of quantity? What sort of number, for instance, is represented by $X = 2/3$? The faithful Pythagoreans multiplied this relation by 3, writing it as $3X = 2$, and were content, but Plato dismisses the stratagem (*Republic*, 525e): "You are doubtless aware that experts in this study, if anyone attempts to cut up the 'one' in argument, laugh at him and refuse to allow it, but if *you* mince it up *they* multiply, always on guard lest the one should appear to be not one but a multiplicity of parts." This is why it was very important—a disaster in fact—that the stratagem did not work for $\sqrt{2}$.

A moment later, Plato is serious again. The One, he understands, is a *mental* One, "which can only be conceived by thought, and which it is not possible to deal with in any other way." As such it entered Plato's theory of ideal forms and became a fit subject for study, but outside the Platonic circle it remained as a problem: In what sense is One really the starting point for an understanding of quantity? Euclid sees the answer and compresses it into a gnomic definition at the start of Book 7 of the *Elements*: "One [or oneness, or unity] is that according to which each existing thing is said to be one." Then the next definition, "Number is a multitude composed of ones." Obviously, for Euclid, numbers are integers. Other quantities are not numbers but can be expressed by the lengths of line segments, and even in those Euclidean proofs that deal purely with the properties of numbers they are always so represented.

Does Euclid's definition really explain what he means? Saint Augustine, in an early work *On Free Will* (1970, II.8.22) clarifies it, pointing out that every object, however small, has a right side and a left side, a top and a bottom and a middle, "and so no object can be called truly one—and yet I cannot count its parts unless I have the idea of one. . . . However I have arrived at my knowledge of what one is, I did not learn it through my bodily senses." It is indeed a mental unity, but whether a Platonic ideal or merely a convention adopted to make discourse possible depends on the needs of the thinker. As Euclid wrote, each existing thing *is said to be* one—whether or not we consider its parts.

Perhaps the epitaph of this discussion should be a remark by the Neoplatonist Plotinus about efforts to understand the One in its broadest sense, but which is applicable here: "The One is central but ineffable—any attempt to think about it leads to multiplicity" (*Enneads*, VI.1).

There has always been more to numbers than just counting. For Pythagoras numbers had shape as well as size; this is why he could think of making the world from them. Even for so subtle a mathematician as Euclid the idea of number as quantity in the abstract simply did not exist; as we shall see, that idea was an invention of the eighteenth century. For Euclid a number represented the length of a line or the size of an area or a volume, and it is much easier to read his book if we keep this in mind. Thus in the definitions of Book 7 we read "when two numbers are multiplied their product is called a plain number [plain in the sense of an area], and the numbers which were multiplied are called the sides of the plain." Is 12 x 18 = 216 a plain number? Who knows? It might be a cube.

Sounds

Pythagoras has something more to tell us. It seems that he and his followers performed experiments in musical harmony. Anyone with a guitar can perform them as well, though it is easier with the laboratory instrument called a monochord, which has a scale of lengths attached below the string and a little slider with which one can vary the length of the vibrating part of the string without changing the tension (with a guitar a pencil will do it). Pluck the open string and hear the note it produces. Now move the slider so that only half the string is vibrating. The pitch is just an octave higher. Set the slider at random and the tone will seem to have no relation to that of the open string; the two would be discordant if sounded together. But let the length of the vibrating segment be some simple fraction of the open length, a ratio with small numerator and denominator, and suddenly the two notes sound well together. Table 1.1, constructed assuming a string of unit length tuned to C, shows some notes defined by these ratios and the intervals measured from the basic C.

To understand the structure of this scale let us look at a few intervals. From C to E$^\flat$ is a minor third, and the ratio of lengths is $5/6$ to $1 = 5/6$. From A to C' is also a minor third, the ratio is $1/2$ to $3/5$, and again this is $5/6$. The sixth does not come out so well: C to A is $3/5$, but the ratio F to D', another sixth, is $4/9$ to $3/4 = 16/27 = 2.96/5$, and the two intervals will sound differently. To avoid this, and to make it possible to play in any key, modern tuning effectively shortens the string by a factor of $\sqrt[12]{2}$ for each semitone, so that only the octave, after twelve semitones, is exact.

The Pythagoreans considered that the law of harmony they had found was a law of nature, and they took it as an example that showed that the universal order is mathematical. Today we would say that if two notes harmonize it is a matter of human perception, but for them, if two tones

TABLE I.I
Notes of the Pythagorean Scale

Length	Note	Interval
I	C	
8/9	D	full tone
5/6	E♭	minor third
4/5	E	major third
3/4	F	fourth
2/3	G	fifth
3/5	A	sixth
8/15	B	seventh
I/2	C'	octave
4/9	D'	ninth

were in harmony that was a property of tones, not ears. The moving bodies of the universe were said to produce pure tones, and the legend of Pythagoras says that he could hear them. It says that as he died, his last words to those gathered around his bed were, "Remember to work with the monochord."

The Pythagoreans were not the first to understand that nature is bound by numbers. That had been done long before by carpenters and engineers. But there is a great difference. The numbers of harmony were expressed in terms of integers, and their relation to the natural world was not obvious; it had to be discovered. The Pythagoreans came to believe that everything immutable in nature, the existence of the world itself, could ultimately be explained in terms of integers. As we shall see, similar convictions flared up from time to time in imaginative minds such as Kepler's, and it is an unquestioned axiom of modern physical theory that the various states of matter and field are distinguished by the integers that characterize them.

In spite of the Pythagoreans' achievement in relating numbers to musical harmony, science was very slow to acquire a mathematical basis. The main current of Greek philosophy was relentlessly qualitative. It is proverbial that over the gateway to the Academy Plato founded was a sign "Let no one ignorant of geometry enter here," but clearly the Academy was seeking proficiency in logical argument, not figuring. It was precisely in these hallowed grounds that the distinction was most eloquently made between the ideal and inerrant world of mathematics and the world of experience and human problems. And of course it was the world of experience that really interested most of the scholars.

Still, the relation of music to arithmetic was a lasting component of Greek thought. Nine centuries after Pythagoras, Boethius's *De institutione musica* (1867) divided the study of music into three branches. *Musica mundana*, the music of the world, is the order in the physical universe, the rhythm of the tides and the seasons. "In astronomy, the very movement of the stars is celebrated in harmonic intervals" (1983, p. 75). This is music not for the ears but for the mind. *Musica humana* harmonizes the music of that great world with the little world of man, discussing Pythagorean harmonic theory and the relations between different rhythms. Finally comes *musica instrumentalis*. In a sense the other two musics are studied in preparation for it, but, because it deals with instruments and fingers and breathing and not with questions reducible to fundamental principles, it is not further discussed. Augustine had already certified that a knowledge of music was essential to understanding scripture and so the *Institutiones* made its way into the curriculum of the liberal arts; as late as the eighteenth century it was still required reading for students of music at Oxford. We shall meet Boethius again in the next chapter.

In Chapter 5 we shall see that in the Renaissance some people celebrated Pythagoras as the greatest philosopher and the teacher of all who came after him. I think this was because it is natural for some people to believe that even if it does not turn out that under the most profound examination the world is governed by mathematical principles of harmony and order, it ought to be.

Parmenides

As readers of Plato will know, Socrates had a poor opinion of most other philosophers, and so we pay attention when he says "there is one being whom I respect above all. Parmenides himself is in my eyes, as Homer says, a 'reverend and awful' figure. I met him when I was quite young and he quite elderly" (*Theatetus*, 183e). Parmenides lived in the colony of Elea, south of Naples, and if he ever traveled as far as Athens it was only toward the end of his life. He is known for a single philosophic poem, so original and so influential that considerable sections of it were copied out by later historians and so have survived. The problems he raised are recognizable in our problems, and the ideal world he describes will endure in some form till catastrophe overwhelms us. Whereas others before him had been content to invent and proclaim some version of universal truth, Parmenides believed that to arrive at any true statement about the world one must argue it, logically. Writing after Heraclitus and Pythagoras, he effectively put an end to the kind of speculation that merely announced to everybody how things are.

I can bring a word into existence by defining it: I can define a sphinx and assert that

$$1 \text{ sphinx} + 1 \text{ sphinx} = 2 \text{ sphinxes.}$$

But what if there are no sphinxes; is the statement true? Suppose you tell me that sphinxes do not exist. I can answer "Of course they exist; you know and I know what they are and we have just been talking about them." If then you insist that you are not talking just of words but of beings that actually have weight and occupy space, the only way to verify your assertion is by a systematic search of the universe, unless it can be proved from some accepted fact that sphinxes cannot exist. Parmenides' uncompromising program involves not a search of the universe but an analysis of what can be known without such a search. What exists? X exists. Either X is, he says, and it is impossible that X is not—this is the way that follows truth—or X is not and it is necessary that X is not—this is a path that cannot be explored, for you could neither recognize nor express it. But what is X, the subject of the verb *is* in this sentence? The subject is not there because to name it is to give it a kind of existence already, as with sphinxes. Rather, the words "X is and it is impossible that X is not" *define* the missing subject. What is, he continues, is eternal. It cannot have come into existence out of nonexistence, since nonexistence is a path that cannot be explored. What does not exist can have no qualities or properties that would allow it to produce anything and thereby be known, and to assume that nonexistence exists or has ever existed is self-contradictory.

Now Parmenides must make a positive statement about the world:

> Nor is it divided, since it all exists alike; nor is it more here and less there, which would prevent it from holding together, but it is all full of being. So it is all continuous: for what is draws near to what is. But changeless within the limits of great bonds it exists without beginning or ceasing, since coming to be and perishing have wandered very far away. (Kirk et al. 1983, p. 250f.)

In plain terms: Let us suppose that what exists is divided in some way. What divides it must be something that does not exist. But this is self-contradictory. . . .

I am sorry to mention that of course the passage is a mass of unstated assumptions and, insofar as it concerns cosmology, the logical program has come to a stop. But the argument continues, and since it is important for what follows, here it is in outline (Park 1980).

1. The object of thought is thought. (That is, insofar as the ideas in our minds are logically arrived at, they have their roots in other ideas, not

in experience. Consider, for example, the theorems of geometry, though this is not an example Parmenides would have used. Considered as approximate statements they apply to the approximate figures geometricians draw, but considered as truths they refer only to figures in the mind.)

2. What exists can be thought about.
3. Nothing can be thought about what does not exist.
4. The world exists. It is what exists. It is One.
5. The world is what it is and does not become something else.
6. Neither the world nor any part of it came into being or will pass away (this follows from nos. 1, 3, and 5).
7. The world is a timeless whole (this follows from 6; if nothing about it comes into being or passes away, it never changes).
8. This world is not the one we perceive with our senses (the perceived world is a world of change).

There are two ways of looking at these deductions. Either they restrict the range of possible discourse to a conceptual world with which we have no sensory contact at all, or they define the world in a new way as the material universe taken together with its history, *sub specie aeternitatis*, the medieval idea of God's view of it. This is the world we perceive, but not represented as we perceive it. Plato founded his theory of ideas on the first interpretation, but it may be that the second is closer to the reading of Parmenides' contemporaries. Plato's world exists in "the heaven above heaven." The world of the second interpretation exists in spacetime.[2] Whatever the case, the world of Parmenides was not so much an unchanging world as a world in which change and fixity are undefined, and thus the problem of what to say about motion (in its general sense of any kind of change) remained to be solved.

Parmenides had no predecessors. Plato and all subsequent philosophers are his followers. He was the first to warn us that the worlds of sense and discourse may be different. For a long time scientists had explained the world without worrying about the distinction. This century's most radical revolution in physical thought, the quantum theory, probably cannot be understood without keeping it in mind—I say "probably" because, as we shall see, there are eminent physicists today who do not think they understand it.

---

[2] That ordinary geometrical space (not spacetime) is the subject of Parmenides's *is* has been argued by Giorgio de Santillana (1968) in a sparkling essay, "Prologue to Parmenides."

CHAPTER 2

How Is It Built?

None of us knows anything, not even whether we know or do
not know, nor do we know whether not knowing and knowing
exist, nor in general whether there is anything or not.
—METRODORUS OF CHIOS

NATURE imposes order on our lives. In farming and in medicine, in
sailing a ship, making wine, glazing pottery, baking bread, rules must be
obeyed. Seasons impose their regularities. It is reasonable to suppose that
nature demands order because it is itself orderly, and in the preceding
chapter we have seen some early efforts to say what its order is. The most
enduring result of that effort was the Pythagoreans' discovery that
musical harmony is mathematical harmony. They also tried to represent
matter mathematically as consisting of aggregates of points. This chapter
deals mostly with attempts to explain nature in terms of material substance, to make sense of what things are and what they do by thinking
about what they are made of. Perhaps it was to be expected that pioneers
like Thales would try to reduce everything to some one basic substance,
but that really explains very little. Nature alternates; there are counter-
acting tendencies at work—heat and cold, moisture and dryness, motion
and rest. First one prevails and then the other. There may be harmony at
the end of the search but there is not uniformity, and it seems more promising to try to understand the world's diversity in terms of diversity on the
material plane. But what is the material plane? In the fifth century there
arose notions as to how the world's material nature is manifest in order

and physical process that can still, after many changes, be recognized at the core of modern science.

Empedocles and the Four Elements

The search for a unifying material principle had turned up first water, then air. These theories, if one may call them that, were unsatisfactory in two ways: they did not explain the properties of the observed world, and they did not even try to explain why anything happens in it. For a while people went on trying to explain the world in terms of elementary substances, but, as we shall see, the ultimate winner was Aristotle, who taught that we learn more by understanding causes than by understanding material properties.

Empedocles was born about 500 B.C. in Acragas, a name that has evolved into Agrigento; its temples still raise their golden columns among almond groves on the south coast of Sicily. Like Parmenides he wrote in hexameter verses, and his works are said to have included forty-three tragedies. They have all disappeared but we know about his teachings from Aristotle, and more than a hundred lines of a poem called *On Nature* are quoted in various works that have survived, the longest text we have from any Presocratic philosopher.

Empedocles explained both substance and process. Apparently he believed like Parmenides that the universe is entirely full, and for him its material contents were combinations of four indestructible elements. In a literal translation (Gershenson and Greenberg 1963) of the passage where Aristotle tells about Empedocles (*Metaphysics*, 985a) they are described as "heat substance on the one hand, and on the other hand dry dust, colorless gas, and clear liquid." Tradition says that he named them after gods: Hades, Hera, Zeus, and the water-nymph Nestis, but soon they were being called Fire, Earth, Air, and Water, and the names have stuck ever since. Natural processes consist in the mixing and separation of these elements under the guidance of two opposing principles, Love, which draws them together, and Strife, which drives them apart. Empedocles has no idea of chemical composition; for him the world's variety arises from a mixing of elements which he compares to a painter's mixing of colors, and although he does not explain Love and Strife, he gives examples to show how they operate.

At no point in the long history of the four elements does anyone claim that they are identical with what we know as earth, air, fire, and water. Fire is usually taken by later writers to be invisible; we shall see that in the Middle Ages the earth's atmosphere was thought to be enclosed in a region of Fire. Ordinary flames are contaminated with other substances:

a knife blade put into a flame becomes sooty with some form of Earth. That element is responsible for heaviness and solidity and it does not (by itself) grow plants.

Empedocles' theory, taken up a century later by Aristotle, started a tremendous development of ideas that dominated physical, chemical, and medical thought for two thousand years. Only after 1661, when Robert Boyle's *Sceptical Chymist* appeared, was the ancient scheme generally abandoned.

The Atomists

Little is known of Leucippus, the reputed inventor of the atomic theory. He is supposed to have been born in Miletus about 475 B.C. and to have written several works of which a single sentence survives, but from the summary Aristotle gives in refuting him we know what philosophical problem he faced and how he tried to solve it.

Parmenides, a generation earlier, had frozen the world into immobility by his logic, but nevertheless it moved. Parmenides' reasoning was not to be doubted. Either, then, the world he describes is not the one we experience (Plato's later choice) or else something is wrong with his hypotheses. Let us start from Parmenides and find the most conservative assumption that will bring theory and observation into agreement. How can change be introduced? The simplest kind of change is movement from one place to another. A thing can change its position without changing in any other way. Why does Parmenides not allow this? Because the world is full; there is no empty place for the thing to move to. This in turn is because emptiness is nonexistence; something that is physically nonexistent is logically nonexistent and so cannot enter an argument concerning things that exist.

A curious corollary of the Greek insistence on the nonexistence of nothing is seen in their number system, which was on the same general plan as the familiar Roman numerals, and was about as convenient for calculation (try dividing MCXXXVII by XLI). The much older Babylonian system was positional, as is ours, and it contained a symbol for zero, that wonderful invention that opens arithmetic to nonspecialists: a circle, a dot, or simply a blank. It is tempting to guess that the Greeks, who of course knew about the Babylonian system, refused to use any such thing because they objected to a symbol that represented nothing at all.

What if emptiness could actually exist in nature as zero exists in the Babylonian number system? It would then be possible to discuss changes of place, for a thing could move from where it was to another place where nothing was. What about other kinds of change? They would be allowed if they could be reduced to changes of place. And here was Leucippus's brilliant invention: make everything of atoms, too small to be seen. Then the different properties of matter can be explained in terms of the atoms' different sizes, shapes, positions, and states of aggregation. For something to change, all they need to do is move around.

All we know of this first atomic theory is handed down in the words of Leucippus's pupil Democritus, born about 460 B.C. He was known to antiquity as the laughing philosopher, for, as Saint Hippolytus wrote long afterward, "he laughed at everything, as if all things among men deserved laughter." Many of his sayings have laughter in them:

One must either be good or imitate a good man. (fr. 39)
To live badly is not to live badly, but to spend a long time dying. (fr. 160)
To a wise man the whole earth is open; for the native land of a good soul is the whole earth. (fr. 246)
Man is a little replica of the universe. (fr. 34)
I would rather discover one cause than gain the kingdom of Persia. (fr. 118)
I alone know that I know nothing. (fr. 304)
The eagle has black bones. (fr. 22)

Diels lists more than three hundred such fragments.

What do atoms do? According to the tradition that stems from Democritus they move around, or else in solid bodies they are hitched together, and in this position they vibrate. Atoms of Fire are spheres (since Fire never adheres to itself or anything else), but since flames deposit soot or moisture not all the atoms in a flame are spheres. The soul also is composed of spherical atoms, so small that they can wander through the body. It was generally believed even before the atomists that soul and body are intermixed—how else can soul move all our muscles?

What are atoms? They are all made of the same primary substance, and this is the unity that ultimately underlies all appearances. Each atom is One, for it cannot be further divided. The four elements are not mentioned in this early atomic theory. In fact there are no elements at all, for corresponding to the continuous gradations of properties we find in nature there is a continuity of atomic properties: no two atoms are exactly alike and they differ in shape as well as in size. Tradition says Democritus taught that atoms are infinite in number; the universe is infinite in extent and contains an infinity of worlds.

The single direct quotation that remains from Anaximander, the one that expresses natural law as cosmic justice, is echoed by the single one that remains from Leucippus: "Nothing happens at random; everything happens out of reason and by necessity (fr. 29)." We are no longer in a law court, but we are at the remote beginning of an idea that law governs not only the general tendency of occurrences on earth but (though it will be centuries before the idea becomes explicit) their exact, necessary sequence, atom by atom, as well. The Greek mind had long been burdened with a sense that the individual was trapped, that justice in the hands of the gods was arbitrary and unevenly applied, that acts had unpredictable consequences. Leucippus simply banished the gods from the world of human affairs to the world of ceremony. There is no record that people took him seriously enough to prosecute him for impiety, but they took him seriously enough to quote him, as an example of an extreme point of view, for centuries afterward.

It is clear that one of Democritus's aims was to understand how we perceive the world, and he seems to have been the first to distinguish between primary and secondary properties. Primary properties are those that inhere in the object, and objects considered as assemblages of atoms cannot have many of them. Secondary properties are the ones that register in our minds. The problem of the relation of one to the other is fundamental. How, looking at a thing's secondary properties, can we form any idea of its primary properties—for example, if all atoms are of the same substance, why do things differ in color?

> It is obvious that we cannot possibly understand how in reality each thing is. (fr. 8)
> Sweet exists by convention, sour by convention, color by convention; atoms and void exist in reality. (fr. 9)
> We know nothing about anything really, but opinion is for everyone an inflowing. (fr. 7)

Probably Democritus means an inflowing of atoms. Taste is explained by the shapes of the atoms tasted; sound, apparently, by passage of atoms of sound through those of air. Sight is more difficult. Light was atomic, but how was an entire image transmitted? There is an explanation (Kirk et al. 1983, p. 429; Diels 1960, vol. 2, pp. 114ff), that is involved and awkward enough to show that Democritus appreciated the question's importance and difficulty. I will discuss it later in the version given by Lucretius. Democritus continues:

> In reality we know nothing, for truth lies in the depths. (fr. 117)

This statement, echoing the "nature likes to hide" (fr. 123) of Heraclitus, is surprising, coming from a man who might have claimed to have at least the outline of a theory of all of nature. But Democritus clearly realized that his theory is terribly incomplete, for it does not explain the world as we sense it. Of course, it was understood that the world we sense is not the world that actually is; this claim, implicit in Thales and perhaps even before, is the central contribution the Greeks made to the world's scientific thought. In all the texts from Egypt, Babylonia, and China, no such statement is to be found. It was taken as fact by many thinkers throughout the Greek period (though not by Aristotle), can be found in the Middle Ages, and remains at the center of today's physics. Theories and language change, but the disjunction is still there. Democritus says clearly that the knowledge we obtain from the senses does not tell the real story, and that the underlying truth about nature, the truth in terms of atoms, is the knowledge that really counts.

The Epicureans

It is usual nowadays that when a new physical or chemical theory is developed someone finds a way to apply it. This is what happened when atomism became the intellectual foundation of a new philosophy of life and ethics. Its founder, Epicurus, was born on the island of Samos in 341 B.C., which makes him a contemporary of Euclid. He went to Athens and, in a garden that became famous as the symbolic as well as the actual center of his teachings, he sought to free his students from the tangle of ethical commands, metaphysical doubts, and superstitious fears in which they had been educated and which stood between them and happiness. Life is short and meant to be beautiful, meant to be enjoyed. His students were Athenian ladies and gentlemen, their existences protected from annoying distractions by the work of slaves. The enjoyment he taught was kind, generous, frugal, and intellectual, the sort that belongs only to one who commands his own life and does not fear the gods or any experience after death: to one who is composed of atoms in a world of atoms that are ethically neutral and governed by law. "The blessed and immortal nature," he wrote, "knows no troubles itself nor causes trouble to any others" (Bailey 1926, p. 95).

As a guide to the conduct of life the theory of Leucippus has a fatal and obvious flaw: lawful atoms allow no room for freedom of choice and therefore no room for ethical considerations. Epicurus thought about this and wrote a book *On Choice and Avoidance*, now lost. Cicero and Lucretius tell us his remedy; Lucretius's version is given below.

The school of Epicurus lasted far longer than such schools usually did, and a century and a half later its teachings were set forth in the long and very fine poem by Lucretius, a Latin poet of whom little is known, called *De rerum natura*, which Rolfe Humphries has translated as *The Way Things Are* (Lucretius 1968). Atomic theory originated in response to philosophical necessities. Here for the first time we find it supported by arguments based on common observation: the wind is not seen but its force results from innumerable impacts of its atoms. Scent is carried by

messengers too small to see. A horn lantern gives light while protecting
its flame from wind and rain because the small atoms of light pass
through the horn window while the larger atoms of wind and rain do not.
If wet clothes dry in the wind it must be because the water in them has
blown away as particles.

Impressions such as color and odor are generated in our sense organs
by the arrival of atoms emitted from the thing we are sensing. Since atoms
are neither created nor destroyed, these messenger atoms must have originally been present in the thing, though not as major parts of it. No atoms,
even the messengers, possess properties like color and odor at all. The
atoms of the sea are not inherently blue, for the color we perceive in it
can change from blue to gray or white according to the influence of cloud
and wind. The primary properties are only the shapes, sizes, weights,
motions, and arrangements of atoms.

Atoms are in ceaseless motion, always with the same speed, whether
going from place to place or vibrating where they are. Atoms of light
move quickly in straight lines; atoms of odor migrate slowly among the
atoms of air; atoms of soul either congregate in the breast, where thought
takes place, or pass very quickly through the body to report sensations
and move muscles. At death these atoms leave the body but they are so
small that we do not notice their going. And here is a point of fundamental importance: their motions are not absolutely governed by law, for
occasionally, and very slightly, atoms swerve from their lawful paths.
Through this little door enters the unpredictable, the undetermined:
freedom of choice enters.

The ideal world may be a world of flawless structure and inviolate
causality, but this is not the one that Lucretius wrote about or that we
live in. Ours is faulty in construction and in function—surely not divine
handiwork—but its faults are essential to it, and make us what we are.
This was Lucretius's solution to the problem of free will, but whether it
came from Epicurus we do not know.

For Lucretius the gods had nothing to do with us on earth. Obviously,
the world with its wild animals and deserts, its droughts and floods, was
not formed for human use, nor is there any evidence of a plan. It was
formed, quite recently, by a fortuitous clustering of atoms that some day
will once again disperse. We are the shapers of our own lives, and when
we die, we die.

As I have said, there is a huge gap in the Democritean scheme. It can
explain how a shirt dries but not how a seed sprouts. As long as the atoms
are doing something simple and easily visualized the mind does not feel a
need to explain further. But if we focus on the question "Why?" we see
at once that we understand almost nothing. Why are atoms of light
emitted from shining bodies? By what power does a chick embryo,

receiving nothing from the outside world but warmth, organize the atoms of yolk and albumen surrounding it so as to develop and grow? To questions like these the atomic theory had no answer. Understanding structure was not enough. The question "Why?" demands an answer in a different style. When an answer issued from Aristotle's vast imagination the atoms of Democritus were blown away and, except as they were preserved by Epicurus and a few practical men, we sight them only fleetingly during the next two thousand years.

The "swerve" of Epicurus and Lucretius is still part of science and perhaps the unpredictability of human thoughts and actions stems at least partly from the unpredictability of the behavior of the smallest particles—the uncertainty principle will be discussed in Chapter 15. But we know of no way, even in principle, to connect the mysteries of consciousness to the behavior of particles. Metrodorus of Chios, whose words began this chapter, was the best of Democritus's students. His words express the skepticism of someone who has tried to think his way from the hypothesis of atoms to even a partial explanation of the world as we experience it. If we do not characterize our ignorance today in such expansive terms it is because we have learned not to ask questions that have no answer, but the gap between physical and mental remains the central question of metaphysics.

CHAPTER 3

How Should We Think About It?

This reality, then, that gives their truth to the objects of knowledge and the power of knowing to the knower, you must say is
the idea of the good, and you must conceive it as being the
cause of knowledge and of truth in so far as known.

—PLATO

THE next two thousand years of the story this book tells are dominated
by the mind of Aristotle. He distinguishes form from substance and provides the words for talking about both; he distinguishes things from our
perceptions of them and tries to show how they are connected. He
explains physical process in terms of four basic types of cause. He says
that we have knowledge of a scientific fact when we can prove that it
could not be otherwise, but since observation never shows whether or not
this is the case he thereby establishes reason rather than observation at
the center of scientific effort. And yet when we look at the body of his
work and see the immense volume of miscellaneous observations it contains we recognize that what separates us from him is largely the word
"knowledge": he equates it with absolute certainty, and we do not.

Plato is also central to our story. He tells us that true knowledge is the
knowledge of ideal forms, but what this knowledge consists of and how
it is acquired we never really understand. Plato and Aristotle will be our
guides through many pages to come. As we approach our own time Aristotle's guidance becomes less sure. The existence of any absolute knowledge becomes more and more problematic until we can hardly believe
that any exists, and at the same time, as our description of nature

becomes more mathematical, the concept of substance withers away until only mathematical form remains. Only mathematical form? Surely not. Everything we know we know through the senses. Everything? Even mathematical form? Surely not. It is not a simple story and we must follow it slowly, realizing that there will be no unusual amount of light at the end of it.

The World as Image: Plato

The way Western philosophy is taught and studied, all paths from thinkers before Plato converge toward him and all paths toward future thinkers diverge from him. Partly this is a convention, an organizing principle based in a widespread conviction as to what problems are central. Partly, though, it comes from the form in which we have received Plato's philosophical ideas. His "esoteric" treatises have all disappeared. What survives is philosophic drama in dialogue form. With his insight into literary psychology Plato realized that a long explanation from the mouth of a character in a story enters the mind through a different portal than when the author speaks for himself. If we read a forced and ultimately unconvincing logical argument presented as the speech of a man trying to change the minds of his listeners, it is not the same as if the author gave it to us directly. Information delivers its message and is silent, but the memory of conversations under trees, at a banquet, at the deathbed of Socrates pour meaning into memory by pathways the senses use, and Socrates, a literary character based on a real man, creates his myths. We should not read Plato without being conscious of three facts about him that make him different from our usual picture of a thinker expounding his ideas.

1. His ways of thinking are rooted in a recent past in which ideas were transmitted orally, with events explained by the motives of men and gods. In these stories the medium was never distinguished from the message. Myth cannot be translated at will into the expository mode; the two express different kinds of thoughts. Plato's theory of Ideas, or Forms, for which he is best remembered, was never expressed as a set of propositions, and I believe it cannot be so expressed without making it ridiculous.
2. Plato was a confident and supremely gifted literary artist. Translation damages him; paraphrase kills him—but what are we to do who write about him? Only warn that a paraphrase is nothing but a paraphrase.
3. Part of the time, in the dialogues, Plato is having fun.

God alone is worthy of supreme seriousness, but man is made God's play-
thing, and that is the best part of him. Therefore every man and woman
should live life accordingly, . . . Life must be lived as play, playing certain
games, making sacrifices, singing and dancing, and then a man will be able
to propitiate the gods, and defend himself against his enemies, and win in the
contest. (*Laws*, 803c, in Huizinga 1949, p. 18)

The *Laws* is Plato's last work, written in his old age, and we should pay
attention when he tells us how to live. His words remind me of Niels
Bohr's remark, "There are some subjects so serious that we can only
laugh about them."

That was a great fuss to make about how Plato should be read, but
perhaps it will save someone from taking the wrong passages literally.

In restricting my account of Plato to what he says about the natural
world, I truncate very sharply. For the problems that concerned Plato
form a single complex—education, government, human relations, and
most of all, how the world ought to be. The *Timaeus*, however, is a dialogue mainly devoted to the natural world, and I will focus most of my
comments on it.

Let us consider the problem of saying something about this world of
ours that is really true. The truth of Parmenides cannot be applied in the
world of human experience; I have said earlier that he has nothing to say
about change. But how is our world related to the one Parmenides writes
about? In the *Timaeus* Plato gives an answer: we believe on the basis of
experience, but we do not know truth: "As being is to becoming, so is
truth to belief." And our world, the world of belief, was made by a Divine
Craftsman as a moving model of the world of timeless being and truth.
What is the status of the tale Timaeus is about to tell about our world?
The same as that of any other statement about it. He says over and over
that it is only a likely story—probability, belief, "beholding as in a
dream"—not that it is truth. In this way Plato's use of myth is not only
for didactic reasons; it is, if you believe Plato, the only appropriate language in which to describe the world we live in. We may then wonder
what the lost treatises were about. Apparently they contained an effort to
reduce the theory of Ideas to mathematical certainty by expressing them
in mathematical form (Findlay 1974). It is unfortunate, but perhaps
hardly surprising, that only the dialogues have survived.

The world existed before the Craftsman touched it. Then "finding the
whole visible sphere not at rest, but moving in an irregular and disorderly
fashion, out of disorder he brought order, considering that this was in
every way better than the other" (*Timaeus*, 30a). "When he was framing
the universe, he put intelligence in soul, and soul in body": the world is
alive. It is One. It is a perfect sphere, without eyes or ears because there is

nothing outside it to see or hear, without feet because there is nowhere for it to go. It is a god. It moves with the only motion consistent with its finite and spherical nature, turning about an axis. That motion expresses the world's rationality; the soul is expressing itself in natural law, which is the law of the cosmos—of stars and planets, whose regularly recurring and mathematically predictable motions had been studied for centuries. Today we would perhaps dismiss these cosmic cycles as mere appearances, surface manifestations of deeper laws that it is our primary business to discover. Plato had no such idea.

> Now the nature of the ideal being was everlasting, but to bestow this attribute in its fullness on a creature was impossible. Wherefore he resolved to have a moving image of eternity, and when he set in order the heaven, he made this image eternal but moving according to number, while eternity itself rests in unity, and this image we call time. (37d)

The timeless eternity is the world of Parmenides.

"The sun and moon, and five other stars which are called the planets, were created by him in order to distinguish and preserve the numbers of time" (38c). This is an interesting point. One might at first think that the sun would be enough for telling time. It goes round and round (the earth being stationary); it tells the hours but one day is like another and one month or year is like another. Therefore let us add the moon, to tell the months. A year does not contain an integral number of months, so the simple periodicity is lost and we begin to have a calendar by which one can distinguish one year from another by noting where the moon was on a given date. But the astronomer Meton of Athens had found that 19 years equals 235 lunar months with an error we now know to be only 2 hours, and so the positions of sun and moon are again essentially identical after 19 years. Now add the planets. There are no integral relations any more, and the celestial motions do indeed serve to "distinguish and preserve the numbers of time." "Preserve" means, of course, that the positions of the heavenly bodies at the moment of an event serve to mark it forever. There is, in principle, a vast interval, the "great year," after which sun, moon, and planets all return to their original positions. Apparently it is at this moment that God rewinds the universe (*Statesman*, 270d) and history begins again.

Form and Matter

Next the Craftsman creates four living species: gods, birds, fishes, and land animals. Most of the gods are stars. He creates souls, one for each

FIGURE 3.1. Plato's five regular solids and the elements assigned to them. From Kepler's *Harmonices mundi* (1619).

star; from time to time soul goes into body for a while and then returns to its star. Finally he creates matter. It starts with a medium called the Receptacle which plays the role of Anaximander's Boundless and seems to be space filled with matter awaiting its form. Then he introduces five different forms, in a diversity of sizes, corresponding to the five regular (or "Platonic") solids (Figure 3.1): these are the only solid shapes that are bounded by regular polygons (polygons all of whose sides and all of whose angles are equal). Four of them are the four elements. Fire is the tetrahedron because its sharp points and angles will sever the connections between other atoms. Earth is a cube because of its structural stability. The elements are

	SIDES	
Fire	4	tetrahedron
Earth	6	cube
Air	8	octahedron
Water	20	icosahedron

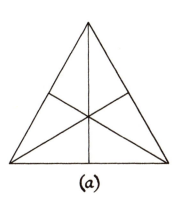
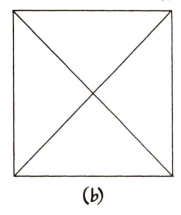

(a) (b)

FIGURE 3.2. Plato divides the faces of the elemental solids into triangles.

The fifth solid, the twelve-faced dodecahedron, decorated "with figures of animals," presumably those of the Zodiac, was used in "the delineation of the universe" (55c).

These bodies are, so to speak, hollow, consisting only of the plane figures that form their faces. The faces of Fire, Air, and Water are equilateral triangles, but Plato does not consider these as primary. Rather, he divides each into six other right triangles as in Figure 3.2(a), and he divides the square faces of Earth into four triangles, Figure 3.2(b). The commonly accepted explanation for this is that the sides of the triangles in (a) are in the ratio of 1 to 2 to $\sqrt{3}$, while those in (b) are in the ratio 1 to $\sqrt{2}$, and that since irrationals are quantities that cannot be constructed from integers Plato wanted to build them into the material structure of the world. Karl Popper (1950, p. 529) mentions the approximation

$$\pi \approx \sqrt{2} + \sqrt{3},$$

with an error of 0.0047, and suggests that Plato may have believed that by doing this he had introduced π and all the other irrational quantities. As to why Plato divided the faces of his elementary bodies into six and four triangles I do not know (but see Cornford 1937, pp. 210ff).

Plato had a different role for the dodecahedron. Some commentators have pointed out that since the constellations of the zodiac lie in a band around the plane of the ecliptic, Plato must be mixing up a dodecagon, which is a twelve-sided plane figure, with a dodecahedron, which is a twelve-sided solid. Possibly they thought that Plato was a stupid man. More than four hundred years later, Plutarch, in the *Platonic Questions* (1976), knew what was meant but did not get it quite right either:

It has been constructed out of twelve equiangular and equilateral pentagons, each of which consists of thirty of the primary scalene triangles, and this is why it seems to represent at once the zodiac and the year in that the divisions are equal in number.

Here Plutarch may have followed Plato (53c) into a slight mistake, the belief that all triangles can be constructed by putting together the two kinds of triangles of Figure 3.2. This is not the case, but the error does little damage to the picture, for the dodecahedron, composed of 360 triangles, still represents the year. Why the lengths of the month and year are not quite the same as those we know is a question to which an answer may already have occurred to the reader; it is discussed below.

Plato's fundamental bodies are not atoms, for *atomos* designates something that cannot be cut, while Plato's elementary bodies easily disassemble into their component triangles. The elements are not indestructible as Empedocles claimed they were; for example, the eight triangular faces of the body of Air can rearrange to form two bodies of Fire, and we know that a fire will not burn without air. Earth, however, cannot participate in these exchanges because its triangles are of the wrong shape. This explains why Earth is incombustible.

Plato's bridge between the ideal and the experienced has two spans. The first stretches from the ideal to the four elements, which are the fundamental constituents of matter but are hypothetical because nobody has ever seen them unmixed; the second is the mixings and dilutions that occur when elements combine to form the world we know. To Pythagoras's claim that all things are numbers Plato answers "obviously not; only ideas are made of numbers." Things are material replicas of ideas, and their materiality warns us not to take too seriously the abstractness of ideas and the remoteness of "the heaven above the heavens" that Plato suggests in poetic passages. Were they so remote, it is hard to imagine them playing the great part in human life and the material world that Plato assigns to them. But even so, as Aristotle later points out, Plato never explains how form makes anything happen (*Metaphysics*, 991a).

The *Timaeus* had a great influence in the Middle Ages; for centuries it was the only Platonic work available in Latin. It seems also to have influenced the history of mathematics. One might think, reading Euclid's *Elements*, that he was simply extending a web of theorems to prove at least the simpler parts of the mathematics of his time. But what of the work's plan, its goal? A work reveals its goal at the end. The *Elements* ends with the construction of the five regular solids, shown in Figure 3.1 ornamented with symbols corresponding to Plato's basic substances. Euclid proves that there are five different regular solids and no more. Is there any significance to their position in the *Elements*?

One of the last great scholars of the Classical world was Proclus Diadochus, born in Byzantium early in the fifth century A.D. By the time he was twenty he was in Athens, professing the old pagan religion in a world dominated by Christians, and in his later years he headed the Academy that Plato had founded 750 years before. Since the Academy had functioned in a reasonably continuous way for this entire time,[1] we may assume that he inherited a mass of written and oral tradition that we no longer possess. In the prologue to his *Commentary on the First Book of Euclid's Elements* (1970), Proclus distinguishes between the work's obvious pedagogical aim to present the elements of mathematics and its grand plan.

> Looking at its subject-matter, we assert that the whole of the geometer's discourse is obviously concerned with the cosmic figures. . . . Hence some have thought it proper to interpret with reference to the cosmos the purposes of individual books and have inscribed above each of them the utility it has for a knowledge of the universe. (p. 58)

Later, Proclus explains how the study of plane figures in Book 1 contributes to understanding the Platonic theory of the world's structure, to be discussed a little further on: "We shall therefore discover how to construct the equilateral triangle and the square. . . . The equilateral triangle is the proximate cause of three of the elements—fire, air, water—and the square the cause of earth." Note how Plato's "likely story" has dried and hardened into fact. Euclid's *Elements*, according to Proclus, is intended to explain the world's cause as well as its composition and structure. There is no mention of such notions in Euclid and Proclus may well be wrong, but he represents an opinion held through many centuries.

The central conceptual device in Plato's dialogues is his theory of Ideas. The Ideas live timelessly in Parmenidean abstraction: Justice, Love, the Good, and so on, which the prepared mind, after much study, can perceive all at once in a single act of understanding. Their influence governs, or ought to govern those areas of life in which choices must be made: how to rule, how to educate, how to live one's life. The Ideas are not quite like the ideal forms of geometry in that there can be many ideal circles or triangles, while the Idea of Justice is One. They are real; they are what is real, the subjects of the verb "exist" and objects of the verb "know." All else is at best opinion. They are the unmoving anchor of the good man's life, and from them radiate the laws of the just republic and its system of education. They are all there is to know; the rest is technique. The purpose of the Ideas was to root ethical values in what actually exists, rather

---

[1] For a lively account of its struggles to preserve and transmit Plato's teachings, see Augustine's *Contra academicos* (1948, 1970).

than in opinion. By the time Plato died the young logicians of the
Academy had shown that his theory contained logical contradictions and
could not be literally true, but at the end of his life (in Letter VI) he still
referred to it as beautiful wisdom, and I think all his students felt that
even if it was false, life should be lived as if it were true.

If it seems strange that even Plato's science (I mean particularly his
theory of elementary geometric structures) is so firmly based on such
abstract considerations, consider its purpose. Plato was first of all a
teacher. He invented the idea of a university, founded the Academy,
defined its subjects, and taught there for years. Probably he supervised
Buildings and Grounds as well as Admissions. His aim was to train the
future leaders of Athens, to teach them to set their course in the storms of
politics by navigating from unchanging principles. The little science there
is in Plato serves only this purpose, and it should not be taken to show
that nobody at that time was interested in getting things right. The purpose of medicine, for example, is to cure sick people, and the case studies
reported by the Hippocratic medical writers of the same period were
acutely observed and were explained as well as possible in terms of physical cause and effect. Evidence from civil and naval architecture shows
that even though little of their writing has survived, the architects based
their principles on the real world.

Ideas have their place in Aristotle's philosophy also, but not as Plato
understood them. For Plato, Justice is real; for Aristotle also it is real, but
its reality resides in the specifics that led us to the idea in the first place.
The distinction between these points of view divided the thinkers of the
Middle Ages; perhaps each of us inclines naturally a little bit one way or
the other.

It was easy for Christians to think platonically. In the Myth of the Cave
(*Republic*, VII), for example, Plato teaches that the light of understanding
emanating from the Good illuminates our minds and the world around
us. For the Neoplatonist this light was absolutely real, more real than
sunlight; it was Plato's light, and we shall find light endowed with this
transcendent reality by thinkers as distant in time as Robert Grosseteste
and Roger Bacon.

Augustine took the path Plato did not take: for him the Ideas exist
within the mind of God and their light illuminates the human mind. It
was in this form that the Platonic conception survived during the Middle
Ages. The only one of Plato's works known during that long period was
Calcidius's Latin version of the first part of the *Timaeus* (Plato 1962).
Isolated from the rest of Plato's work and from all relevant Greek commentary it was perceived as the scientific report of a leading expert, but
since Plato was a pagan it did not deeply enter the medieval mind before
the twelfth century when the recovery of other Greek texts began.

The World of Opinion and the World of Ideas

The Greek texts invite us to choose between two atomic theories of matter. One, invented by Leucippus, imagines atoms as little hard pieces of primal matter, while it is vague on the questions how the physical properties of atoms determine those of the perceived world and what makes the atoms move and combine as they do. Plato's theory has no such material substrate and bases everything on geometry, but it is vague on the same two points. We shall follow the first theory as it stumbles through the Christian centuries, opposed by the church for the Epicurean philosophy that had adopted it. It never died. In the seventeenth century it enjoyed a renaissance but not a reformation, for it is scarcely an exaggeration to say that even in 1900 the only new idea added to Leucippus's theory was that each chemical element was identified with a separate atomic species. The Platonic theory warns us that insights based on ordinary experience do not help to understand atoms, and that our explanation of the material world must ultimately flow from an understanding of mathematical forms. We shall follow the fortunes of these two ideas but I will say now that at the end of the twentieth century the second is triumphant. In no sense is an atom, or any of the particles that we now say compose it, a hard little piece of anything. Asked to explain it at the most fundamental atomic level, a physicist starts writing down equations, but honesty demands that we not forget the old question, what is the connection between the world as mathematically represented and the world we experience? The answer is simple: we do not really know. Knowing so much, and I shall try in later chapters to show what we do know, we cannot answer that question. In laboratories and factories we can make things happen, but we cannot explain to our own satisfaction, at the most fundamental level, the relation between what is conceived and what is observed.

The real world is a world of ideas, said Plato, while the world of opinion depends on it; it is a copy of it with flaws inherent in the nature of the copy. The real world is the one we experience and nothing else, said Aristotle. There was no clear winner, and I think that for many centuries educated people with no special bias tended to think both ways, at least some of the time. Consider the atoms of Lucretius, swerving slightly from their ordained paths so as to free mankind from the chains of determinism. The real world can be seen as an imperfect copy of an ideal one, whether or not the ideal is thought to have the transcendent existence that Plato proclaimed. According to the representation by Timaeus's dodecahedron there ought to be 30 days in a month, but there aren't; there ought to be 360 days in a year but there aren't. The ideal world exists in order to make clear to our minds the basic tendencies of things that the real world so often submerges in details. In modern physics, the world is portrayed in terms of mathematical forms. Plato's mathematical forms, invented out of so little knowledge of the physical world, seem to us naive. Our own are incomparably more complex and make contact with experience at many more points. It must be understood that the existence of such mathematical structures is validated by experience rather than by a priori considerations; we know no reason why nature *ought* to be amenable to such representations. It is hard to see how these structures can be derived from experiment, and history suggests that they are not, even by the little slow steps that scientific progress normally follows. As Einstein said, "It seems that the human mind has first to construct forms independently before we can find them in things" (1954, p. 266). Later in this book, when we have more theories to talk about, I shall give examples to illustrate these generalities.

The Rational World: Aristotle

They say that one morning when Plato arrived to begin his lecture he looked around and did not see Aristotle. "The room is empty," he said, "the intelligence isn't here." He might also have said the energy. Glance through Aristotle's *Collected Works* (1984) or the index to them (Organ 1949); you see logic, physics, astronomy, meteorology, medicine, anatomy, physiology, psychology, politics, ethics, literary criticism, and an immense amount on natural history. Every bit of it has influenced twenty centuries of thought. The influence was not in every case deserved—the physics, to name but one subject, could have been handled better—but considering the entire range of his work we must say that he rendered an incomparable service to humanity. He insisted that every theory, no matter what its subject, be logically adequate, constructed without errors on clearly stated assumptions of a fundamental character; he set forth methods for rigorous argument and issued warnings on the ways in which they could be misused. Where he fell into error we can often trace it to the assumptions from which he started and which he admitted are open to question. More than anyone else, Aristotle is responsible for our knowing how to distinguish between observation and theory, between the specific and the universal, and for our being able to say something sensible about the relation between them.

What most recommends Aristotle's work are its vigor and optimism, its huge scope, and its unfailing attention to consistency. He expounds a program, not a system; more than once he calls it "the science we are seeking." In the biological sciences he reports experiments, not merely observations. The description of the day-by-day development of a chick embryo (*History of Animals*, 561a) is based on very careful dissections. On the other hand, he describes fictitious differences between men and women in size of brain and number of sutures in the skull, and a remark that an actual count shows women to have fewer teeth than men (*History of Animals*, 501b) shows that the sample studied, presumably by his students at the Lyceum, must have been very small. If experiments on physics were performed they leave no trace in his writings. In general his physical theories are explanations of the external aspects of things that ignore the warning of Democritus that the truth is hidden in the depths.[2] Aristotle set the style of our thinking, and if at certain moments his work exerted a baneful influence on the development of science it was because readers denied it the criticism it deserved. They ignored its virtues and canonized its defects, and until people began to take a fresh look at the world around them it dulled their eyes even as it sharpened their minds. Today we admire it for its heroic attempt to show that considerable parts of human knowledge can be arranged so as to follow from a few qualitative axioms, and in order to understand anything about almost two millennia of scientific thought that followed, we shall have to get our hands into it a little.

Let us start with one of his definitions of nature (he gives several, not the same):

> Of the things that exist, some exist by nature and some from other causes. By nature the animals and their parts exist, and the plants and the simple bodies [Earth, Fire, Air, and Water]—for we say that these and the like exist by nature. All the things mentioned plainly differ from the things which are *not* constituted by nature. For each of them has within itself a principle of motion and of stationariness. . . . On the other hand, a bed or a coat or anything of that sort, . . . products of art, have no innate impulse to change. (*Physics*, 192b)

There are several ideas to be noted in these sentences. First, motion and change. The words are used synonymously here and throughout. Motion is any kind of change, qualitative or quantitative, including the change of position that we call motion and that Aristotle calls local motion. Second, one seeks the sources of motion and rest within a thing: physical processes and animal behavior are explained in much the same way. Third, a bed or a coat exists by nature insofar as it is made of wood or wool, and for that reason it will burn if ignited or fall if dropped, but there is nothing that it *does* specifically because it is a bed or a coat. Finally, note the Aristotelian term "bodies" for the four elements.

---

[2] For the development of experiment in this period see Lloyd (1979, ch. 3).

If we experience a stone, say, what are we experiencing? The thing itself, or our sensory perception of it? Aristotle defines the *substance* of a thing as its existence as a concrete, specific object without regard to its properties. *This* stone is a substance, and it is what we experience.

The word *matter* means almost the same as substance: the thing stripped of all properties save those that are inherent, but matter may refer to what we would call matter in general, while a substance is always particular.

The *essence*, or *being* of a thing or a class of things is the hardest of these terms to define, precisely because it is meant to be the most fundamental. It is what the thing or class is by its very nature.

A *property* of a thing is a quality that belongs to it by nature. We know a substance by learning about its properties.

An *accident* is a quality that a thing happens to have. Redness is a property of a cherry, but if the cherry is small or cold or unripe these are accidents.

The sum total of the properties of a thing is its *form*, "by reason of which matter is some definite thing" (*Metaphysics*, 1041b). The form of a thing is all we can know about it, but generally, matter is necessary for form. (God is pure form, Saint Thomas thought that angels are also, and of course mathematical objects are pure form.) Form and matter are distinguished in the process of sensation, for sense organs receive the imprint of form and not matter. Perception is the process in which the form of a thing enters the soul—not some representation of the form but the form itself (*On the Soul*, 431b), and this is why Aristotle disdained Democritus's theory of perception in which, for example, we see a thing by means of particles of light passing from it to us. We shall see in Chapter 8 that Aristotle's theory explained vision quite differently. It is important to realize that the account of perception just given, so different from our own, was taken as the exact, literal truth by almost every educated person down to the sixteenth century.

Armed with these concepts, Aristotle must now solve the riddle of Parmenides: how is change possible? Parmenides and Plato solved it by avoiding it; they distinguished two separate realms of existence, the ideal and the experienced. In the ideal world, as in the familiar world of geometry, change is not defined and rational discourse is possible. Aristotle declines this gambit and looks more closely at change. It is possible for the material in an egg to turn into a bird but not for it to turn into a sapphire. There is something about the matter composing the egg that allows one change and not the other. Even though we cannot know matter we know that the potentiality for change inheres in it, as part of the matter, and change occurs when what was potential becomes actual. The form of a thing can change but its substance does not change. Substance is what corresponds, in the material world, to Plato's timeless Idea. Aristotle summarizes this doctrine, a bit too roughly, in six words: matter is potentiality; form is actuality (*On the Soul*, 412a). In the Middle Ages it became clear that one cannot really tell from Aristotle's definitions where matter leaves off and form begins. Form in its manifold variety can be at least partially known, while the nature of matter can only be inferred from what is known of its forms. A human body consists of flesh and bone of a certain shape, but it is more than that: it is alive, it has a soul that pervades every part of the matter composing it. The soul is the form that produces the unique substance of a particular man. Such a form, organizing matter into a particular substance, was called a *substantial form*.

With these fundamentals established we can begin to talk about Aristotelian physics and its later history.

All matter is composed of a mixture of the four elements, each of which shares two of the four qualities:

FIRE

dry   hot

EARTH   AIR

cold   wet

WATER

As in Plato's account, the elements are not permanent but one can change into another. This is especially simple if the two share a quality: Water cannot easily change to Fire but it can easily, by evaporation, become Air.

This theory explains the observable properties of things by the qualities associated with the elements: heat, cold, moisture, dryness, rather than in terms of their elemental content, and the explanations are apt to be both general and complicated. Consider, for example, Aristotle's account of things that change from liquid to solid state:

> Whatever solidifies is either water or a mixture of earth and water, and the agent is either dry heat or cold. Hence those of the bodies solidified by heat or cold which are soluble at all are dissolved by their opposites. Bodies solidified by dry heat are dissolved by water, which is moist cold, while bodies solidified by cold are dissolved by fire, which is hot. Some things seem to be solidified by water, e.g. boiled honey, but really it is not the water but the cold in the water which effects the solidification. (*Meteorology*, 382b)

The confusion arises from deducing the constituent elements of a thing from the properties associated with them. Because cold will solidify water and heat will solidify mud the explanation must be elastic enough to include both, and it ends up explaining nothing.

A little reflection shows that if one is going to catalogue separately all the properties of a thing, the list grows very long. To its properties add its accidents. All may be apprehended by our senses, and Aristotle taught that they are the direct causes of our sensations. Contrast him with Democritus, for whom qualities like color and temperature are only conventions adopted to describe impressions arising from atomic properties whose true nature lies hidden. Democritus distinguishes the message our senses receive from the content of the message (say, "red"). Aristotle makes no such distinction.

I have said that the great weakness of Democritean atomism is that, except for phenomena at the most elementary level, it does not provide causal explanations; one of Aristotle's criticisms of Plato is that he does not explain how form makes anything happen. Cause, on the other hand, is at the center of Aristotle's theory of nature. For a first orientation, let us look at a simple example. Suppose someone makes a bowl out of silver. Four causes are said to operate here:

The *formal cause* is the design for the bowl that determines its final form.

The *material cause* is the silver.

The *efficient cause* is the silversmith.

The *final cause* is the purpose for which the bowl was made.[3]

Aristotle claims that *every* process can be explained in terms of these four causes, that no more are needed, and that we have scientific knowledge when we know the necessary connection between a thing and its cause. It is a science of why. As we shall see, questions of how are of little importance to it.

In explaining natural change the causes must be defined more abstractly. The formal cause is the one whose operation we recognize when we see the new form after the change is complete. The material cause, even though we often cannot know it except by analogies, must exist in order that there can be a formal cause; the efficient cause is whatever initiates the change; and the final cause is the reason why the efficient cause operates, guiding the world one step further toward a final, divinely determined actuality. The material and formal causes define the "this" that changes; the efficient and final causes produce the change. Now it is clear what is wrong about atoms: by mechanizing the world Democritus omits the final cause and therefore reduces all the operations of nature to

---

[3] Aristotle's sense of cause is not quite the same as ours. Aristotle's words *aitia* and *aition* (related to our *etiology*) were otherwise used mostly in a legal sense to denote someone's responsibility for an event. Once again, as in writers mentioned earlier, the terms are those of the law court.

necessity. For Aristotle the final cause is a device for escaping from necessity, or at least, since the final cause is hidden from our eyes, it removes the appearance of necessity.

Aristotle's doctrine, which I have outlined in a superficial way, does not sound much like the principles of today's science, and indeed it had to be swept away before that science could develop. But look around you for a while, notice what is there and what happens, and see how easily and naturally Aristotle explains it. Of course no numbers emerge, and when experiment replaced simple observation as the criterion of fact, that was one of the limitations that led to the theory's downfall. But most people do not think about the world quantitatively unless they are forced to. It is not hard to see why, when the Aristotelian books were recovered in the twelfth century, people looked around them as I have asked you to do and said with newly opened eyes "Yes, that is how it is; I never really understood anything before." In particular, the final cause stirred their imaginations. Now for the first time they could sense how each leaf that falls brings the world a little closer to that final and not distant day when "the heavens depart as a scroll when it is rolled together," and the earth is united with God.

The Living World

Aristotle wrote far more about biology than about any other science, and biological thinking is evident in the four causes. For example, it is very easy to think of final causes in biology if we are not careful—we may find ourselves saying that the zebra developed stripes in order not to be seen against a background of reeds, even though we know that in explaining the fact we should not imply that the zebra had a plan.

Aristotle looks carefully at the causes that produce local motion. The fundamental hypothesis, with which we shall have to deal again and again, is that everything that moves is moved either by itself or by something else (*Physics*, 243a). Let us first see how a thing can be moved by itself. Here the inspiration is clearly biological. All of an animal's voluntary movements are by definition natural. The flow of a river is also natural. The river's material is mostly Water, and Water's natural motion is downward, carrying the river along its course to the sea. Similarly, the natural motion of Fire is upward, as we see in a flame, which is mostly Fire. Earth falls, Air rises. Cause inheres in substance. There is another kind of natural motion that is explained in the same way. One might at first incline to say that the stars and planets are made of fire, but fires burn out after a while, and besides, planetary motions are circular or compounded of circular motions. Such motion is neither up nor down, and "therefore we may infer with confidence that there is something beyond the bodies that are about us on this earth, different and separate from them; and that the superior glory of its nature is proportionate to its distance from this world of ours" (*On the Heavens*, 269b). The higher the better. Aristotle names this body beyond those we know *Ether*, a word usually, and probably correctly, derived from *aithein*, to burn, but which, following Plato (*Cratylus*, 410b), he prefers to derive from *aei thein*, continually in motion. Since its natural motion is not vertical but circular we must not think of it as a form of matter.

Figure 3.3 shows two schemes for assigning the elements their places in the world: in ascending order they are Earth, Water, Air, Fire, and Ether, but the last four are assigned either to concentric shells or to successively larger complete spheres. There does not seem to have been a canonical opinion on the question. The sphere of Earth, with Water on and beneath its surface, is the abode of man. Next is the sphere of Air, which is invisible even though we feel it. Above that is the sphere of Fire, which is also invisible; Ether fills all space beyond the level of the moon. Because, except for their local motion, the heavens are never seen to change, it is assumed that change occurs only in the region below the moon that is ruled by the four elements (the moon appears to change, but that is only an effect of illumination). There is no vacuum anywhere.

Not all motion is natural. If a thing is not moved by itself it is moved by something else. Such motion is called violent. Violent motion persists only as long as its cause persists; afterward, natural motion resumes. On this account it is not obvious why a thrown ball keeps traveling for a while after it leaves the hand; Aristotle ties himself in knots (*On the Heavens*, 301b) to explain that the throw imparts motion to the air, which in turn pushes the ball with diminishing force until natural motion dominates and the ball falls to earth. The experiments that Galileo performed two thousand years later to refute these ideas could have been done by Aristotle, had he cared to, using available technology, and much trouble would have been saved. Why didn't he? If we trouble to put ourselves in frame of mind of those for whom natural and violent motion were very different things, we see at once that an experimenter deals only in violent motion; it is his only tactic. If, as the Aristotelians thought for two thousand years, the overwhelming majority of all the world's motion (remembering that the word refers to change in general) is natural, experiment can touch only an irrelevant fringe of everything that happens. The world was to be observed, not tinkered with. Only in a few areas like optics, where there is no motion, was it considered that experiment could provide insight into the ways of nature.

If the motion of a thing is caused by something else, what is the cause of that cause? Does the chain of causes ever end? Aristotle says yes: it

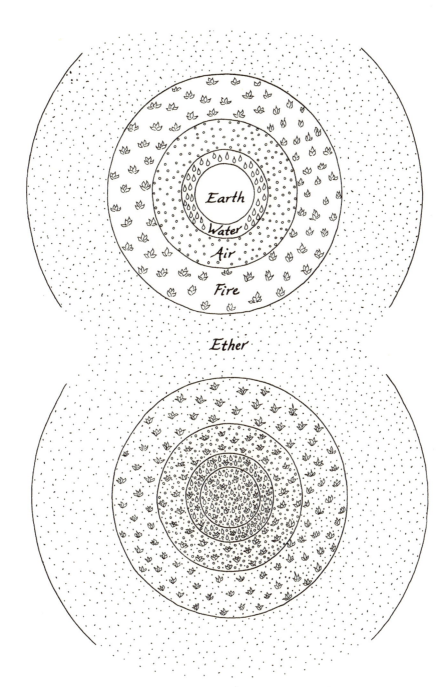

FIGURE 3.3. Two schemes for distributing the four elements in space.

ends in a first mover, or prime mover, or *primum mobile*, one and eternal, which causes other motions by virtue of its own natural motion: "There is a mover which moves without being moved, being eternal, substance, and actuality. And the object of desire and the object of thought move [i.e., cause motion] in this way; they move without being moved" (*Metaphysics*, 1072a). The prime mover lies outside the universe. "There is neither place, nor void, nor time outside the heaven. Hence whatever is there, is of such a nature as not to occupy space, nor void, nor time, outside the heaven; nor does time age it" (*Movement of Animals*, 699a). It is divine (*On the Heavens*, 279a); "it thinks that which is most divine and precious; and it does not change, for change would be change for the worse, and this would already be a movement" (*Metaphysics*, 1074b). That something can cause motion without itself moving is clear to Aristotle—it can cause it, for example, by being loved (*Metaphysics*, 1072b). Aristotle nowhere identifies the prime mover with God, but his discussion of the divinity of the prime mover leads him to a meditation on God that I will quote here to counteract any impression I may have conveyed of Aristotle as a reason-machine.

> And life also belongs to God; for the actuality of thought is life, and God is that actuality; and God's essential actuality is life most good and eternal. We say therefore that God is a living being, eternal, most good, so that life and duration continuous and eternal belong to God; for this *is* God. (*Metaphysics*, 1072b)

In most science after Aristotle, until recently, God dominates every model and the model's reasons are ultimately his reasons, but in the Christian centuries Aristotle's remark about love was taken to reveal the central cause of all motion. In the *Consolation of Philosophy* (Book II, poem 8), Boethius writes of the concord and stability and lawfulness that make the earth a proper home for man:

> This harmonious order of the world
> Proclaims the love
> That rules the earth and seas and steers the sky.

Dante expresses the same idea in the *Divine Comedy*. The *primum mobile* is a thing, the outermost of the nine celestial spheres, and all motion derives from it. How that actually happens is not clear, in Dante, Aristotle, or anywhere else. The fixed stars are on the eighth sphere. God, outside the *primum mobile*, is the first cause, but he does not move anything. The *primum mobile* moves slowly, by itself, and the cause of its motion is its burning love of God. This is why Dante speaks, in the closing words of the Divine Comedy, of "the love that moves the sun and the other stars."

Aristotle spent twenty years in the Academy under the daily influence of its loved and revered founder, and during this time he developed and sharpened the tools of logic. Probably his original purpose was to see what happened to Plato's theory of Ideas when it was subjected to logical scrutiny, for much of the great collection of speculative papers published after his death as the *Metaphysics* focuses on this question. His conception of science is founded not on observation of nature but on logical demonstration. Science deals in general statements, while observation yields only particular instances. No matter how many times you throw a stone into the air and observe its motion, you can never, from doing that, find out why it falls as it does. That knowledge can come from reason alone. Logic is the tool that creates science. Clearly, he writes, "it is impossible by perceiving to understand anything demonstrable—unless someone calls this perceiving: having understanding through demonstration" (*Posterior Analytics*, 88a).

To see how Aristotle's program worked in practice let us look at a sample proposition: "The shape of heaven is of necessity spherical: for that is the shape most appropriate to its substance and also by nature primary" (*Physics*, 286b). The argument goes on to explain that a circle is drawn from a single line and is therefore primary among plane figures, and similarly for the sphere, which is covered by a single surface. How are we to consider such statements? Not as the result of ignorance or carelessness, but of conscious choice, for a deductive argument must start somewhere, and here is where Aristotle chose to start. How was the choice made? Since the same situation exists more transparently in mathematics, let us consider that first.

Euclid begins his *Elements* with a group of definitions, such as "a point is that which has no parts," and "a straight line is that which lies evenly between its extreme points." Let us not worry about whether these words uniquely define what they are supposed to define; we know what they mean. Then Euclid asks us to assume certain postulates, statements about the objects that have just been defined: "between any two points a straight line may be drawn" is the first. It is possible to imagine worlds in which the various postulates do not hold, but they are assumed to hold for the world we inhabit, in that not only they but the consequences drawn from them fit with our experience. Finally, there is a group of "common notions," later dignified by the name of axioms. These are supposed to be statements assumed not merely for the sake of argument but because it would be folly to deny them. Who would dispute that "things equal to the same thing are equal to each other," or that "the whole is greater than its part"? Euclid did not create his system as a seamless web; many of these statements are older than the *Elements* and one can learn about the early history of mathematics from studying how they have been fitted together (Szabó 1964, 1978).

Aristotle's logical works also are concerned with starting points of arguments, and the beginning of the Topics shows how Aristotle considered statements such as the one relating to sphericity of the world:

> Now a deduction is an argument in which, certain things being laid down, something other than these necessarily comes about through them. It is a demonstration, when the premises from which the deduction starts are true and primitive. . . . Things are true and primitive which are convincing on the strength of not anything else but of themselves, for with regard to the first principles of science it is improper to ask any further for the why and wherefore of them; each of the first principles should command belief in and by itself. (*Topics*, I.1)

Clearly these premises are analogous to Euclid's postulates. It seems to me that the word "true" arises from wishful thinking and is inconsistent with Aristotle's usual modesty about "the science we are seeking." If Aristotle had not been so positive about the premise of the spheres astronomy would have moved faster.

The logical machinery most used by Aristotle and his successors consisted of the analysis and construction of several kinds of syllogisms in which necessary conclusions are drawn from premises of several kinds, carefully distinguished in the *Prior Analytics*. Let us look at the prototypical example:

> All men are mortal. This is a hypothesis, supported by experience. It is not contradicted by experience, but as long as there is anyone left alive it could still be wrong. It assumes tacitly that there are men.
>
> Socrates is a man. Assuming that Socrates exists, this follows from definitions.
>
> Therefore Socrates is mortal. This follows from the premises.

Now another,

> Some winged horses can fly.
> All winged horses are horses.
> Therefore some horses can fly.

The first term is, as before, a hypothesis. It seems to make a modest and reasonable (though possibly false) statement about winged horses. In fact, without saying so, it is making the much stronger claim that there are winged horses. The second term follows from definitions as before. The unlikely conclusion follows from an assumption that was never stated and might be overlooked. To help us toward understanding, a logical implement must be more than a pattern of words. Aristotle knew all this and catalogued the possible modes of error in his logical books. His scholastic successors in the Christian centuries knew it too, but they were obliged to assume so many propositions without proof that they were often not skeptical enough concerning the rest. It is not hard to spot scholastics among us today, even though some, who consider their tendencies advanced, might be surprised at being called that. I suppose the rest of us believe that we will never know the truth, or even whether it is really useful to talk about such a thing, but that perhaps we ought to dig around in the available information for a long time before we start uttering postulates.

Plato's world of heavenly ideas with counterparts in the world of experience is a myth, as I understand the word (the use of myth as a synonym for untruth is a sordid modern joke). Aristotle's version comes to us in language that is expository rather than narrative, language that for pages and pages is laboriously exact as to shades of meaning and the legitimacy of conclusions to be drawn. Is it any less a myth?

Aristotle makes it clear that the applications of his program to natural science are limited by the rules he follows. If the propositions from which we argue are universal, our conclusions will also be eternal verities, but concerning matters of a transitory nature there can be no proof and no certainty. There can be at best a science of accidents, properties a thing has for a while and under some circumstances. To formulate true statements about time-dependent phenomena is very hard. The laws of optics are timeless and universal and so are the laws of motion—anything that can be summarized in a mathematical statement—but Aristotelian science cannot explain such transitory happenings as eclipses, aurorae, rain, or the development of a chick's eye except perhaps as effects of some very generally stated causes. In no sense does Aristotle disparage descriptive science; in fact that is most of what he writes, and the situation is not very different today, for scientists who deal with the properties of matter tend not to claim they really understand a thing unless they can trace a path that leads, at least in principle, back to fundamental laws. It may be a century before we can explain the development of a chick's eye in such terms. What then is the difference? We are separated from Aristotelian science by more than a mere disagreement concerning postulates. The difference is that whereas the circles and spheres on which Aristotle based his arguments floated down from Plato's heaven, our heaven is much smaller and its gifts are regarded with suspicion. It contains beautiful mathematical structures, but which of them are realized in some process or structure and which are not must be guessed from what we know; and when that is done there remains a question just as difficult. Sensory experience is one thing and mathematics is another. What is the connection between them? As we shall see, something is known about this, but not too much.

CHAPTER 4

The Sky Is a Machine

In order that its motion may continue the same, the moved
must not change in relation to the mover. So, from first principles, it must lie either at the center or the circumference. But
things nearest the mover move the fastest, and in this case the
circumference moves the fastest; therefore the mover is at the
circumference.

— ARISTOTLE

A thousand years before Thales and Anaximander, the Babylonians wrote
how to interpret celestial omens, for stars and planets act upon us, or they
reflect the intentions of gods who do so. The night sky was seen as a
message to be read and pondered upon and understood. Stars move in a
predictable way and the sun moves among them predictably. Planetary
motions are more complex and follow diverse rhythms. Eclipses were
unpredictable, and so they, like unusual winds and rains, were interpreted
as signs of battles in heaven, of events on earth that would one day be
known about, of the gods' anger. By about the fourth century B.C. Babylonian astronomers had learned enough about the moon's motion that
they could predict lunar eclipses, but solar eclipses are more difficult.
When the earth's shadow falls on the moon the eclipse is evident to
anyone who can see the moon at all, whereas the moon's shadow on the
earth occupies so small an area that a solar eclipse that might terrify the
inhabitants of Cyprus could miss Alexandria entirely. The prediction of
eclipses seems to have been a powerful motive of early astronomers. Perhaps one of the reasons why people began to explain celestial phenomena

in terms of mechanisms is that by doing so they diminished the area in which divine power could be seen as acting capriciously. The astronomical theories to be discussed in this chapter go beyond the mere search for numerical regularities that would enable astronomical configurations to be predicted; their makers wanted to know what was actually up there, what was happening. For many of them, though not for Aristotle, the aim was to understand celestial events in terms of materials and processes that had some similarity to things seen on earth. One can argue that astronomy would have progressed faster had they been less successful.

This chapter will follow the two intertwined strands of astronomy: positional, for the purpose of making forecasts and interpretations of astronomical events; and physical, to tell what is really in the sky, how it is constructed, and how it functions to create the appearances we see. We shall look at the Eudoxan model that mounts the planets on concentric spheres and then, after a brief visit with the Roman encyclopedist Pliny the Elder, at the model that became known as Ptolemaic. The chapter ends with some remarks on models in general intended to be useful later.

What's Up There?

In the year 467 B.C. news came to Athens that a large meteorite had fallen near the shore of the Dardanelles, and perhaps because this happened during the daytime it encouraged one of the local professors, an Ionian named Anaxagoras, to declare that the stone had fallen from the sun and that the sun was a ball of glowing rock larger than the Peloponnesus. After a public trial on charges of impiety, Anaxagoras was exiled from Athens. He died at home, much honored.

Many years later Socrates was on trial for his life, and one of the charges brought against him was that he had spread this idea. "You surprise me, Meletus." Socrates turns to his accuser. "Why do you say that? Are you suggesting I don't believe the sun and moon are gods, as all mankind believes?" (*Apology*, 26d). That was exactly what Meletus thought, and the judges may have agreed, for impiety was among the crimes of which Socrates was convicted. Perhaps from the popular point of view Meletus was right. The last words of the *Apology* echo in the mind: "It is time to leave and go our ways—I to die, and you to live. Which is better, only God knows." God: *ho theos*, the God. For Socrates, the Divine, manifesting itself in nature as well as in the gods of the pantheon, is One. It is shared, in appropriate measure, by a tree or a stream or Apollo or the sun, and the believer worships it where it is perceived.

Anaxagoras had added to his legal difficulties by teaching that the moon is not self-luminous but shines with the light of the sun. He also gave the right explanation for the moon's phases and told how the sun gets to be eclipsed. Although he believed that the earth is suspended in space he thought it was flat, and so was unable to give a convincing explanation of lunar eclipses. That the earth is round must have become clear a little later to many people, since otherwise tradition would single out a first discoverer. Diogenes Laertius (1925, VIII.48) credits Pythagoras, but by then almost any discovery could be attributed to him. Both Parmenides and Plato write of a round earth, in mythic language that permits but does not require a literal interpretation. I suspect that in their days the idea was a commonplace. A little later Aristotle knows it. He remarks that the earth is "of no great size" and mentions the circumference, but it is almost 100 percent too large (*On the Heavens*, 297b).

In reading the Greek writers of this period we can see how the old pantheism was beginning to make way for a philosophy that emphasized the role of matter. By assuming that the sun and moon are material objects Anaxagoras explained familiar astronomical appearances. Matter does not exclude the divine, but if Anaxagoras thought his astronomical objects were divine he never mentioned it in any writings that survive. Obviously his fellow citizens in the fifth century thought they were and people continued to think so for a long time. They also continued to think of the heavens in material terms.

The Bowl of Night

If the earth is a sphere suspended in space it is natural to suppose that the stars are lights mounted on another sphere with the same center, and if one wants to explain their nightly motion one may suppose that the sphere turns. This is because the stars do not alter their relative positions as they move—the shapes of the constellations stay the same. If the stars were at different distances from the earth, the ones that are further away would have to move faster than the nearer ones in order to explain the appearances. Things could of course be arranged that way, but the simplest plan is to have them all at the same distance, permanently attached to the inside of a revolving spherical surface. As the mind focuses, imagination starts to move. This means that the stars do not suddenly assemble at sunset like actors taking their places before the curtain goes up; they are there above us all the time, exerting their influences by day as well as by night, and only invisible by day because the sun's light erases them.

Day by day, against invisible stars, the sun moves through the sky. The motion cannot be watched, but it can be plotted with a little effort. Make sketches of the stellar patterns visible during the course of a year. On a clear evening, note where the sun is an hour before sunset with respect to trees and houses; then note where it sets. Now you can easily estimate where it is an hour after sunset. By then the stars are out. Consult your sky chart and put a dot on it representing the position of the sun among the stars. Do this again a week later and you will find the sun has moved quite far eastward against the stars, about fourteen times its own diameter. Continue and you will find that it moves at an almost uniform rate in a circle through the sky, ending, after a year, exactly where it started. Astronomical custom divides this circle into twelve equal parts where it passes through twelve constellations: Aries, Taurus, Gemini and the rest, the constellations of the zodiac. Followers of newspaper astrology think, if they think about it at all, that late in April, for example, the sun is in Aries. It isn't; anyone who makes the little chart will find that it's in Pisces. The data in the newspapers are wrong because they ignore a phenomenon called the precession of the equinoxes that has been known for two thousand years (of course, educated astrologers know this even if they do not write about it). The vernal equinox, which we define as the first day of spring, used to occur as the sun moved into Aries. The position of the sun at equinox, however, crawls backward around the zodiac, moving about one degree in the course of a lifetime. Now it is approaching the edge of Pisces on its way toward Aquarius. The equinox marks, roughly, the beginning of spring. In order that spring occur in the same part of the calendar each year we peg the calendar to it, and so the constellations of the zodiac appear to drift past.

There is another circle that can be drawn in the sky but has nothing to do with the sun. The sphere of fixed stars, which rotates once a day, has an equator midway between its fixed poles. This celestial equator lies directly above the earth's, and the plane containing them is called the equatorial plane. The plane on which the sun moves is called the plane of the ecliptic. The two are not parallel (Figure 4.1), so that the sun spends half the year to the north of the equatorial plane and the other half to the south. This fact was known very early—even the untrained eye sees that the sun is more northerly in summer and more southerly in winter. Winter is colder because the days are shorter and the sun is lower in the sky, but long ago another reason was given: the sun is under water. This was because the spherical earth, which has to be supported somehow, was often thought to be floating on water. Aristotle tells us that Thales thought so. In the first century A.D. the Chinese astronomer Chang Hêng describes the heavens as a sphere half full of water with the earth floating in the middle (Needham 1954, vol. 3, p. 217), and even as late as the fourteenth century the Arab historian Ibn Khaldun compares the earth to a floating grape, though perhaps he does not want to be taken literally (1958, I.I). This may explain why three of the winter zodiacal signs are aquatic: Capricorn the goat-fish, Aquarius the water-carrier, and Pisces

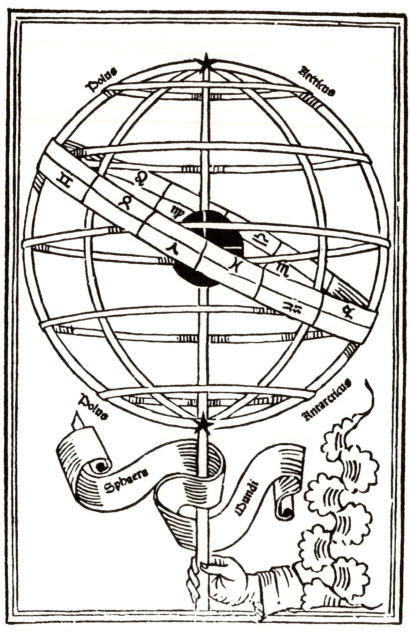

FIGURE 4.1. The sphere of the world, showing the equatorial (horizontal) and ecliptic (oblique) circles. From John Sacrobosco's *De sphaera* (1485).

the fish. For the great geographers of the classical era, Herodotus, Eratosthenes, and Strabo, the earth was not afloat, but even so there was nothing but ocean south of the equator.

The idea that the fixed stars are mounted on a sphere may well be prehistoric, but after people began to measure and record the positions of stars they began to realize that this sphere must be vastly larger than the earth. Figure 4.2 shows why. Go out at night onto a level space, the sea perhaps, and compare the constellations that can be seen with your sky map. It turns out that you can see just half the sphere: on the average, six of the twelve zodiacal signs. If the earth's size were appreciable, you would see fewer. The ancients were not in the habit of making numerical estimates, but the stars on that immense sphere, making a full circuit in twenty-four hours, must move at gigantic speeds, and, if you imagine things that way, you think, as they did, of noise and rushing winds.

Because they move against the stellar background, sun and moon were called *planetes*, wanderers. There were also five others, Mercury, Venus, Mars, Jupiter, and Saturn: seven altogether. Seven is a number that tells us that a list has been completed: it counts the days of the week, the wonders of the world, the stars of the Pleiades and the Great Bear, the gates of Thebes, the openings of the head, the ages of Man—Cicero, taking its continual occurrence as evidence of divine planning, writes that seven "might almost be called the key to the universe," (*De republica*, VI.18, in Cicero 1928) and Macrobius devotes a whole chapter to it (1952, 1.6). (In Christian times there were added the cardinal virtues, the deadly sins, the sacraments, the acts of mercy, the liberal arts.) But apart from their role in divine numerology the planets make it very hard to understand what one sees from week to week because of the strange ways in which they move.

The sun travels slowly inside the celestial sphere, making a complete circuit once a year. The sphere itself rotates once a day and the sun shares in that motion also. The moon revolves around the earth in the course of a month. You can see it move against the stars from night to night, and if it happens to be near a bright planet you may even notice that it moves across its own diameter in the course of an hour. You know it is below the fixed stars and the other planets because occasionally it moves in front of one and hides it for a while.

The motions of sun and moon are simple to describe only at first glance. In fact the sun does not move against the stars at a perfectly uniform rate, for the times of the vernal and autumnal equinoxes do not cut the year exactly in two. The plane of the moon's orbit wobbles with a period of about eighteen years the way a spinning coin wobbles as it settles down on a table. And the five remaining planets have their own peculiarities of motion. Figure 4.3 plots the path of Mars during nine months

FIGURE 4.2. If the earth were comparable with the celestial sphere, less than half of the sphere would be visible at night.

of 1928–29. The westward loop in its generally eastward motion is called retrograde motion, and all the planets except the sun and moon show similar behavior. How is this to be explained?

There is no program for explaining planetary motion in Plato's surviving works but there is a tradition that he proposed one. One of the last members of the Academy that Plato founded was Simplicius of Cilicia. About A.D. 530 he wrote in his *Commentary on the Four Books of Aristotle's Book on the Heavens,*

Plato lays down the principle that the heavenly bodies' motion is circular, uniform, and constantly regular. Thereupon he sets mathematicians the following problem: What circular motions, uniform and perfectly regular, are to be admitted as hypotheses so that it might be possible to save the appearances presented by the planets? (Duhem 1969, p. 5)

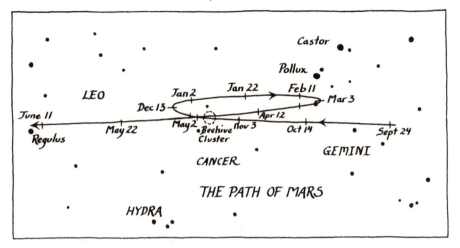

FIGURE 4.3. The retrograde motion of Mars.

Simplicius lived nine centuries after Plato and philosophical orientations had changed. The phrase "save the appearances" recurs in the writings of his time. An argument that explains some observed fact was said to save the appearances, or save the phenomena. It did not need to state facts about the universe; even if it was only a geometrical construction and gave the right answer, the appearances were saved. But Plato's works chronicle a search for truth, and there is no reason to think he would have been satisfied with a mathematical trick. I find it hard to believe Simplicius's story, but the problem he describes, whoever set it, controlled astronomical thinking through the time of Galileo. As it turned out, the appearances were never saved by Plato's hypotheses, for the paths of planets are ellipses, not circles. Still, the ellipses are only slightly different from circles, and though the motions cannot be represented in terms of uniform rotations they are only slightly nonuniform. Of course it was in everybody's mind that the earth lay at the center of these circles. Once more, Nature had been dressed in ill-fitting clothes cut according to philosophical fashion.

Celestial Spheres

Eudoxus, mathematician and astronomer, was born on the island of Cnidos about 408 B.C. He was the first to work out a theory of the planets that saved the phenomena reasonably well. The theory depends on a conceptual model requiring an acrobatic geometrical analysis using tricks invented for the purpose, many of which later showed up in Euclid's *Ele-*

ments. Figure 4.4 shows the first step in a representation that saves the sun's apparent motion. We are to imagine that the stars are fixed to an immense dark spherical shell (these shells are called spheres), rotating once a day around an axis, which in the figure is vertical. The line representing the axis of rotation passes through the center of the earth, cutting its surface at two points now (but not then) called the north and south poles. To the inner surface of this sphere are fixed two pivots that serve as axes of a second, smaller sphere, made of transparent substance, that rotates within the first. The sun is mounted on the equatorial plane of this inner sphere. Now let us try the machine. First, turn the outer sphere on its axis without touching the inner one. The stars execute their daily motion through the sky, the sun along with them. Now hold the outer sphere still and slowly turn the inner one. The sun will move slowly against the background of stars, along a path that is not parallel to the celestial equator. This represents the sun's annual motion through the constellations of the zodiac. Finally combine the two motions, turning the outer sphere uniformly at one revolution per day and the inner one uniformly at one revolution per year with respect to the outer. The phenomena are saved, but not exactly, since as I have mentioned, the sun does not move quite uniformly along its path. That was taken care of later.

The moon's motion is similarly explained, although because the plane of its orbit wobbles a third sphere is required. Again the principal appearances are saved, with small errors.

The motions of the other five planets, including their fits of retrograde motion, were explained by Eudoxus with the same mechanism, using four spheres for each. Add the sphere of the fixed stars and, for some reason, one extra sphere for the sun and you have twenty-seven spheres. Kuhn (1957) gives a particularly clear and helpful account for anyone who wants to learn some of the details.

In order to aid the imagination I have described this apparatus as if it were capable of being constructed by a model-maker, but a little thought shows that it is not, for each three- or four-sphere system assigned to a planet must be separately fixed to the outermost sphere, and Eudoxus proposes no mechanical way to do that. Apparently there was no need. "It seems probable," writes Dreyer in his *History of the Planetary Systems from Thales to Kepler* (1906), to which all who study the history of astronomy are indebted, "that he only regarded them as geometrical constructions suitable for computing the apparent paths of the planets."

To reduce the residual errors of the Eudoxan scheme his successor Callippus brought the number of spheres up to thirty-three, and this is the system that Aristotle inherited. It saved the phenomena and assumed uniform circular motion but it did not satisfy Aristotle. As a geometrical

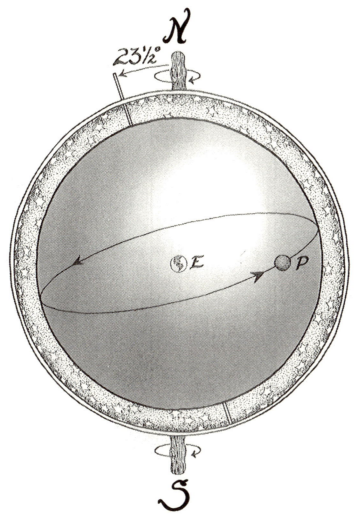

FIGURE 4.4. Planetary motion modeled by Eudoxan spheres.

construction it was very clever and clearly has at least some relation to the true state of affairs, but it was good only for calculation, and as long as the model had nothing to say about causes, for Aristotle it explained nothing. He proceeded to treat the idea literally, as if there were a real machine, and he found a way to make it work. By adding twenty-two more spheres that did not carry any planets but served only to adjust the motions of the planetary spheres with respect to the outermost one and to convey its motion inward, he designed a mechanism that could explain why the planets move and how the outer sphere moves all the rest (*Metaphysics*, 1074a). The spheres are made of ether, which, as we have seen, is not a material substance; the planets are glowing masses of ether. The whole is immersed in ethereal vapor so that there is no empty space anywhere. Since the planets are made of a substance whose nature is to move in circles they may need to be guided but they do not have to be propelled, and indeed the immaterial substance of the shells could do no such thing. I think that for Aristotle the essential purpose of this vast complexity was to be material and efficient cause of events everywhere, emanating from final causes residing in the prime mover. Without this unifying control the universe would function at random, a machine serving no purpose at all. Much later it was widely believed that the planetary spheres are solid and impenetrable, but such a model, if one tries to imagine how it could actually be driven, is absurd, and very few serious astronomers committed themselves to it.

Aristotle's effort was heroic but it contained a tragic flaw, since there was one set of appearances that it could not possibly save. If all the planets are mounted on spheres concentric with the earth their distances from the earth can never change, no matter how they move. This means that unless their sizes or brightnesses change from day to day—unthinkable to Aristotle—they should always appear the same. But they do not. Everybody knows that Venus and Jupiter shine more brightly at some times than others, and that the fit between sun and moon in an eclipse varies: sometimes the moon more than covers the sun and the eclipse is total, while sometimes it does not cover it all and we see a ring of the sun's surface. The moon's apparent diameter varies visibly from month to month; the total fluctuation is almost one part in seven. Therefore a model based on spheres concentric with the earth cannot possibly save all the appearances. Nevertheless, this is the system of ideas that Europe inherited in the thirteenth century when Latin translations of the *Metaphysics* and *On the Heavens*, as well as Arabic works based on them, became available for the first time. It was modified to include Ptolemy's improvements and a few later ones, and received its final revision from Copernicus.

Later Versions

Heraclides of Pontus was a little younger than Eudoxus and probably survived Aristotle, but Aristotle does not mention him. This is regrettable, since he had two good ideas for simplifying the Eudoxan model (Dreyer 1906, ch. 6). The first is that the earth is not stationary but turns on its axis. This simplifies the entire scheme because the sphere of fixed stars no longer needs to rotate. In fact it no longer needs to exist at all, for its only purpose was to explain why the stars keep their relative positions as they whirl around the earth. If they do not whirl they can be at any distances, and Heraclides is said to have believed in an infinite universe. Aristotle dismisses the idea of a rotating earth: an object thrown straight up would not fall straight down.[1] And of course, the idea conflicts with his conviction that there is a divine prime mover at the periphery of the system. The prime mover would now have to stand with his feet on the muddy and corrupted earth. Perhaps that is the real reason Aristotle rejected it.

Heraclides' second suggestion simplifies both the mathematical structure and the mental picture that supports it. One of the remarkable planetary phenomena is that whereas Mars, Jupiter, and Saturn travel through the entire zodiac independently of where the sun is, Mercury and Venus always accompany the sun, both of them alternating as morning and evening stars. Mercury is never more than 28° away from the sun, Venus never more than 47°. These facts were incorporated into the Eudoxan construction and of course they come back out of it, as the result of a miraculous compensation of the motions of two immense spheres. The new suggestion was that the spheres of the first two planets have the sun as their center. It is now obvious why they are never found far from the sun, and the quantitative details work out just as well as in the older scheme. Heraclides' works are lost and the tradition is uncertain, but such a useful simplification was not forgotten. More than three hundred years later Cicero mentions quite casually (*De republica*, VI.17, in Cicero 1928) that the two interior planets follow the sun, and one encounters the idea in astronomical texts and diagrams throughout the Middle Ages, though it never gained authority (Duhem 1913, vol. 3, Jones 1936). It was usually called the Egyptian system.

Among the essays collected as Plutarch's *Moralia* (ca., A.D. 100) is one in dialogue form called "On the face in the moon's disc." Here one of the speakers mentions that Aristarchus of Samos was accused of impiety "on the ground that he was disturbing the hearth of the universe because he sought to save the phenomena by assuming that the heaven is at rest while the earth is revolving along the ecliptic and at the same time rotating about its own axis" (1957, p. 55). The term "hearth" may go back to the Pythagoreans, for according to a fragment from one of them named Philolaus, "The first composite thing, the One, which is at the center of the sphere, is called the hearth" (fr. 7). An astronomical text from Aristarchus (Heath 1913) describes a very clever way to find the distance to the sun in terms of the distance to the moon, but does not mention such a

---

[1] He is right, of course. The effect was found in 1833 by Ferdinand Reich, who dropped little balls down a 188-meter mine shaft in Saxony (Dugas 1955, p. 380; Armitage 1947).

hypothesis. Archimedes, the great mathematician and physicist, mentions it but is skeptical, for it requires the cosmos to be very much bigger than anyone had previously supposed. Figure 4.5 shows why. Whereas a proper construction of the planetary spheres determines their relative sizes, it says nothing at all about the sphere of the stars. Nevertheless, most people seem to have thought it is not far beyond the sphere of Saturn. Suppose that the earth in such a system moves around the sun. Then any given star in the sky should be seen in a different direction in July from where it was in December. But this is not the case. The old argument of Figure 4.2 required that the outer sphere be very much larger than the earth; the new one requires that it be very much larger than the distance from the earth to the sun. Nobody knew how big that would be, but clearly it would be huge in comparison with the earth. Therefore either the stars are at unimaginable distances or the hypothesis is false. This is the great observational objection to the heliocentric theory, raised again and again by reasonable people when Copernicus proposed the same model. Yes, the distances are unimaginably large.

The Skies Over Rome

Pliny the Elder, Roman soldier, statesman, and student was born in about A.D. 23 and died on his galley in the Bay of Naples during the great eruption of Vesuvius that buried Pompeii and Herculaneum in 79. Instead of seeking safety he had ordered his ship near the shore so he could watch the eruption and perhaps save some friends. During his career he collected facts the way some people collect money. He was never without a book in his hand, and he claimed that to write the thirty-seven books of his great encyclopedia, *Historia naturalis* (1938), he and his secretaries combed 20,000 facts out of 2000 volumes by 100 carefully chosen authors. His table of contents and list of authorities occupies 143 printed pages.

I mention Pliny because it is natural to wonder whether Rome, the great Mediterranean power, had a school of natural philosophy. Pliny's curiosity is factual rather than speculative, and this is generally true of Roman thought. It gave rise to some competent medicine and mathematics, but never to a deep idea. Still, the *Historia* was of immense importance for perhaps 1500 years, for it contained so many singular facts, found nowhere else but in books copied from Pliny, that medieval scholars tended to regard it as the ultimate authority. It was widely distributed—about two hundred manuscript copies still survive, and its first printed edition came only a few years after the Gutenberg Bible.

The first books of the *Historia* concern the earth and its geography;

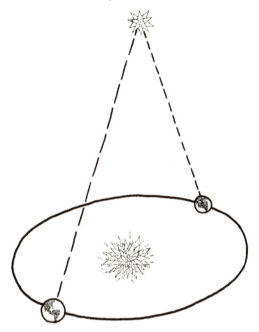

FIGURE 4.5. Stellar parallax expected if the earth orbits the sun.

later ones tell of animals, plants and agriculture, fish, birds, storms, earth-quakes, monstrous births, magical cures, a strange race of men with one enormous foot that served as a sunshade in times of need, and a talking rooster. We are told that

> it rained iron in the district of Lucania the year before Marcus Crassus was killed by the Parthians. . . . the shape of the iron that fell resembled sponges; the augurs prophesied wounds from above. But in the consulship of Lucius Paullus and Caius Marcellus it rained wool in the vicinity of Compsa Castle, near which Titus Annius Milo was killed a year later. It is recorded in the annals of that year that while Milo was pleading a case in court it rained baked bricks. (*Historia naturalis* II.57)

According to Pliny (Book II), the starry sphere is filled with air and the spherical earth rests at its center, kept from falling by the contrary tendencies of its light and heavy elements (Pliny has not solved the problem of down): "Supported by air between earth and heaven, at definite spaces apart, hang the seven stars which owing to their motion we call planets, although no stars wander less than they do."[2] (II.4)

---

[2] This curious remark refers to the fact that while with respect to us the starry sphere

It seems that educated Romans believed the sun's rays cause the planets to move irregularly; we find a rather confused account in Pliny (II.70, 71) and a little earlier in the *De architectura* of the Roman architect Vitruvius (IX.1). I find no mention in Pliny of planetary spheres—not that there really are any planetary spheres, but a Greek encyclopedist would not have considered his account complete without them.

The Romans were not lazy or stupid people and they excelled the Greeks in areas as diverse as bridge-building and love poetry. We can see in Pliny, though, that they did not share the Greek passion for explanations based on the orderly development of ideas, and if we look around at the great old civilizations of Babylonia and Egypt and China we do not find it in them either. It seems the Greeks were different.

Many reasons might be suggested for the "Greek miracle." In the city-state of Greece or Asia Minor a citizen (who was by definition a man) had certain rights and duties that required him to think for himself. The Greek religion did not encourage intellectual subservience. The gods knew things that mortals did not, but nothing in early Greek literature (mostly Homer) suggests that the all-powerful Zeus ever thought very much, had any interesting opinions, or was in the slightest degree more intelligent than the average man in the street. If nature was to be understood in terms other than those given by mythology it was up to humans to make the attempt, and they were free to do so. When writing was introduced the Greeks had created a system that was easy to read and easy to write, so that no group had a monopoly in literacy. But why, sitting in the prosperous Ionian city-state of Miletus, did Thales undertake to expose the unity of things by saying that the whole god-filled world was made of water? Why did it happen, this effort to replace the experienced world by an ideal?

Consider the statue of a young man (a *kouros*) shown in Figure 4.6, dating from about 525 B.C. when Pythagoras was alive. Its form is so schematic that one cannot regard it as a portrait. I think we should see it almost as a diagram that tries, in a formal language full of geometry, to explain "This is what a young man *is*." Essence before substance, even if there were no words yet to express the distinction. A trip through a book on art history shows us that the progression from schematic representations like this one to the forms of later Greek sculpture took about two centuries. As Emanuel Loewy argues in a classic study (1907), the abstract and general concept of humanity expressed in the earliest works was successively modified and nudged toward realism by comparison

---

turns from east to west, with respect to that sphere the planets move from west to east, and therefore with respect to us they move a little more slowly than the stars.

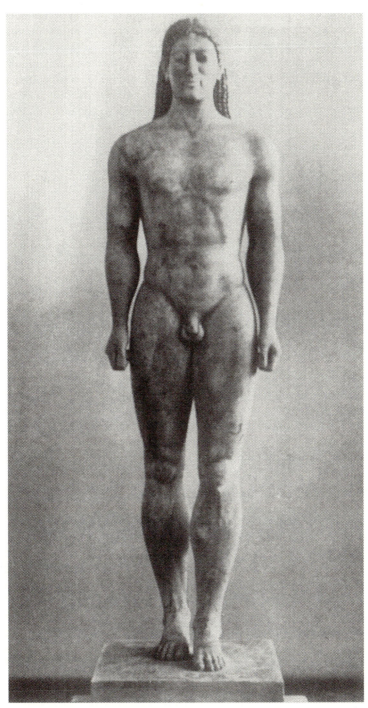

FIGURE 4.6. *Kouros*, Greek, ca. 525 B.C., Athens.

with specific examples, but even the lifelike images of later sculpture seldom look like portraits.

In two centuries the abstract and nonrepresentational physics of Thales and Anaximander had evolved into the observational science that produced pages on the day-by-day development of a chick embryo, and the primitive notion of the "bowl of night" could be replaced by complex hypotheses for explaining celestial motions. In literature, over a longer time, there is a similar development. Homer places his characters in a thousand situations among friends and enemies to show how one ought and ought not to behave, while in Euripides we see how they do behave. But before we give ourselves over to facile generalities we should remember that one of the pleasures of Homer is that his people are so real. Their functions in the story may be schematic but their natures are not. If the Greek miracle was a miracle at all, it was not a simple one.

Almagest

The model of concentric spheres was impressive if one enjoys intellectual constructions, but it did not save the phenomena. Consider how many phenomena there are. First, the orbits of the seven planets. Second, the planets' positions in their orbits at a given date (as marked, for example, on Figure 4.3). Third, though they could not be measured exactly, the planets' obvious changes in brightness. Fourth, conspicuous changes in the relative sizes of sun and moon. Obviously, if the moon rides on a sphere it is not a sphere centered on the earth, and so astronomers in the generations after Aristotle gradually placed the whole Eudoxan system on hold and tried again with a new program: never mind how the system actually works; is there just a quantitative description that accounts for its motions and changes and can be used to calculate how they will occur in the future?

Why did anyone want to predict how the planets would move? An immediate reason was the establishment of the civil calendar, in which many dates were defined in astronomical terms. But why were they defined in this way? Why could they not simply be assigned to particular days of the year? For at least two reasons. First, no calendar exactly fits the solar year. Suppose you assume that the year has 360 days. Then in about 35 years you will be celebrating the Fourth of July in midwinter. If you use 365 days you will do better, but ancient history is very long and the Egyptians had to grapple with feasts that slid through the seasons because they neglected leap years. Even when Hipparchus had measured the year with great accuracy the problem was not solved, for feasts have purposes. They are connected with human actions, and these, in turn, are influenced by the stars, the planets, and especially by the moon. Everybody knew this. Ptolemy, the great astronomer, was better known during the late Middle Ages for his *Tetrabiblos* (1940), a book on astrology, than for his astronomical writings. Tides rise and fall according to where the moon is; storms were thought to occur at the equinoxes, crops were planted and harvested under celestial signs; the menstrual cycle and the agitation of insane people during the full moon (which is why they were called lunatics) leave no room for doubt that the moon affects the human organism.

Once it is established that effects occur, the human tendency to elaborate things takes over, and Ptolemy devotes two hundred pages to recipes for assessing the meaning of celestial happenings. An event occurring at the time when one of the planets reverses its motion in the sky will have effects of long duration; what effects and upon what class of beings is determined by which planet it is and in what zodiacal sign it stops. Similarly, with eclipses of sun and moon, the duration and degree of totality determining the duration of their effects. Agriculture, medicine, shipping, the conception of children and the signing of contracts, every action depended for its success not only on how but under what stars it was undertaken. Passover and Easter are lunar feasts, the calendar of Islam is controlled by the moon, and the United States today finds work for 20,000 professional astrologers as against 2000 professional astronomers. Astrology belongs in this book, but there is not enough space. Even these brief remarks, however, show why it was necessary to know well in advance of an act where the planets would be and whether there was even the slightest risk of an eclipse.

So astronomy, as it became more accurate, became more of an applied science. That made it simpler, though it was still complicated, because the system no longer needed to work as a machine does. The models of Eudoxus and Aristotle needed to be constructed out of spheres because the axis of one sphere had to be fixed to another, but the rotation of a sphere carried its planet in a circle, and for practical purposes of calculation it was simpler to focus on the circle. The spheres were still occasionally mentioned on ceremonial occasions and in literary texts, and as we shall see, the Arabic astronomers did not give them up, but in Europe they were almost forgotten until the Aristotelian books were rediscovered, translated, and enthroned in the twelfth century.

The pioneers of the new applied astronomy were Apollonius of Perga in the third century B.C., Hipparchus of Rhodes a century later, and finally, in the second century A.D., Ptolemy of Alexandria. In order to allow planetary magnitudes to vary they introduced new devices: they mounted the centers of the planetary orbits on rotating circles called eccentrics, centered on the earth, and on the planetary circles themselves they mounted other rotating circles called epicycles. Figure 4.7 shows these, and how simply epicycles describe the retrograde motions of the planets, but the whole picture is much more complicated. The system now contained dozens of numbers chosen purely in order to fit the data in astronomical tables, assembled by Hipparchus and modified by Ptolemy, which summarized the results of many Greek observations and a few from Babylonia.

Ptolemy's works include tables of planetary positions, books on geography and optics, the *Tetrabiblos* and a great treatise, the *Almagest* (1984).[3] It shows exactly how to calculate planetary positions using the machinery of the spheres, but it pays little attention to the question of how the machinery is made or how it works. This work, lost and rediscovered, copied and recopied, was the standard reference on astronomy until the seventeenth century, though because it was too mathematical for the average reader other texts such as John Sacrobosco's *Sphere* (1485), written early in the thirteenth century, were more widely read.

But in the midst of all this empiricism sat the ghost of Plato, legislating that the curves drawn must be circles and nothing else, and that the planets and the various connecting points must move along them uniformly and in no other way. Therefore it is not correct to say that Ptolemy's constructions are purely empirical. The planets were really up there, embodying divine perfection and executing motions representable by combinations of Plato's circular paths. Ptolemy's was an empirically derived picture of this reality.

The Ptolemaic tables were so accurate that they were used with only minor numerical changes until Copernicus, who computed most of his numbers from these same tables, showed how essentially the same numbers can be obtained more simply if the sun replaces the earth at the center of the diagram. Was this just another, more convenient geometrical construction in the same spirit as Ptolemy's? For a hundred years after Copernicus most people thought so, but Copernicus knew better. Recognizing the substantive nature of the Platonic postulates enshrined in Ptolemy's geometrical constructions, Copernicus understood that to change even one of them is to make a new hypothesis about the nature of the universe, to adopt a new model.

I have written about Ptolemy's astronomy as if his only purpose were to calculate astronomical tables, but there is a late work, the *Planetary Hypotheses* (See Neugebauer 1975, V.B.7), which shows his concern to establish how things really are. It is based on what Neugebauer calls an

[3] Originally it was called *Mathematike syntaxis*, The Mathematical Compilation. Later it was called *Syntaxis magiste*, The Greater Compilation (there was also a lesser one). The Arabs shortened this to *Al Magiste*, The Greatest, and suitable corruptions produced its Latin name.

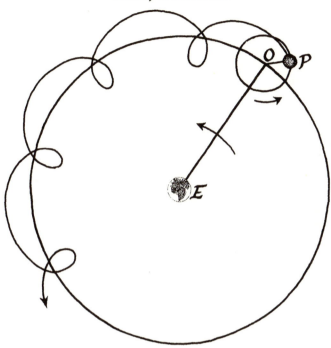

FIGURE 4.7. Ptolemy explains retrograde motion with an epicycle.

incredible numerical accident: Ptolemy's orbital calculations had resulted in figures (entirely incorrect) for the least and greatest distances from the earth's center of the orbits of the moon, Mercury, Venus, and the sun, and it turned out that they fit so that the moon's greatest distance equals Mercury's least, Mercury's greatest equals Venus's least, and Venus's greatest equals the sun's least. The same construction was then continued to Mars, Jupiter, and Saturn beyond the sun's orbit. Thus originated the "Ptolemaic system" of planetary spheres considered as thick spherical shells with the planetary mechanisms inside them, lubricated in between by thin layers of ether, with the stars not far beyond the outermost sphere (Lindberg 1978). This model dominated astronomical theory during the later Classical phase, the Arabic period, the Middle Ages, and in the seventeenth century Johannes Kepler's first book was an elaboration of it. It sheltered the extravagant numerical assumptions that planetary theorists were obliged to make within a smooth and simple system of nested revolving spheres, stupendous in size and ultimately simple in structure. Thus the model served two purposes: it saved the phenomena and it satisfied the soul's thirst for order. Perhaps it is time to think about models for a while.

Models

A model is a mental image, or sometimes even a physical mechanism, that explains how something works, that saves the appearances. The phrase implies a separation between the thing observed and the appearance it presents to us, between the thing and a representation of it. The thing itself does not save the appearances; it produces them, but we do not necessarily know how, for nature likes to hide. A model does not explain everything but it represents a vital intermediate step: it summarizes a number of observations that may be vast, it summarizes our ideas as to cause and effect, and usually it involves some mechanism we can understand. The rotating sphere of fixed stars is an example. It represents every observation of stellar positions ever made before modern telescopes made it possible to see that a few stars move with respect to the rest. Nothing could be simpler. Why the stars are on the sphere is not explained.

If I think about my pencil I am creating a model. Considerations of wood, graphite, paint, and so on, precisely reproduce the appearances of the pencil, but this is a path not to be followed, for at the end of it each of us is a model surrounded by models, seated in the Garden of Philosophy, where people discuss the problems of appearance and reality. Great events may happen there but it is not the place for us. My pencil lives in my hand and I do not need a model for it; at present we are concerned with stars and atoms and other such constructions, forced on us because we have no direct experience of the things they represent.

A word of caution. "Model" is perhaps an unfortunate term, since it connotes something that can be seen, at least in the mind's eye. But to every such model belong also the theoretical principles that make it work: the movements and the principles that guide and govern it. They are part of it.

What makes a good model? It should be comprehensible, as uniformly rotating spheres are. It should save the appearances. It should be able to exist alongside other models invented for other purposes. Another way of saying this is that in the light of what we already think, it should be plausible. A model is best if it also suggests ideas for further research. Are the celestial spheres plausible? Today few would think so; there are too many ideas quietly but firmly established in our minds that conflict with them.

A model must satisfy other criteria also: it should be economical. Suppose all you want to do is to explain a single observed fact, say an eclipse of the sun. There are models that use shadows, smoke, dragons; how to select among them? But if Anaxagoras's model explains the phases of the moon at the same time as it explains eclipses, it acquires a certain authority because we cannot think of another model that so easily explains both. Its author was exiled not because his model failed to save the appearances but because it conflicted with other widely accepted models, involving the actions of gods, that had been developed for a complex of purposes, some of them undoubtedly scientific, in the remote past.

In a model one hopes for simplicity but is often disappointed, for nature is not simple. More to the point is intelligibility, but what is intelligible is a question of personal judgment. Gaston Bachelard (1949) wrote that "scientific explanation does not consist in replacing what is concrete but confused with what is theoretical and simple, but rather in replacing confusion with an intelligible complexity." This remark will be useful to think about later.

Suppose we model the moon by mounting it on a sphere that rotates uniformly. Then we have to supply two pieces of information: the sphere's rate of rotation and the direction of the axis around which the rotation occurs (because the eye does not tell us how far away the moon is, we need not specify any sizes). But the moon, when we observe it carefully, does not seem to progress evenly at constant distance from the earth. Suppose we decide to save the phenomena by mounting the sphere off-center and making it rotate unevenly. Many more numbers must be supplied: the direction and relative size of the offset and the particulars of the varying motion. If you assume enough details selected ad hoc you can fit anything.

This was the route taken, but only after many others had been tried. A hypothesis that assumes a profusion of numbers is not as attractive as one that assumes only two. Plato called for a theory that fell from the world of ideas, and that is not a world full of empirically determined numbers. In the *Timaeus* (35–36) the planets are mounted on seven circles but the ratios of their sizes are not arbitrary; they are determined from the numbers of Pythagorean harmony. Circles plus harmonic intervals plus uniform rotation are reasonably parsimonious assumptions for a planetary theory, but of course they could not save the phenomena and no such theory was found. Instead, as we have seen, circles became spheres, one inside another, and their sizes and modes of suspension had to be adjusted to fit the data; then they became circles again. How did the whole development wander so far from its Platonic program? People who study politics know how, though they may not know why. Since the time of Thucydides, historians explaining the beginning of a war trace a series of steps, each one a logical or at least a comprehensible consequence of preceding steps, and the result of each of these steps is generally to carry the parties further from their goals. The more extensive a theoretical structure is, in science, politics, or anything else, the harder it is to think about it in a new way.

ASTRONOMERS

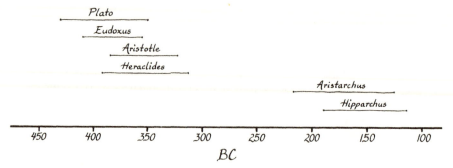

FIGURE 4.8. Time chart: The ancient astronomers.

There are good models and bad ones. The foregoing remarks suggest ways to distinguish them, but physicists and astronomers use also a much stricter criterion: a model must lead to an intelligible theory that in turn must lead to precise, quantitative predictions that can be checked later with observations. It must not only save appearances; it must predict new ones. A theory must expose itself to destruction, as Pavlov says. How to know when a theory has been destroyed is a delicate question, but scrupulous workers seem to agree that it is the proposer of a theory and not its critic who is responsible for specifying the experiments that could show that it is wrong.

Finally, one would like to feel about a model that it is not just a strategem for saving some appearances and predicting others, but that it also contains a measure of what, for want of a better word, we might call truth. But how can we know whether or not our model expresses in some way what lies hidden in the abyss? Ibn Rushd (1126–98) was a physician from Cordova whose commentaries on Aristotle were so widely used in the West that under the name of Averroes he became known as the Commentator, just as Aristotle was the Philosopher. He criticized the artificial nature of Ptolemy's orbits:

> For mathematicians propose the existence of these orbits as if they were principles and then deduce conclusions from them which the senses can ascertain. In no way do they demonstrate by such results that the assumptions they have employed as principles are, conversely, necessities. (quoted in Duhem 1969, p. 30, from Averroes' *Commentary on Aristotle's On the Heavens*)

Averroes' ideal model is one that is not merely suggested but forced on us by the facts, one from which there is no escape, and such a model could tell us something about how things really are. I know of no model or theory in fundamental science that would satisfy him. Even where we have but one model, we never know that it is the only one. To require that given a large but necessarily finite amount of experience our concepts will be uniquely determined is to ask too much. It reckons without the inner states of the mind, hints and glimpses and inclinations rooted in past history and not derived through any rational process. These imponderables sit at the center of every scientific theory. In Chapter 19 I shall have a little more to say about them.

CHAPTER 5

The Christian Cosmos

For they were ignorant of the figure of the earth, and were una-
ware that the heavenly bodies are moved in the air by angels.
—COSMAS INDICOPLEUSTES

THE fourth and fifth centuries saw the intellectual triumph of Christianity
in Europe. What happened was far more than one state religion sup-
planting another: for the first time a state religion had a coherent philosophy to go with it. Rome, as the center of ecclesiastical power, imposed
its language, and since Greek was the language of a literature whose most
famous works expressed a pagan culture there was little reason to make
translations, and knowledge of it began to fade away. Western Europe
lost most of Aristotle and almost all of Plato, and except as summarized
in a few encyclopedias the works of the mathematicians and astronomers
disappeared entirely. Latin speakers were not interested in the old questions of natural philosophy and that is just as well, since they would have
had to start again from the beginning. They were, however, interested in
the earth, the home of man, the footstool of God, the stage on which the
drama of salvation was being played. It was when they considered the
earth that people began once again to pose basic scientific questions,
slowly and cautiously at first, but as time went on mankind's natural
tendency to dig deeper led to radical questions. Do we really know that
the world is enclosed within a finite sphere? Do we know there is only
one world? Do we know that the earth is at the center and does not move?
These were not questions to which observation held the key, but in the

long run they turned out to be the ones that led to the scientific renaissance of the seventeenth century.

The Survival of Greek Knowledge

As the Christian centuries passed, Latin became the learned language. Most church fathers of the fourth century wrote in Latin or were quickly translated; by 425 Saint Jerome's Vulgate Bible was being copied and distributed and Western scholars no longer needed Hebrew or Greek. In 389 Christian monks sacked the great Greek library in Alexandria that had been reassembled after Caesar's general destroyed the first one, and in a few years what remained of the Greek texts had either been burned for their associations with paganism or had come to rest in dark places. Tastes differed however, and among the pious Christians were some who distinguished pagan learning from pagan gods and sought to preserve the knowledge that was fast disappearing.

If there was actually a last philosopher in the classical tradition it was Anicius Manlius Severinus Boethius (ca. 480–524), born while Proclus was still alive, and whose book on music I have already mentioned. He was a statesman and scholar who came into conflict with his sovereign, the Emperor Theodoric, and spent his last years in prison where he composed manuals for the liberal arts and a moving volume of ethical reflections, the *Consolation of Philosophy*, which was probably more widely read than any other book throughout the Middle Ages. In prison Boethius started a grandiose project of translating all of Plato and Aristotle into Latin, which, if he had been allowed to continue, would have changed the history of the world, but he had finished only Aristotle's logical books when the executioner came. That was in Pavia, where this last of the ancients lies near Augustine, the first of the moderns, in the solemn church of San Pietro in Ciel' d'Oro. Until the rediscovery of Aristotle in the twelfth century his translations were the basic texts for all students of logic.

Boethius wrote a book on arithmetic (1867, 1983), but it looks strange to a modern reader because except for a 10-by-10 multiplication table (in Roman numerals) it doesn't mention calculation. Until the Renaissance, that belonged to another study called logistics. Arithmetic was the study of the different kinds of numbers (odd, even, prime, composite, perfect, imperfect, etc.), the relations between them, and the harmonic and moral qualities of each. A perfect number is one that is equal to the sum of its factors; for example, 6 has factors 1, 2, and 3 and is equal to their sum. Boethius mentions 6 and also gives the next three, which are 28, 496, and 8128, and he gives the rule, from Euclid's Book 7, for finding them (1.20).

300 – 700 A.D.

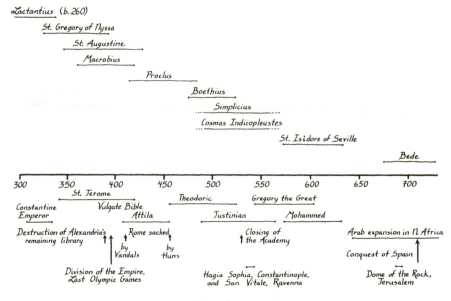

FIGURE 5.1. Time chart, A.D. 300–700.

It is one of the few nontrivial mathematical facts he quotes. Perfect numbers are associated with virtue because, like virtue, they occur infrequently.

The book *De geometria* (1867) may have originated with Boethius but not in the form that has come down to us, since that contains an account of the Hindu numerals (we call them Arabic), which in Boethius's time had not yet arrived in the West. It has been dismissed as a useless work, a list of theorems without proofs, but I think this misses the point. There were no mathematicians any more—the last one, Hypatia, had been pulled from her carriage and torn to pieces by an Alexandrian mob a century earlier—but under the Ostrogoths a gentleman still needed some geometry as part of his education, and gentlemen do not have to concern themselves with proof. The popularization seeks to convey the content of Euclid without involving the reader in technicalities he would never need.

For us, so deeply imbued with Greek intellectual values, the idea of mathematics without proofs is silly or even self-contradictory, and yet mathematicians in the other great cultures do not seem to have thought so. Egyptian and Babylonian mathematical texts give no proofs at all, and while the Chinese wrote down a few they erected no structure in which one proof stands on another. That was a Greek idea, and it now disappeared for a long time.

The flame of divine curiosity burned low in those centuries, but it did not go out. From time to time someone took up the task of making sure people still knew that there was a world and they were in it. After Boethius died the Senator Cassiodorus (ca. 487–ca. 580) retired at the end of a long and distinguished career in public service under Theodoric and his successors and founded a monastery in southernmost Italy to preserve learning as the Western Empire lurched towards its end. In his *Institutiones* (principles of education), written in about 562, he sets down brief sketches of arithmetic, geometry, music, and astronomy, most of which reappear in Isidore's *Etymologiae* sixty years later and so need no further comment here (Cassiodorus 1946). He believes that the knowledge embodied in the liberal arts is scattered through the scriptures and that the great scholars are those who dig it out and arrange it according to discipline.

Isidore of Seville (ca. 560–636), historian, bishop, and Saint, was an ecclesiastical statesman, but he is best known for the encyclopedia, the *Etymologiae*, in which he gathered together as much of the world's knowledge as had survived until his time. The strategy is original: first he explains the derivation of a word; then he tells briefly what is known about the topic referred to. A sample entry will convey the flavor:

> *Nox* (night) is derived from *nocere* (to injure), because it injures the eyes. . . .
> Night is caused either because the sun is tired from its long journey . . . or
> because the force with which it carried its light over the earth now drives it
> downward, and thus the shadow of the earth makes night. (*Etymologiae*,
> v.31)

Readers uncertain about their Latin may be assured that the derivation is wrong, and those he draws from Greek are even worse.

The *Etymologiae* follows Boethius in discussing numbers arithmetically but not logistically. Geometry is represented only by definitions, of which the following will serve as examples: "A sphere is a figure of rounded form equal in all its parts" and "A cube is a solid figure which is contained by length, breadth, and thickness." These sad mutterings are mentioned to suggest how great must have been the astonishment of scholars in the Middle Ages when Greek mathematical texts came out from hibernation.

The World and the Bible

"I would know God and the soul," Augustine's reason says to his soul. "Nothing else?" "Nothing else at all" (1948, vol. 1, *Soliloquies*, 1.6). The

saint was still young. As the years went on his finely honed intellect sliced
into a wide range of problems, but in those days the belief that salvation
involves some tremendous mystery that must be unraveled with all our
powers began to push aside the old philosophies and finally buried much
of the learning that encouraged them. This rejection in its extreme form
is typified by what Bishop Lactantius Firmianus (ca. 250–ca. 325), the
"Christian Cicero," wrote about studies of nature:

> For to investigate or wish to know the causes of natural things—whether the
> sun is as great as it appears to be, or is many times greater than the whole of
> this earth; also whether the moon be spherical or concave; and whether the
> stars are fixed to the heavens, or are borne with free course through the air;
> of what magnitude the heaven itself is, of what material it is composed . . .
> to wish to comprehend these things, I say, by disputation and conjectures, is
> as though we should wish to discuss what we may suppose to be the char-
> acter of a city in some very remote country, which we have never seen, and
> of which we have heard nothing more than the name. (Lactantius 1871, III.3)

Of course he was partly right—results achieved by brute speculation do
not stand up very well, but it is for speculation itself, the activity of speculation, that Aristotle is more remembered than Lactantius. And even if
he did not believe in speculation, Lactantius held a definite idea as to the
shape of the earth. Obviously, it is flat (III.24), and there were only a few
church fathers, among them Augustine, who did not think so.

The works of the old geographers lay largely forgotten and their maps
were lost. Except perhaps in a few quiet monastic cells, no branch of
study was esteemed or investigated for itself; everything it revealed was
taken as a symbol of some theological reality, and all that was required
of a map was that it should locate the continents and peoples of the world
so as to illustrate eternal truths. Figure 5.2 shows a ninth-century map
from Strasbourg. East, the direction of Paradise and Jerusalem, is at the
top. The vertical stripe represents the Mediterranean; *Flumen Tanai* is the
River Don, and the copyist has mistakenly extended the label onto the
right-hand branch, which should be the Nile. The flowing loop of
Oceanus, mentioned in Homer and Hesiod, surrounds the whole, and
while Friesia is mentioned, Britain is not. The three great land divisions
are as they were distributed to Noah's descendants: Asia to the sons of
Shem, "Affrica" to those of Ham, and Europe to those of Japheth. The
waters unite to form the cross that symbolizes dominion over the world.
Please consider the significance of this: When God created the world he
fashioned it in the form of a cross in anticipation of Christ's suffering and
the mode of his death.

Saint Isidore thinks the earth is a flat disk at the center of a spherical
heaven. "The sphere of heaven is a certain surface, spherical in shape. Its

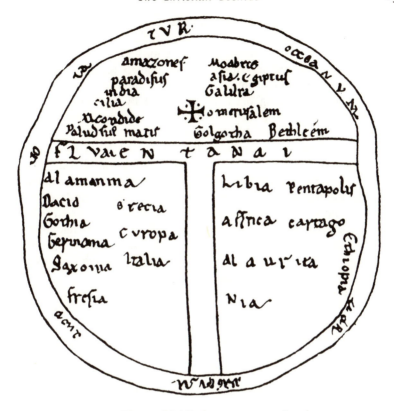

FIGURE 5.2. The world: Ninth-century map, Strasbourg.

center is the earth and it is shut in equally on all sides. They say that the sphere has neither beginning nor end; since it is round like a circle its beginning and end cannot easily be seen. . . . The sphere revolves on two poles . . . and with its motion the stars fixed in it circulate from east to west." It is the voice of a child reciting something it does not understand. But then: "Heaven has two gates, east and west, for the sun issues from the one and retires into the other" (III.32, English in Brehault 1912; see also Kimble 1938). Collecting the old-time learning did him no good. It meant nothing to him, if he did not understand that the revolving sphere explains the rising and setting of the sun and that gates belong to a different story altogether.

In the early Middle Ages it was generally agreed that the larger universe consists of ten spheres: those of the earth, the seven planets, the fixed stars, and the Prime Mover. The whole is illuminated by the sun; even the stars shine only by reflected light (e.g., Isidore III.61). The universe is full of the sun's light all the way out to the eighth sphere, and it must therefore not be so big that the light would be faint at that distance. The darkness of the night sky is an optical illusion: we look out into the cone of the earth's shadow where only stars and and planets reflect the sun's light. Everything else is dark. "Since the sun moves and the earth is stationary, we must picture this long, black finger perpetually revolving like the hand of a clock; that is why Milton calls it 'the circling canopy of night's extended shade' (*Paradise Lost*, III.556)."[1]

From the heavenly mechanism, influences rained down upon the world (*in fluere*: to flow inward; hence, for example, *influenza*). Nobody doubted that the stars influenced physical objects: metals, plants, animals, the human body. A doctor chose and timed his procedures with an eye on planetary positions. Any claim that stars really control human actions contradicted the postulate of free will on which religion based its view of moral responsibility, and religion did not allow it. Nevertheless the influences, just like a bad cold, could affect one's judgment; the wise person knew this and acted accordingly.

I have mentioned that each element had its sphere: Earth, Water, Air, Fire, Ether. The sphere of fire extended as far as the moon's orbit. Its lower boundary was not very high, for in the poetic exaggeration of Ariosto's *Orlando Furioso* (published in 1516), when Ruggiero and the Tartar King Mandricardo leveled their spears at each other and charged,

> As Turpin tells, the splinters upward flew.
> They soared into the air and even higher,
> And two or three dropped burning from the blue,
> For they had risen to the sphere of fire.
> *Orlando Furioso*, XXX.49

The whole huge space inside the starry sphere was filled with conscious and active beings: angels (*angelos* means messenger) and demons and countless other spirits. Demons started out as good spirits, but by the late Middle Ages almost all were malevolent, messengers of Satan just as angels were messengers of God. Every one of these beings had functions to perform. Arthur Lovejoy (1936) stated what he called the *principle of plenitude* to describe this assumption of the medieval world: that every ecological niche is occupied. Water has fish in it, air has birds, and the ether above the moon, filling a vast region of space, was inhabited by a correspondingly vast hierarchy of beings, all active, all with some part to

---

[1] Lewis (1964, p. 112). This splendid book vividly recreates the medieval world as it is reconstructed from literary sources. It should be remembered, though, that at this time and to an even greater degree later, philosophers were debating and contesting every point of this description (see also Bernardus Sylvestris 1973, Tillyard 1945, Lovejoy 1936).

play in the divinely ordered scheme (Bernardus Sylvestris 1973). Compared with this cosmic world mankind fails to impress, and commenting on the scriptural text, "Behold, the nations are as a drop of a bucket," Saint Jerome remarks that "the number of people in the world is as nothing compared to the angels and celestial agents" (1963, p. 461), and we shall see presently that Sir Isaac Newton thought likewise.

In this vast cosmos nothing, absolutely nothing happened by chance. A leaf lying on the ground was there for a reason, playing its tiny part in God's plan for the world. "Even the very hairs on thy head are numbered" (Luke 12:7) was in no way an exaggeration. The world was the stage on which the battle between God and Satan was being fought, and in the last days, which no one thought very far off, after the final accounts had been rendered and paid, it would cease to exist.

The High Middle Ages

So much for the surrounding cosmos, but what about the earth itself? Of course, it was the center of the spheres and, metaphorically, God's footstool. With respect to the heavenly spheres it was the lowest point, made of the grossest matter, the abode of change, misery and sin—and yet, paradoxically, the entire system had been created for the sake of its inhabitants. The sun gave it light; the constellations provided seasonal signs for agriculture and bearings for navigation; the planets kept time and their influences exerted God's pressure on human events. But to the people who lived in it, what was in the world and what did it all mean? José Gaos (1973) suggests that we look around the cathedral of Chartres, for the world is there, and he enumerates: Christ in majesty, his sixteen forebears, the twenty-four elders of the Apocalypse, the twelve Apostles, the four Evangelists and their symbols, the twelve signs of the zodiac, the occupations of the months, the seven liberal arts represented by ancient worthies (Aristotle as Dialectic, Cicero as Rhetoric, Euclid as Geometry, Ptolemy as Astronomy, and so on). There are a chorus of angels, the Virgin and child, histories from the Old and New Testaments, the life of the Virgin, thirty kings and queens of Judea, Joseph, Solomon, the Queen of Sheba, and Balaam on his ass. There are twenty-four martyrs and their martyrdoms. On the capitals there are reliefs representing agriculture, metallurgy, medicine, architecture, music, painting, philosophy, and magic. There are the three Theological Virtues, the four Cardinal Virtues, and the seven deadly sins. There is God the Father, the Creation of the World, the Fall of Man, the Crucifixion, the Last Judgment, and the Resurrection. The entire cathedral is in the shape of a cross.

There is more, but why list so much? Because this, for ordinary people, was the world. Note what is not there: secular history, for example. History concerned the Old Testament, Christ and his Mother and his church, the acts of God, the life of Man, and nothing else. There is nothing about the stars or the earth and its geography, nothing about the physical world. There is no representation of the heavenly spheres, God's largest handiwork (and it must have seemed puzzling that their construction was not mentioned in Genesis). The universe represented in Chartres required immense intellectual efforts to comprehend its nature and function, the connection of its parts, and (to the extent of human abilities) its purpose, in order that the church, the state, and other human ordinances could cooperate in the divine plan. A knowledge of natural science might possibly help in one or two details of this task, but it was of small importance overall. As long as the world was seen as a system planned and constructed with a single purpose, it was the general scheme and not the specific details that lived in the imagination, and all the wearying dialectic of the schools was finally aimed at elucidating as much of it as human beings could grasp. Since every detail of the created world was understood as relating to the divine ordering of things, layer after layer of meaning could be discovered even in its smallest parts. Emile Mâle quotes from Hugh of Saint Victor (d. 1141):

> The dove has two wings, just as for the Christian there are two sorts of life, the active and the contemplative. The blue feathers are the thoughts of heaven. The subtle gradations of the rest of the body, the changing colors that remind us of a rough sea, symbolize the ocean of human passions in which the church sails. Why does the dove have eyes of a beautiful golden yellow? Because yellow, the color of ripe fruits, is the very color of experience and maturity. The yellow eyes of the dove symbolize the wisdom with which the Church regards the future. And finally, the dove has red feet because the Church proceeds through the world, its feet in the blood of the martyrs. (Mâle 1984, p. 34)

The dove further represented the church as the visible symbol of the Holy Spirit and, having brought to Noah the olive leaf, symbolized safety and salvation and a better world to come.

We have seen how the encyclopedists of the early Middle Ages scraped together some pieces of what their Classical predecessors knew and wrote. The thirteenth century brought a flood of translations of Greek and Arabic philosophy, science, and mathematics, to be understood and adapted to Christian ways of thinking, and from this effort emerged a great encyclopedia that displayed the breadth and richness of the highest European culture of its time. Vincent of Beauvais (ca. 1190–ca. 1264) was a French Dominican scholar, eminent enough so that the royal library was open to him, and out of it he pulled a compilation of six thousand folio pages that he called *Speculum majus*, the Greater Mirror (Vincent 1624, 1964). It was divided into the Mirrors of Nature, Knowledge, and History, and someone later added a fourth, the Mirror of Morals, which is essentially a condensation of Saint Thomas's *Summa theologiae*. The only earlier work remotely comparable to it is Pliny's. Vincent's focus is the continuity and interconnectedness of the Christian world view, but there are hundreds of pages on science and history, carefully documented, and he quotes the latest developments in philosophy from Hebrew and Arabic sources as well as what suits him from the Romans and the Greeks.

The *Mirror of Nature* is organized around the Creation, and each day's work provides the subject for books in which the vast pastiche of summaries and quotations is arranged. Vincent starts with the first actions of the Trinity and the powers and hierarchy of the angels. Then, with "Let there be light" and the creation of the sensible world, there begins a twenty-chapter discussion of the nature and qualities of light. Aristotle says it cannot be a substance because if it were, when light fills a room there would be two substances, light and air, occupying the same space at the same time. But Augustine says that this is not a real difficulty—one can mix two liquids together in a bottle. Albert argues that light is a substance that descends from the sun (its natural motion being downward) and bounces off the surface it hits, just as a material object would. But John Damascene says that light is the quality of fire and Vincent is inclined to agree that the best opinion places light among the accidents, not the substances. We do not in fact ever see light move, and (physicists please note) there is no reason to think that the light reflected from a surface is the same as that which arrived at it.

After the long debate between authorities Vincent goes on to discuss the formation of images by mirrors, the phenomena of color, and the four elements; then he is ready for the work of the second day. The creation of the sun and stars introduces a discussion of time that reaches a conclusion not very different from Augustine's, time began at that moment. Natural phenomena of thunder, wind, and rain lead on to plants and animals and finally Man, whose anatomy, physiology, psychology, and illnesses are set forth in several chapters. Book xv deals with astronomy and presents a competent summary of the late medieval cosmology which, as we have seen, mixed Ptolemy's circles with Aristotle's spheres.

History in the *Speculum historiae* is mostly religious, but it transcends the history in Chartres by listing some early kings of England, and Book xxiv relates the exploits of Charlemagne. He appears as a mixture of fact and legend—eight feet high, great in arm, thigh, and belly, with beard a foot long and nose half as much, and able to raise an armed knight above his head on his open palm; yet withal gentle, temperate, devout, and something of a scholar. A rich and wonderful book; there ought to be an English version.

The cathedral of Chartres spoke its message to all who came in, mostly the unlettered poor. Vincent spoke to those who knew Latin, most of them clerics. A third level of instruction arose also, books compiled in the vernacular languages for the literate of the towns, aware of the ferment of new knowledge in intellectual circles and hoping to broaden their own horizons. Of these, an interesting example has recently been published (Anon. 1980). It is in the form of dialogues between a master, Timéo, and his chosen student, Placides. Placides is the son of a minor ruler somewhere, and we are told that his choleric disposition makes him cheerful and eager to learn. Timéo is not further identified, and there is no suggestion of an era. The dialogues fall into four main sections: the world, its creator, and man's place in it; human and animal reproduction (subjects of immense curiosity and interest in those days); meteorology; and human laws and civilization.

In explaining the physical world Timéo carefully distinguishes substance from accident: substance can exist by itself but accident cannot; the distinction is essentially Aristotle's. Substance has body in the four elements and the living and inanimate things formed from them, while substance without body belongs to celestial powers, spirits, and souls. Incorporeal substance does not occupy space, so that four thousand souls, for example, can fit into a small pail.

"Master, in the name of God," said Placides, "You who speak of body, tell me what is body." "Willingly, my son. Body is everything that can support something and can fill a vessel and if it strikes something else makes a sound." Earth and water clearly pass these tests; air supports birds and if one strikes it with a long rod it makes a sound. Fire is less clearly dealt with.

Anything that moves by its own volition has a soul. Placides asks whether the elements have souls. Timéo answers that they do not, but Placides protests that water flows toward the sea. "Placides," says Timéo, "he who asks well wants to know well. As Aristotle says, it is never unprofitable to doubt, but always profitable. . . . Just as I have told you that the earth on which we stand is not the element earth, so also the springs and rivers and the fire we use are not elemental but take their nature from the element." The sea is purer in the element than is the fresh water of springs and rivers, and the fire of the sphere overhead is purer than the fire on the hearth. "Earth, air, water, and fire do not move of themselves but each tends toward its nature and its being." The downward flow of rivers and the upward motion of a fire inhere in the nature of elements seeking their right places and are not matters of volition. Thus Timéo distinguishes natural motion from motion produced by will and begins to show his pupil that the various physical and vital processes going on around him all manifest God's great plan, complex in its working out beyond our capacity to understand, but simple and logical in its central conceptions. Faith is a tree with many roots, but this was to become one of the solidest.

A Thousand Years

After Aristotle, philosophy began to turn its searchlight away from nature toward religion—first pagan, then Christian. The Academy kept Plato's teachings alive until it was closed by an edict of the Emperor Justinian in 529, which forbade pagans to teach. Invited by King Chosroes I of Persia, most of its members went off to Ctesiphon to learn a new language and translate the old texts. The enterprise was soon swallowed up in a conflict of cultures. The historians Procopius (1935) and Agathias (1975) describe the bloodthirsty tyrant who fancied himself an intellectual. Simplicius of Cilicia was one of the philosophers who made the trip. Agathias (II.31) tells the story and describes the barbarity and licentiousness that made them long for home. "They felt that merely to set foot on Roman territory, even if it meant instant death, was preferable to a life of distinction in Persia." In the event, however, it did not mean instant death, for Chosroes wrote into a treaty with the Romans, and the Romans agreed, that the philosophers should be allowed to return to Athens and finish their lives in peace. So ended the Academy, the world's first university, its oldest, and its most distinguished. Its influence, however, continued to spread, for alumni went on to Constantinople and opened schools there from which in due time Platonic doctrines came to the Arabs and thence to Europe, so that even medieval scholars who had never read a word of Plato had a sense of his reputation and what he taught.

But Platonic teachings were only a small part of the flow of knowledge into the Islamic world. In the ninth century Harun al-Rashid established a brilliant court at Baghdad that began to attract scholars and continued for two hundred years, while the outward rush of Islam carried Syrian learning to India and in the opposite direction into Arab domains as far as Cordoba and Toledo. The main philosophic texts were translated from Greek to Syrian, and thence to Arabic. In the twelfth century European scholars traveling in Spain were astonished to find people reading books of which they had barely heard. Now, after learning Arabic, they could own and study them. Adelard of Bath rendered Euclid into Latin in about 1120, and a few years later Gerard of Cremona established a workshop that turned out one translation after another. By 1175 it had produced Latin versions of most of Aristotle as well as an *Almagest*. For many of these texts this was their fourth or fifth time on the translator's desk and much of their meaning had become obscure, but the excitement of their discovery sent students rummaging in the monastic libraries of Europe where a few Greek manuscripts were found, and during the fourteenth century many more began to arrive from Constantinople. After mid-century there were professors of Greek in Italy and soon the study of Greek became the absorbing passion of hundreds, perhaps thousands of literate Italians. There was never any shortage of people who knew the language—after all, there were Greek merchants in the streets and Greek monasteries in Palermo and Rome; there were (and are) Greek-speaking towns in Calabria, and Greek was the living language of millions at the distance of a few days' sail. What was new was the interest in the classical past. Before Constantinople fell to the Turks in 1453 a great part of its textual treasures had been bodily moved to Italy and were being edited and studied. By 1498 a complete Aristotle was in print, and Plato followed in 1513.

What had the Syrians and Arabs done with Greek scientific knowledge during the centuries when they were its sole possessors? They made great progress in medicine and physiology and some in physics and mathematics, and they developed the logical works with care and exactness. They admired Plato, idolized Aristotle, and despised the astronomers who had tried to smash his spheres. In the eleventh century the philosopher Ibn al-Haitam (ca. 965–1038), known in the West as Alhazen, labored to produce a constructible cosmic model free from the difficulties of Aristotle's. Others followed him, but it is not necessary to describe the structures used, for they originated not in any new conception of the world but in uncritical admiration of the philosopher who, Averroes wrote, "founded and completed logic, physics, and metaphysics. . . . I say that he completed them because no one who has come after him up to our own time, that is, for nearly fifteen hundred years, has been able to add anything to his writings or to find any error of any importance in them" (*Commentary on Aristotle's Physics*, quoted in Duhem 1969, p. 29).

Even with the newly translated texts, the European establishment was no more ready than the Arabs had been to look for new ideas. Still, it was good at comparing and combining verbal formulations, and soon it had produced a version of astronomy that combined Plato, Eudoxus, Aristotle, and Ptolemy, a structure of spheres and epicycles that has become known as the Ptolemaic system, though Ptolemy would scarcely have recognized it. Few people thought critically about it. Thomas Aquinas, in his commentary on Aristotle's *Metaphysics* (XII.10), objects to the system on doctrinal grounds. Aristotle defined three classes of natural motion: toward the center of the earth, away from it, and in circles around it, but epicyclic motion is none of these. Later, commenting on *On the Heavens* (1.17), he complains that since a planet mounted on an epicycle mounted on a sphere would have to pass through the sphere every time the epicycle turned, the spheres cannot be continuous, but that when Aristotle described the spheres he mentioned no discontinuities. For Thomas and his followers astronomical truth issued from Aristotle in three streams: the principles of physics, those of cosmology, and the logic that makes deductions from them. This does not mean that arguments based on observation or plausibility are to be dismissed. They have a dignity of their own and may even be necessary to guide one's steps toward truth, but such knowledge, however much of it one has, is not the same as truth. Truth is not made by man; it is the shining palace in the sky toward which we climb with patient dialectic.

Where Is the Earth?

It was Copernicus and his followers who made the most direct, specific, and publicized refutation of the Aristotelian picture of nature, but he did not come like a bolt of lightning. For two hundred years people had been publishing speculations that went beyond Aristotle, questioning the hypotheses on which Aristotle and all who followed him had based their theories of nature. In the next chapter we shall see the church striking out against doctrines that attacked its essential beliefs and policies, but orthodoxy in questions of physics and cosmology was considered far less important. When we remember Galileo's trials and punishment for teaching that the earth moves, we might suppose that the church savagely opposed the idea. This was true in the seventeenth century, when Europe was torn by the Reformation and the Thirty Years' War, but earlier, before the troubles began, the idea could be discussed freely, even in front of a pope.

In the fourteenth century Nicolas Oresme (ca. 1330–82), Bishop of Lisieux, mathematician, economist, translator and commentator on Aristotle, questioned whether the earth stands still. He wrote: "If a man in the heavens, moved and carried along by their daily motion, could see the earth distinctly and its mountains, valleys, rivers, cities, and castles, it would appear to him that the earth is moving in daily rotation, just as to us on earth it seems as though the heavens were moving" (Oresme 1968, p. 523). Oresme shows with careful arguments that absolute motion cannot be inferred from observations of relative motion and he disposes of the classic arguments against a rotating earth. But he mentions these considerations only to exercise reason in defense of faith, for we know from scripture that Joshua caused the sun to stand still, and the psalmist

The Renaissance In Science

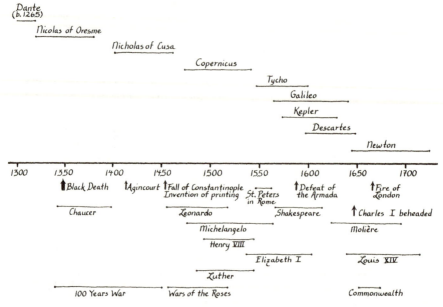

FIGURE 5.3. Time chart: The renaissance in science.

wrote "the world also shall be established that it shall not be moved."
The cat was still in the bag, but it was looking out.

In about 1440 the German bishop Nicholas of Cusa (1401–64) wrote
a remarkable book called *On Learned Ignorance*. The theme of this original but difficult text is that most of so-called knowledge is only conjecture, and that to admit it is to be wise. I use the word "knowledge" in its
scholastic sense, as conclusions reached through correct logic from
undoubted premises. Learned ignorance is Nicholas's name for his brave
attempt to escape that swamp, not by clever footwork inside it, but by
avoiding it altogether. In his discussion of cosmology (Cusanus 1981,
II.11, 12) he writes that the earth cannot possibly be at the center of the
universe, for the universe is unbounded and therefore has no center. (It is
not infinite; only God is infinite. How the universe can be unbounded but
not infinite Nicholas does not explain, though such a geometry is not
impossible and in fact is popular among cosmologists today.) "Hence the
world-machine will have its center everywhere and its circumference
nowhere, so to speak; for God, who is everywhere and nowhere, is its
circumference and center." It was a famous medieval metaphor (Harries
1975) that God is a circle whose center is everywhere and circumference
nowhere. Here Nicholas surprises us by turning it into a literal statement

---

about the universe. If there is no center, then (if I understand his argument) there is no rotation about a center, and so one cannot meaningfully say either that the starry sphere rotates or that the earth stands still.

Nicholas was not the first to point out that an infinite universe can have no center. In Plutarch's *Moralia*, which Cusanus may have known, the narrator in the dialogue "On the Face in the Moon" makes the same point: "After all, in what sense is the earth in the middle, and in the middle of what? The sum of things is infinite; and the infinite, having neither beginning nor limit, cannot properly have a middle; for the middle is a kind of limit too, but infinity is a negation of limits" (Plutarch 1957, vol. 12, p. 77).

Rushing from topic to topic, Nicholas impatiently questions the unargued assumptions of two millennia regarding the universe and our place in it. Like Oresme, he emphasizes that absolute motion cannot be detected, for the appearances are the same in either case. "Thereupon you will see—through the intellect, to which only learned ignorance is of help—that the world and its motion and shape cannot be apprehended." Apprehended by finite minds, that is; but for himself, he prefers to think that the earth is moved. One can see that distrusting the dialectic of the schools he is at every moment eager to think his way toward new conclusions; he is eager to say new words but except for his epistemology he has not found in natural philosophy anything new to talk about. He echoes the old formulas, "The most nearly perfect motion is circular; and the most nearly perfect corporeal shape is therefore spherical." He still tries to explain what things do in terms of properties that inhere in them, but we are approaching the end of that kind of talk. In the next chapters we shall see how scientific thinkers tried a new kind of explanation in terms of mechanical actions.

Nicholas is beginning to think of the earth in cosmic terms, as one of a company of similar objects and not as the center of everything. He does not qualitatively distinguish stars from planets, and he calls the earth "a noble star which possesses light and heat." If there is this one earth there may be many, and he is ready to consider the possibility that they are inhabited by beings whose natures correspond to the astrological and elemental qualities of the places they live.

"It may be conjectured that in the region of the sun there exist solar beings, bright and enlightened intellectual denizens, and by nature more spiritual than such as may inhabit the moon—who are possibly lunatics— while those on earth are more gross and material" (1981, II.12). This and many other passages in the book show the author at play. Needless to say, such speculations were considered unnecessary in ecclesiastical circles on account of the questions they raised. Were these other beings also descended from Adam, and if not, who were they? Did Christ save only this world or must he offer the redemptive sacrifice in each? Nevertheless, the brilliant son of a poor Moselle boatman became and remained a cardinal of the church.

Like the infinite universe, the plurality of worlds was even then an old theme. According to Diogenes Laertius (1925, IX) and other commentators, Leucippus and Democritus believed in it, and Lucretius in *De rerum naturae* (II.1045) writes of other worlds formed like our own from random aggregations of atoms and inhabited by races of animals and men. Today many people find the suggestion plausible and there are systematic efforts by astronomers, who realize the immense odds against them, to intercept radio signals that extraterrestrials may be sending in our direction.

CHAPTER 6

What Are These Things I See?

But seven and three make ten not only now but always. In no circumstances have seven and three ever made anything else than ten, and they never will. So I maintain that the unchanging science of number is common to me and to every reasoning being.

—AUGUSTINE

THIS chapter will trace the survival of Greek ideas into the Middle Ages with special attention to mathematics and atomic theory. We have seen that the atomic theory may have arisen from attempts to devise a logically sound way of talking about the world just as much as from a desire to explain natural phenomena. Aristotle's physics also is a carefully argued theory that starts from explicit assumptions and accounts for most of the phenomena we notice in the course of a day. As such it helped people make sense of the world, but since it confined its attention to the kinds of things and processes we ordinarily perceive, it had no place for atoms. Therefore the philosophical doctrine of atoms survived mostly among antiquaries and encyclopedists, people who saw the intellectual treasure of Classical times evaporating and hoped to preserve some of it. We shall see Isidore and Bede writing about atoms in terms that suggest they had some knowledge of Epicurean philosophy but no interest in the Classical arguments either for or against it.

As the years went by, the Christian world developed a strong philosophic tradition of its own. Guided by Aristotle's logical books and by his conviction that knowledge is built of necessary conclusions, the phi-

losophers of the Middle Ages found it very hard to build a natural philosophy. They knew some mathematics but not how to apply it scientifically.
Tight reasoning required syllogisms, and though geometry could be presented that way, numerical calculation could not, and its role in the construction of truth therefore remained obscure. Not until Chapter 8 will
we see the beginning of a just appreciation of its uses.

Atoms in the Dark Ages

Saint Isidore knows about atoms: "The philosophers call 'atom' certain
parts of the bodies in the universe which are so very small that they
cannot be seen nor do they admit of *tomen*, which means division,
whence they are called atoms." Excellent, and the etymology is essentially
correct. Then:

> These atoms are said to flit continually through the emptiness of the universe,
> carried here and there like the fine dust that is seen when the rays of the sun
> pour through a window. Certain heathen philosophers maintained that these
> produce trees and plants and all fruits and fire and water—everything is made
> from them. (Isidore 1911, XIII.22)

If a reader finds it hard to visualize a tree as made from particles like fine
dust moving here and there, it is probably because Isidore does not try to
do it. As when he described the earth in Chapter 5, he is repeating verbal
formulas. But note that Isidore has no trouble with "the emptiness of the
universe." Either the philosophical arguments pro and con empty space
have not reached him or he does not think they are important. The passage continues in terms that do not merely echo ancient words:

> There are atoms in a material body and in time and in numbers and in letters.
> You can divide a body like a stone into parts, and the parts into grains like
> sand, and again the grains of sand into finest dust, continuing, if you could,
> until you come to some little particle that you cannot divide or cut. This is
> an atom of the body.
>   An atom of time means this: You divide a year, for example, into months,
> the months into days, the days into hours, and you can still divide the hours
> until you come to an instant, a droplet of time as it were, so short that it
> cannot be even slightly lengthened [*sic*], and therefore cannot be divided.
> This is an atom of time.
>   There are atoms of numbers: for example, eight is divided into fours, four
> into twos, and twos into ones. One is an atom because it cannot be divided.
> It is the same with written language, for one can divide speech into words,
> words into syllables, and syllables into letters. The letter, the smallest part, is

the atom and cannot be divided. The atom is therefore what cannot be divided, like the point in geometry. (XIII.2.2–4)

Isidore briefly mentions the four elements, which

are said to be related in a natural way . . . so that Fire fades into Air, Air condenses into Water, Water solidifies into Earth—and again, Earth dissolves into Water, Water rarifies into Air, and Air into Fire.

Thus all the elements are present in everything, and each receives its name from the substance that contains most of it. And divine Providence has assigned appropriate living beings to each element; for the Creator himself filled the heaven with angels, the air with birds, the sea with fish, and the earth with men and other creatures. (XIII.3.2–3)

In Isidore's mind the atomic theory is tarnished by the church's opinion of the atomists:

Epicurus, a certain philosopher, [was] a lover of vanity and not of wisdom, whom even the philosophers themselves called a swine because he wallowed in carnal filth and declared that bodily pleasures were the highest good. . . . He assigned the origins of things to atoms, . . . from whose chance combinations all things arise and have arisen. He said that God does nothing, that everything is made of particles, and that in this respect the soul is no different from the body. And so he said "I shall not exist after I die." . . . The same material is used and the same errors are repeated over and over by heretics and philosophers. (VIII.6.15–23)

Saint Isidore was a member of the Visigothic nobility and one of the most powerful men in Spain. If his ideas seem simple we might ponder what some of today's statesmen might have to say on these same subjects.

We know very little of the life of Saint Bede, or Beda, or Baeda Venerabilis (ca. 673–735), as he was called, except for the few lines of autobiography he appended to his *Ecclesiastical History of the English Nation*. Born in the territory of the monastery of Jarrow, near the mouth of the River Tyne, "When I reached the age of seven I was given by my family first to the most reverend Abbot Benedict and later to Abbot Ceolfrid for my education. I have spent the rest of my life in this monastery and devoted myself entirely to studying the Scriptures. And while I have kept the regular discipline, singing every day in the choir, it has been sweet for me to study, teach, and write" (Bede 1930, vol. 2, p. 382).

The *History* is a beautiful book, thoughtful, judicious, and full of faith, and from it one can sense the tranquility of the man who wrote it and of the place where it was written. Bede knew Greek and perhaps some Hebrew, and the endless copying of manuscripts that went on in the monasteries of the time had provided his library with texts of Vitruvius's *De architectura*, Pliny's *Historia naturalis*, the *Etymologiae*, English historical records, some fragments of the classical authors, and of course mountains of theology. For almost a thousand years Bede's works were at the center of English education, though the age in which he lived is commonly called dark.

Bede wrote little on science, but a few lines will suggest the scientific thought of his age. In his short pedagogical dialogue *On the Divisions of Time* he cites Isidore as his authority on corporeal atoms but goes further.

> STUDENT. How many kinds of atoms are there?
> TEACHER. Five.
> STUDENT. What are they?
> TEACHER. Atoms in matter, in the sun, in speech, in number, and in time.

The teacher explains atoms in matter. Then,

> TEACHER. Atoms in the sun are the little motes of dust that we say are moved by the sun's rays.

The atoms of speech are letters, and the atom of number is unity.

> STUDENT. What is the atom of time?
> TEACHER. It is thus: Divide a moment into 12 parts and each of these into 47 parts. The 564th part of a moment is an atom of time. (Bede 1563, vol. 1)

There follows a long discussion relating the duration of a moment to the motions of the sun and moon, which we need not follow. As far as I can make out, an atom of time is about 1/6 of our second, and therefore of the order of the briefest sounds that we distinguish in speech.

All this is very speculative, but Bede also contributed some fresh observations concerning tides. Living at a river mouth, he knew that it is important for a ship's captain approaching a port to know the state of the tide there. For a long time it had been known that the periodicity of the tides is the same as that of the moon, but that is not enough. One needs to know how much time elapses between the moment the moon passes the meridian, the imaginary line in the sky that runs north-south, and the moment of high tide. Bede found that this interval varies widely from port to port but is essentially fixed for each. Thus by observing the moon and knowing this single number for the port he is approaching, the captain knows when he can make land. Collecting these numbers is known as the establishment of ports, and Bede seems to have begun it.

Isidore's life spanned that of Mohammed. By Bede's time the Moslem empire stretched from China to the Pyrenees, having overthrown a hundred ancient sovereignties, including the one that Isidore had served. Writing the history of England in his peaceful valley, Bede mentions none of this. As the centuries rolled on, the outside world figured less and less often in the works that came out of the monasteries. Less time was devoted to arranging old knowledge in new ways and more to simple copying. I shall mention only a few of the brightest figures of this period.

The Early Middle Ages

Plato's Ideas, Aristotle's logical physics and metaphysics—I have called them myths because each appeals to the imagination and opens up a way of thinking and a vision of understanding that is not rigorously justified by anything in either. Then came the Christian myth which, starting from a literal reading of scripture, offered to make sense of everything if only the seeker would first believe. In the second century, Tertullian wrote "I believe because it is absurd," meaning of course that where everything is perfectly plain there is no exercise for the muscles of belief. Saint Anselm, nine centuries later, wrote "I believe in order that I may understand." A new myth containing such paradoxes was bound to collide with the older ones that claimed to be lighted by reason.

The situations of Plato and Aristotle in this great game were somewhat different, since Plato's works were well known to Augustine and his immediate successors, while most of Aristotle's had already dropped out of sight. After Bede, almost no one in the Latin monasteries knew Greek until the thirteenth century. A handful of Greek texts were available in Latin translation: the first part of the *Timaeus*, Aristotle's *On Interpretation* and *Categories*, and the *Eisagoge* (Introduction), a commentary on the *Categories* by Porphyry, a Greek scholar of the third century who was probably born in Tyre. At the start of the *Isagoge* (as it was known in the Middle Ages) Porphyry makes an innocent comment that focuses the mind for a moment on the question of universals, terms or concepts referring to classes of things and the qualities by which things are included in such classes. In the Middle Ages, when belief was making great demands on reason, the study of logic had become especially urgent, and people were being required to understand and accept as true a variety of propositions, such as those relating to the nature of God and the Trinity and Original Sin, that were based neither on reason nor on sensory experience.

Porphyry formulates the problem of universals in terms of genera and species. (Roughly translated into English, a genus is the form or idea that defines a class of things and a species is its visible manifestation. Two members of the same species have the same formal cause but different material causes.)

> For the present I shall not speak about genera and species, as to whether they really exist or are simply posited in an intellectual sense—and if they do exist, whether it is corporeally or incorporeally, and whether they are separate from or related to objects of sensation. These are very deep questions. (*Migne* 1844, vol. 64, Boethius I, p. 82)

Suppose we are confronted with a child behaving in such a way that we say it is happy. On reflection, we see that we cannot make this judgment unless we have some prior idea of happiness. Now what kind of idea is it, and where did we get it? Did we get it from experience or did we learn about it when we learned the word? Do you and I each have such an idea, or is there just one of it? Does it cease to exist when we sleep? When we die? Or has it an independent existence? Porphyry offers three choices: the universal may exist *in experience*, as a word used to express the qualities of something we are thinking about at the moment, or else it may be said to have a more independent kind of existence. In that case we have two choices: the universal may exist *prior to experience*, with an existence of its own, even if no one is thinking about it or has ever thought about it (Plato's view), or else it may exist *posterior to experience*, deduced from our experience of specific things and events (Aristotle). During the Middle Ages, in general, the devout inclined toward universals prior to experience and were called realists because they believed that the universals are real, and in fact the ultimate reality, while more skeptical people considered that universals were posterior to experience. This view was called nominalist, meaning that for them the universal, while still real, was identified by names derived from real experience.

As it turned out, the passage from Porphyry was quoted a thousand times, and the question of universals fermented for centuries in the philosophy that was taught in universities and theological schools and is called scholastic philosophy. The question was never settled; of course it could not be settled. It relates to the axioms of thought that are assumed in order that reasoning may begin, and so it cannot be argued true or false, nor is there any experience that decides it. Such questions usually die after a while. The reason this one did not is, I think, that it happens to be important, and in our day it is just as important as it ever was in the Middle Ages, though in other contexts.

Consider, for example, a question whose answer today affects both the foreign and domestic politics of the United States: is a fetus a human being? Is there something called humanity that exists independently of any particular fetus and inheres in every one, however immature it may be, or is humanity a name we give to a collection of characteristics gradually acquired during gestation and childhood? To say that this is a medieval question is not to disparage it; simply the Medievals were there first, and they discussed it with rather more care than one perceives among either "lifers" or "choicers." To suggest that the question can be settled, once for all, by adopting appropriate definitions would be absurd, since we may not fiddle with the word "humanity."

The Proper Use of Mathematics

Aristotle's physics is a science of causes built on the axiom that when the causes of a phenomenon have been established the phenomenon itself can be explained by a syllogistic argument as a necessary effect of its causes. In this pattern of explanation there is no use for the mathematician,

> for in his investigation he eliminates all the sensible qualities, e.g. weight and lightness, hardness and its contrary, and also heat and cold and the other sensible contrarities, and leaves only the . . . attributes of things *qua* quantitative and continuous, and does not consider them in any other respect. (*Metaphysics*, 1061a)

A remark in the *Posterior Analytics* shows what Aristotle thought was missing from mathematics. One cannot prove from geometry, for example, whether the straight line or the curve is the most beautiful of lines (75b). Again, in the *Physics*, he mentions a fact fundamental to his theory of planetary motions: "rotatory motion is prior to rectilinear motion, because it is more simple and complete" (265a). The properties of beauty, simplicity, and completeness belong to his science, for they are fundamental to further propositions that explain the causes of things.

Mathematics has nothing to say about causes and explains only facts. If we start with the optical law of reflection, which states that when light strikes a plane mirror the angle of reflection equals the angle of incidence, mathematics enables us to explain complicated phenomena of reflection, but it cannot tell the reason behind the law. Astronomy also depends on mathematics to tell where Venus will be tonight, but the sphere-haunted axioms of astronomy were not mathematical.

A hundred years after Aristotle, Archimedes of Syracuse tried to formulate a mathematical theory of levers. His postulates are innocuous enough. The first one is that if equal weights are suspended at equal distances from a fulcrum they balance each other. This is obviously true; we know it without trying it. If the two weights were absolutely identical yet one went down and the other went up, what principle could possibly determine which went down? If they are not identical, how could their equality be established except by weighing them on an equal-arm balance? There are seven more postulates, such as "If two weights are in equilibrium and a weight is added on one side, that side of the balance will go down." These are all verifiable by experience, and all of them are qualitative, not quantitative. None of them depends on beauty or fitness.

Archimedes then sets out to derive the law of the lever, which states that a weight W_1 at a distance D_1 from the fulcrum is balanced by W_2 at a distance D_2 provided that $W_1 D_1 = W_2 D_2$. The result is true but it does not and cannot follow from these assumptions. An assumption about the real physical world had to be made, but Archimedes' proof is cumbersome enough so that one does not at once notice where it was smuggled in (Archimedes 1895, Book I; see also Dugas 1955, ch. 1). The situation is exactly as Aristotle had said it is: to derive even a simple physical law, mathematics is not enough. This is because mathematics talks only about itself, and its propositions can be used to discuss the physical world only with the help of two other kinds of statements: nonmathematical statements, expressed in words, that tell how the mathematical symbols are to be related to our sensory experience of the world around us, and mathematical propositions derived from experience as to how the world actually seems to behave, as opposed to other ways in which it might behave without contradicting what we understand of logic or mathematics. I know of no derivation of the law of the lever that deduces it from clearly defined physical assumptions before one given by Simon Stevin in 1608; this will be mentioned in Chapter 10.

Because the question of the relation of mathematics to our experience of nature is so very fundamental, allow me a trivial and boring example. I put an orange into a paper bag; then I put in another one. I have long known that $1 + 1 = 2$. In the present situation this tells me that there are two oranges in the bag, as I can verify at leisure. Next I take a test tube and put in a drop of water, then another. It is still true that $1 + 1 = 2$, but where are two drops of water? No, there is a larger drop. To apply mathematics to the world we must understand a little about the nature of oranges, paper bags, drops of water, test tubes, and many other things, and this necessity cannot be avoided by using more mathematics.

Even if mathematical proof cannot provide physical understanding, it has more to say than it says in Aristotelian physics, and the thirteenth century brought a torrent of new ideas. I shall mention only a few people, though the period was one of great intellectual activity and there are many more I would like to write about. During the preceding century the rest of Aristotle, including some works wrongly attributed to him, had been rediscovered and translated. Up until this time, Aristotelian logic had been used mostly in the service of theology, but the new books showed how to apply it to the world. Of course, every topic was studied within the Aristotelian rules. A proposition was considered scientifically true if it could be given a formal syllogistic demonstration from premises found in Aristotle, and in science the logical process was used to explain effects as necessary consequences of the four kinds of causes. Perhaps this seems to us an arbitrary and crippling method of approach, but in the beginning it swept away accumulated cobwebs and opened paths to new knowledge.

Robert Grosseteste (ca. 1168–1253)—Robert Big-Head in the Norman French used by literate English—became chancellor of Oxford University, probably the first one, and later Bishop of Lincoln. He dominated English scholarship in his time, and his main contribution to science was a long analysis, contained in commentaries on the new Aristotelian texts, of the ways in which we know about nature. The following sketch can be filled out by reading Crombie (1953).

Leaving aside for a moment the question as to whether we can have any true knowledge of nature, there are three paths to what one might call well-founded opinion: logical deduction, experimental observation, and mathematics. Of the three, Aristotelian theory gave absolute primacy to the first. It provided what was called the *demonstratio propter quid*, the demonstration that something is necessarily so, while the others merely give the *demonstratio quia*, which proves that something is so but does not prove that it is necessarily so.[1] Any credit given to experiment and observation detracts a little from that given to logical necessity, and so if they were to be advocated before an Aristotelian audience it had to be with care. In his *Commentary on the Posterior Analytics* (1.8; there is no modern edition but see Crombie 1953, ch. 5), Grosseteste argues the radical thesis that under favorable circumstances mathematics can provide a demonstration *propter quid* that has as much authority and explanatory power as a syllogism. That is, while argument can prove that a motion (i.e., a change) happens necessarily, one can also prove the same thing mathematically:

> There is an immense usefulness in the consideration of lines, angles, and figures, because without them natural philosophy cannot be understood. They are applicable to the universe as a whole and in its parts, without restriction. . . . For all causes of natural effects can be discovered by lines, angles, and figures, and in no other way can the reason for their action possibly be known. (Grosseteste 1912, p. 59, quoted in McEvoy 1982, p. 168)

[1] Aristotle sets out the distinction clearly and illustrates it with examples in *Posterior Analytics* (1.13.)

Two points can be made about this passage. First, the words "cause" and "reason" are here used in the strong sense of *propter quid*, and second, mathematics means Euclidean geometry, not numbers. In a civilization in which most people reckoned by counting on their fingers and long division required a specialist using an abacus, it was hard to see any future for calculation except in astronomical drudgery, in which its function was surely not explanatory. In a few pages I will copy a passage from Galileo that echoes Robert's words almost literally.

Later, perhaps reflecting Augustine's statement that God reasons and acts mathematically, Grosseteste makes an even stronger claim: the underlying mathematical structure is *causally* responsible for the motion that is observed (*Commentary on the Posterior Analytics*, 1.11). Even today, when we know something of the mathematical structure of the laws that govern the universe, this would be a radical statement. In Robert's time none of these laws were known except two: the law of equal angles governing the reflection of light, and the law of the lever.

On the structure of matter Grosseteste had little to say. Aristotle's arguments in favor of a continuum and against a vacuum were so strong that although some of Robert's contemporaries called him an atomist his atoms are logical rather than physical—points of zero size, infinitely numerous, everywhere touching each other. In this way he could enjoy the advantages of the Democritean answer to questions of material change and permanence without having to challenge accepted beliefs concerning the impossibility of a vacuum.

Many moderns dismiss this kind of labor and reasoning as metaphysics and empty talk. In our own time people make a little progress in science every day without worrying about problems of epistemology, whereas in the Middle Ages science was a unit and there was no question of splitting it up. In the recent past, puzzles associated with relativity and the quantum theory have driven thinkers like Einstein and Bohr and Heisenberg to immerse themselves in the philosophical tradition and write of scientific knowledge in new terms. There is no sign that it is now or will ever be possible to reduce the content of physical science to the recital of facts and laws.

Reason and Faith

In the later Middle Ages, the ultimate concerns of the schoolmen were largely theological. The conceptual apparatus that was developed to deal with them has little bearing on our subject and so I can be mercifully brief. The great event was the arrival during the twelfth century of the first wave of translations of Aristotle into Latin. Plato came more slowly. The first part of the *Timaeus*, stopping just before the theory of the elementary solids, had long been available in the fourth-century version of Calcidius (Plato 1962), who also provided a long and poorly informed commentary that included tantalizing quotations from other Platonic dialogues. For scholars who did not know how to read the *Timaeus* in the context of Platonic realism it represented the cutting edge of scientific research, and not surprisingly, they did not make much sense of it. Augustine often refers to Plato but rarely quotes him. It seems that few texts were available, he did not read Greek, and he shows no signs of knowing the Latin *Timaeus* (Klibansky 1939). Scholars sensed that the surviving fragments of Plato had a quality of exaltation that was lacking in the Aristotle they had plowed through in school, and they opened the first Latin versions of Plato with wonder and veneration. The *Meno* and the *Phaedo* were translated directly from Greek at Palermo in about 1156; other books came more slowly, and there was no complete corpus of first-rate versions before Marsilio Ficino's printed edition late in the fifteenth century.

Medieval scholars happily welcomed Greek science, hoping it would clear up questions that had puzzled people for generations. Imagine a scholar reading Genesis to find out exactly what happened at the Creation. Hard questions arise: God created light and divided it from darkness, so that there were mornings and evenings before he created the sun. Was this light physical or spiritual? If physical, what physical source was there other than the sun, and what action other than sunset made light give way to darkness at the end of the day; if spiritual, why did any darkness at all follow the first light? The early Fathers produced a series of works on the first six days, notably Augustine's *Literal Meaning of Genesis* (1982) and Gregory of Nyssa's *Explanation of the Six Days* (1960), but they left the deep questions unanswered. The puzzles remained as puzzles, but Greek science at least put them into a larger context of cosmology.

The new material gave scholars an immense amount to absorb into the fabric of their ideas, since Christian ideology had developed for a thousand years with no serious competition, and here, suddenly, was a body of highly persuasive philosophy that was not Christian at all. The Christian world was organized, outwardly at least, as a commonwealth of faith; to turn men's attention toward the weak points of theology, and to give them the intellectual weapons to do it, was seen at first as the Devil's work. In 1215 the doctors of Paris were forbidden to lecture on certain propositions in Aristotle. The ban was repeated in several times, but in the end it was futile, since between them Albert the Great (ca. 1200–80) and his pupil Thomas Aquinas (1225–74) seemed to have solved the problem of bringing Christian and Aristotelian philosophy into accord. A few points of difference remained: for example, Aristotle said that the world has existed forever, but on this he was simply overruled.

In the new synthesis there are two forms of cognition, sensible and intellectual. Though the sensible is second in importance it is first in time, for only through sensory experience do we learn enough about the world to be able to think at all. Even our understanding of logical principles is based in sensory experience. Again and again Thomas's categorical statement is quoted: "Nihil in intellectu quod prius non fuit in sensu," There is nothing in the understanding that was not first in the senses. I see a vehicle and I see that it is red. I have seen such vehicles before and I know what they do. I *understand* that I am looking at a fire engine.

To have yoked Aristotle to Christian ideology was an immense achievement on the intellectual plane, but in the resulting theory-of-everything it was hard to find the simple faith that sustained the church and most of its members, while Aristotle's immense corpus was cramped and constricted when forced to serve a foreign doctrine. As soon as the work was finished the demolition began as critics like Duns Scotus and William of Occam in the fourteenth century sought to disconnect the theory of knowledge from theology and draw clear distinctions between the knowledge that comes from faith and that which comes from reason and experience. But the scientific influence of these great theoreticians was small and we need not follow them further.

The Condemnations of 1277

It would be a great mistake to assume that the enthusiasm of scholars for the new learning extended throughout the church. In Rome sat Pope John XXI and his cardinals, and they had more to think about than species and genera. Their responsibility was to preserve and strengthen the faith, and they saw it wavering before the new ideas. Early in the year 1277 Pope John wrote to Etienne Tempier, Bishop of Paris, expressing his concern over reports of errors circulating at the great university from which pure Catholic doctrine had once flowed to the edges of the world. The Bishop was told to find out what had been taught that was damaging to the faith and who was teaching it, but apparently he exceeded his instructions, for he got together a council of "doctors of Sacred Scripture and other prudent men" and before the end of the year, with the Archbishop of Canterbury sitting beside him, he anathematized a long and disorganized list of propositions. "We excommunicate all those who shall have taught the said errors or any one of them, or shall have dared in any way to defend or uphold them, or even listen to them, unless they choose to reveal themselves to us or to the Chancery of Paris within seven days" (Hyman and

Walsh 1967, p. 542). Of the 219 propositions listed I mention only a few of those that touch our story. The first three condemnations strike down limitations on God's power inferred from Aristotelian physics and cosmology; while the last two uphold the church's central doctrine that God has given man freedom to choose between good and evil.

27A. That the First Cause cannot make more than one world.

42A. That God cannot multiply individuals of the same species without matter.

66. That God could not move the heaven in a straight line, the reason being that he would then leave a vacuum.

92. That with all the heavenly bodies coming back to the same point after a period of 36,000 years, the same effects as now exist will reappear.

154. That our will is subject to the power of the heavenly bodies.

Clearly, in the first three of these propositions the decisive point is the distinction between what God can and cannot do. It was not expected that God could do anything regarded as self-contradictory. It is impossible to imagine that 7 and 3 could ever equal 9 and nobody thought that God could make them do so, but it is easy to imagine multiple worlds or the creation of matter out of nothing or a one-foot displacement of the outermost sphere of the universe with respect to the earth. Even though nobody thought that such things would happen, there is nothing self-contradictory about them, and to say that God could not bring them about was to impose arbitrary limits on his power. That he was able to deliver Mishach, Shadrach, and Abednego from Nebuchadnezzar's fiery furnace was proof enough that he could set aside his own natural law when he chose to.

Actually, propositions 27A and 66 sprang from a misinterpretation of Aristotle's physics. If there could be several worlds, i.e., several complete systems each enclosed in its own sphere, the question arose as to the nature of the region between them. As I have mentioned, Aristotle was clear about this: outside the sphere of our world there is neither space nor time and so nothing need be said about the region between or the locations of the several worlds. Apparently this did not satisfy some scholars and they played with the idea that the region would have to be a void, which the logic of physics did not allow, and that this is why there is necessarily only one world. The argument leading to proposition 66 was similar. It follows from Aristotle's postulates that there cannot be a void but there is nothing absurd or self-contradictory about the idea, and God could surely make one, even an infinite void (*On the Heavens*, 274b) if he chose to. It was understood that in condemning these two propositions Bishop Tempier condemned the arguments as well as the conclusions. The crushing force of the injunctions was felt not only in Paris but in other universities that looked to Paris for guidance on theological matters.

In imposing the ban Tempier exceeded his instructions. Not only that, but at many points his reasoning was successfully attacked, and in 1325 the ban was revoked. Nevertheless an entire generation of scholars had been compelled to think critically about the relation between Christian doctrine and the Aristotelian philosophy of nature. Natural law decreed that there is no vacuum, but God in his irresistible majesty is not bound by it: he could make one if he wanted to. What then is natural law? It comes from God, that is why he can break it; it is part of his plan for an orderly universe, and it is not broken otherwise. Natural law, considered as ordinances that hold everywhere with perfect exactness, had hardly been thought of in early science; from now on the suspicion grew that such ordinances might exist, and that to know them would bring mankind closer to a knowledge of God. Very slowly, people began to look for them (see Oakley 1984, O'Connor and Oakley 1969, ch. 3).

These were some episodes of medieval philosophy that dealt with substantive matters that arose in trying to understand the material world. Limitations of space and knowledge have reduced my account to a sketch. In doing so I may have created a false impression, for I have not mentioned the endless debate about the nature of language and its relation to experience. Questions in this area have never been settled and they cannot be, because in using language to discuss language one is creating a tempest in a closed world that excludes direct experience. But scientific thinking has progressed nevertheless, for science has an intellectual life of its own.

From the scientific point of view the great synthesis was a regression, since during the battle between faith and reason the claims of observation as a path to truth were scarcely heard. The mere fact that some notion arrives in the intellect via the senses does not mean that it is true. Mathematics, a science of pure form, was generally thought to deal with form in nature but not with substance, explaining *quia* but not *propter quid*. In spite of this, physics continued to grow in the direction of mathematics as a plant grows toward the light, but there were other sources of ideas which, in their various ways, contributed to a broader understanding of the natural world, and in the next chapter we shall look at some of them.

CHAPTER 7

The Wider Shores of Knowledge

The spirit of the world . . . is a very subtle body, almost not a
body but a soul. Or almost not a soul but a body. Its power
contains very little earthy nature, more watery nature, still
more aerial, and most of all, fiery and starry nature. . . . It vivi-
fies everything everywhere and is the immediate cause of all
generation and motion.

—MARSILIO FICINO

AT the beginning of the *Metaphysics*, and again in the *Ethics* (1139b),
Aristotle distinguishes between science and art. Science deals "with the
first causes and the principles of things" and draws its conclusions by
exact reasoning. "Art arises when from many notions gained by experience one universal judgement about a class of objects is produced."
Nothing here about principles. Architecture and politics are arts, and
when Hippocrates says "The art is long, life is short," he is talking
about medicine. Art tells how but science tells why, and that is what's
fundamental. This chapter is mostly concerned with the arts. Reading
about the intellectual history of the Middle Ages we are liable to forget
that there was life outside the universities of Paris and Oxford where
the voices were so loud. The arts developed almost independently of
the schools and of Latin, and although there were university courses in
medicine, most doctors entered the profession through apprenticeship,
studying their masters' books. We shall look at some of the activities and
ways of thought that developed outside the doctrines taught in the European schools. We shall see how atomistic ideas survived and were trans-

mitted by two writers on practical subjects, how old theories were extended into the practical domains of magic and alchemy, and how the discovery of the supposed writings of the Egyptian sage Hermes Trismegistus brought magic into the highest intellectual circles. In Chapter 8 we will have to go back to school.

The Survival of Atomism

Hero of Alexandria, who probably lived in the first century A.D., is known for mechanical contrivances such as his device designed to use the expansion of heated air to open the doors of a temple automatically after someone lights a fire on the altar outside. He also wrote on mathematics, performed experiments in physics (something that Greek philosophers before him had rarely done), and theorized about the results. His writings have none of Plato's poetry or Aristotle's insistence on the primacy of reason; instead, he asks nature to answer his questions.

At first glance Hero's treatise on pneumatics (1851) seems to be a book about steam engines and other such devices, prefaced with some remarks about the science of air and steam, but a closer look shows that the machines are more like lecture demonstrations, and that as machines they are fit for little more than opening doors. For us the introductory remarks are the most important part, since here the author speculates on the nature of air.

Hero explains that the four elements consist of atoms, and that substances such as atmospheric air are mixtures of them. Between atoms, allowing them to move about, are small empty spaces. Nature avoids a vacuum as she avoids stones that fly upward, but, just as one can throw a stone against its natural tendency, one can produce a partial vacuum. A bottle that is said to be empty is not really empty, for it cannot be filled with water unless one allows the air inside it to escape. To show that the particles of air have some empty space between them, take a metal globe holding two quarts or so and solder a short metal tube so that you can blow into the globe. Then,

> if anyone, inserting the [tube] into his mouth, shall blow into the globe, he will introduce much wind without any of the previously contained air giving way. . . . Again, if we draw out the air in the globe by suction through the [tube], it will follow abundantly, though no other substance takes its place. . . . By this experiment it is conclusively proved that an accumulation of vacuum goes on in the globe; for the particles of air left behind in the globe cannot grow larger in the interval so as to occupy the space left by the particles driven out. (Hero 1851, 1)

From this experiment the existence of empty space around the particles is argued plausibly enough. Hero also points out that wine poured into water spreads through it in a short time and that the light from one lamp passes freely through light from another; these phenomena would be hard to explain if there were no spaces. Hero's works were available in Latin from the twelfth century on, from translations made in Sicily.

The ideas of Epicurean atomism survived also in another very respectable text, the *De architectura* of Vitruvius, in which he explains (II.2) the properties of building materials by talking about their atoms. Bede knew this book and it was the standard architectural treatise until the Renaissance.

Magic and the Occult

Some discussion of magic is necessary to this account, but the subject is so vast and its history so long that a historical summary would be pointless. Therefore I will give only a sketch—a little background and a little theory.[1]

The word "magic" derives from *magus*, a wise man, going back to the Persian *magu*. To understand the role of magic in medieval thought we must first think of the kinds of questions that true science could not answer at that time. How do the planets act so as to influence our lives? By what means does an infusion of willow tips act to relieve pain even though there is nothing in the manifest form of the willow that suggests any curative power? How does a magnet act on a piece of iron? True science is the knowledge of causes. What kind of science is it when we know nothing but the effects? In this dilemma one spoke of occult causes, meaning hidden causes; they are no less real than the causes one knows, but of a different kind. They were the causes of effects learned by experience and they belonged to art, not science. Inevitably people tried to invoke them by supernatural operations, by the black arts involving commerce with demonic intelligences, or by less dangerous forms of magic; the common meaning of "occult" has hardly changed since. The superstitious precautions we still sometimes notice show that for many of our fellow citizens there are still occult causes that affect their lives.

As time passed the range of the occult broadened, and by the seventeenth century it included spot removers, cough medicines, complexion remedies, fish lures, kitchen procedures, and of course poisons and other ways of modifying people. All are examples of what was loosely called

[1] For those whose interest goes further I recommend Lynn Thorndike's *A History of Magic and Experimental Science*, in eight volumes (1923–58).

magic, and Giambattista della Porta in his *Natural Magick* (1658) gives four hundred pages of procedures for making the natural world obey one's will.

I will begin the story at the dawn of history. The oldest magical text is the Ebers papyrus in the collection of Columbia University. It dates from about 1550 B.C. and contains material probably much older. In it are more than eight hundred recipes, of which I mention only one: for falling hair, treat with a mixture of the fats of ibex, horse, crocodile, cat, snake, and hippopotamus. From the beginning, magical literature has consisted largely of such recipes, but over the centuries magic also developed a theoretical structure which, though less involved with philosophical questions than that of science, was not in conflict with it.

One of the first extended accounts of magic is in Pliny's *Historia Naturalis*, which he begins with the words: "In the earlier part of my work I have very often refuted the foolishness of the Magi, whenever the subject and the occasion demanded, and I shall continue to argue against them and expose them" (1938, XXX.1). Pliny then gives hundreds of what we would call magical recipes and remedies, many of them credited to the Magi. Apparently, what he dismisses with the word magic is any claim of action at a distance through the agency of gods and spirits, but he seems to think there is no limit to what can be accomplished when one thing touches another. For epilepsy, "The Magi recommend the tail of a python attached as an amulet in gazelle skin by deer sinews, or the bits of stone from the crops of baby swallows fastened to the left upper arm" (XXX.27).

If you don't need that one, try one of Pliny's own: "I find that a heavy cold clears up if the sufferer kisses a mule's muzzle" (XXX.10). Evidently the cold is supposed to pass along to the mule, but the technique is carried too far in a Magian treatment that requires us to "take the parings of a patient's finger nails and toe nails, mix with wax, and say that a cure is sought . . . fasten them before sunrise on another man's door as a cure for these diseases. What a fraud if they lie! What wickedness if they pass the disease on!" (XXVIII.23).

Pliny tells us that magic originated with Zoroaster in Persia, about six thousand years before the death of Plato. The tradition was apparently oral, for there were no treatises. "Empedocles, Democritus, and Plato went abroad to learn it, . . . and upon their return taught some of it openly and cherished the rest among their secrets" (XXX.2).

Astrology was another occult science that was widely practiced in Rome, but Pliny has little to say about it. Early in his work he mentions Hipparchus, "who can never be sufficiently praised, no one having done more to prove that man is related to the stars and that our souls are part of heaven." But when Pliny mentions astronomical portents that were followed by catastrophic events, he says that contrary to the usual interpretation he thinks the portents were caused by the catastrophes that were about to occur. In general, except for the immense scope of his knowledge, Pliny held much the same opinions as most other Romans of his time; we must think of him and his fellow-citizens anxiously looking out the door in the morning to make sure that there was not a flock of birds on the left as they started off for work, and assembling little piles of lizard tails, weasel feet, and doves' dung at home in case someone got sick. It is not necessary to identify magic with the Magi or with special practitioners; it was everywhere.

Precisely because magic was everywhere it is not much mentioned in Classical literature except in extreme or comic manifestations, but to see how it entered the lives of practical men we can go to the doctors. The greatest of these was Galen, born in the city of Pergamon in Asia Minor in A.D. 130. Most of his active years were spent in Rome, where he had a large practice and wrote prolifically. I mention here (mostly from Thorndike 1923, ch. 4) only some details of treatments that to us suggest magic. To treat warts, he touched them with a gem he possessed called *myrmecia*, which had little bumps on its surface. Clearly its influence is occult, for no matter can have passed between the hard stone and the wart. In other treatments he used an immense variety of substances, including sweat, crocodile's blood (expensive in Rome), mouse dung, viper's flesh, and, for toothache, the tooth of a dog, burnt, pulverized, and boiled in vinegar. For some of them it is hard to see whether they were used because of magic affinity or because they were supposed to produce some physical effect by heating or irritation. When he can, Galen tries to relate their efficacy to manifest causes. For example, a piece of peony root worn around the neck was found very effective in preventing epileptic seizures, and he suggests that this was either because the patient was inhaling particles of the root or because the root somehow changed the air around it. Galen rarely mentions astrological considerations, but he is aware of them and tells how to take the moon's phases into account in prescribing certain remedies. He is reincarnated in the prologue to the *Canterbury Tales*:

> With us ther was a Doctour of Physick;
> In al this world ne was ther noon hym lik
> To speke of physik and of surgerye
> For he was grounded in astronomie.

Today as always, we try to distinguish magic from science by looking at the causal principles they invoke. If a scientific theory asserts that one event causes another it assumes that cause and effect are linked by some physical process. Perhaps we do not know what the process is, but we assume it exists. When people claim to find a cause for which such a process is unlikely or impossible, we speak of magic. Perhaps, for example, some effects of climate and diet might operate to produce different characteristics in people born in different months of the year, but in the prevailing scientific view, to ascribe effects to the positions of the sun and planets relative to the earth and stars is to talk magic. Modern surveys show that in the minds of millions of our fellow citizens magic and science live peacefully side by side. It is possible to follow the latest scientific developments with interest and still believe that the weather is influenced by underground atomic tests that only shake the ground, and some readers of Immanuel Velikovsky believe with him that the close passage of a comet caused the earth to stop rotating for a few hours and then start up again, perhaps in the opposite direction. There are books, magazines, and newspapers devoted to the "paranormal," most of which must be classed as magic. As I write, the morning paper tells of a psychic to whom a Philadelphia jury has awarded over a million dollars in damages because her special powers were destroyed by a CAT-scan. It is easy to deride such credulity, but we should realize that it springs from a profound and widespread conviction that there are occult connections running through all of nature, that the world is One.

Alchemy

There have been chemists since remotest antiquity. The smelting and purification of metals are chemical processes. There is fine gold in the royal graves at Ur from the third millennium B.C., and in Egypt the development of ceramics, metals, paints, and dyes shows that since the Bronze Age extensive research must have gone into learning how materials behave and how to control them. There are no texts. Secrecy in the service of economic profit, the existence of an unlettered artisan class—one can find many reasons, but an immense body of knowledge not formally written down existed alongside philosophers' knowledge. In about the eighth century, in North Africa, a fusion of the two began to produce books, and they were prolix, varied, and confused. The alchemical tradition in Europe dates from the twelfth and thirteenth centuries when translations from this Arabic literature began to appear.

From the beginning the alchemists set themselves the task of making gold. Gold was precious but not just because it was rare, for it was the king of metals in the universal hierarchies of things just as the dolphin was the king of fish, the lion was the king of beasts, and Man the king of earthly creatures. Alchemists aimed to produce the best of all material substances. The situation was like that of modern physics, in which special effort is concentrated on the attempt to find the fundamental constituent of matter. Perhaps that thing, whatever it is, is the king of the physical world.

The theory of transmutation that guided alchemists was grounded in Aristotle. He taught that all matter is formed from mixtures of the four elements but that the elements are not permanent. They can be transformed into one another—Water can form Air or Earth rather easily and Fire with more difficulty, and so on according to the properties they share. To change base metals[2] into gold was to change the proportions and mode of mixture of their elemental constituents, and this could be done either by combining one substance with another or by operating on a single substance so as to change the elements in it.

The first of the great chemists was Jabir Ibn Hayian, later known as Geber, who was attached for some years to the court of Harun al-Rashid at Baghdad. While there, Jabir studied texts that were being translated from the Greek and organized further translations. The enormous mass of works bearing his name cannot have been written by him, but it appears that he was master of all known techniques of practical chemistry as well as the inventor of the theory that, more than any other, guided the alchemists who followed him.

As we saw in Chapter 2, each element has two qualities that it shares with two other elements: Earth is cold and dry and so on, and these, together with its other properties, constitute its form. Jabir chose to base his chemical system on two substances: sulfur, which on obvious grounds is hot and dry, and mercury, which is cold and wet. Since each contains all four elements, any other material can be formed by proper combination of these two, and since we cannot know substance but only form, our search must aim at the form of the desired product, gold. The inner nature of gold is cold and dry, and Jabir was able to measure these qualities in numbers that furnished some guidance for the practitioner.

In Europe before the Renaissance, alchemists, later called chemists, were little thought of. Many were poor men and women, often forlorn or comic figures as in Chaucer's Canon Yeoman's Tale, who ranked socially little higher than artisans. In English, "chemistry" derives from "chemist" as "sophistry" and "bigotry" derive from their roots: the sense is derogatory. As time went on educated people in the merchant class and the nobility became interested, and by the Renaissance there were adepts in all levels of society. Since the transformations that occupied them all took place, as we would say now, at the molecular level, the causes were necessarily occult, and practitioners sensed unseen intelligences all about them as they watched mysterious changes of color and texture in their

---

[2] Note that our common usage of "metal" is of recent origin. Traditionally it referred to any solid substance at all, and astronomers still apply the word to all chemical elements other than hydrogen and helium.

flasks and retorts. By the late Middle Ages there was a widely shared doctrine which, if one allows it to vary a little from person to person, was this:

The act of turning base metals into gold, known as the magnum opus, the Great Work, was to be accomplished by means of the Philosopher's Stone, a substance, often symbolized as a cube, that would do it easily, once it was made or found. Perhaps the Stone is scattered all around us if we only know where to look for it—perhaps children play with it and the maid sweeps it out of the house, but if we cannot find it we must make it. The magnum opus was usually thought to proceed in two stages: the Work in White, making silver, and then the Work in Red, making gold.

Just as the four elements of philosophy are not identical with the materials whose names they bear, the material known as gold is not the impure metal we know but rather an idealized material known as "sophic gold," or "our gold," and similarly for silver, sulfur, and mercury. Sophic gold and silver are represented in the pictures as sun and moon, king and queen, and their union in the course of the Great Work is represented as sexual union of the king and queen or else the formation of a hermaphrodite, or androgyne, that combines the two.[3] These representations should not be considered as mere analogies. It is exactly because a man and a woman can produce a child, or grass can be converted into beef, that people believed in the possibility, at first glance much simpler, of transmuting one metal into another. Of course we could now tell them in their own language that child and beef exist *in potentia* in the parent substances, while gold is not potential in sulfur and mercury, but how could anyone have known that until everything had been tried? If one merely heats sulfur and mercury together until they react, they form a black, smelly mass. They fight instead of uniting. The procedures of alchemy were designed to produce fruitful unions.

The Philosopher's Stone was usually imagined as a yellow powder. The literature is not clear as to what it is: perhaps a compound of gold, silver, and mercury, perhaps sophic gold; there are many guesses.

The Great Work was to be accomplished in a fixed series of steps, determined by astrological correspondences. The process started at Aries, the first zodiacal sign:

ARIES: calcination (heating a metal so that it turns to powder),
TAURUS: congelation (crystallization),
GEMINI: fixation (making a material nonvolatile, usually by heating),
CANCER: solution,

and so on (Holmyard 1957, p. 154), each operation carried out at the right astrological moment.

[3] Mircea Eliade's book on alchemy, *The Forge and the Crucible* (1962) contains an interesting chapter on the sexualization of nature.

Alchemists tended to agree as to the early stages of the Work; as it progressed they went their own ways. Here is a typical scheme that illustrates the mode of thought (Read 1937, p. 32):

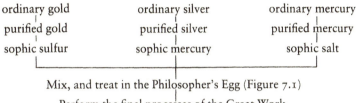

Mix, and treat in the Philosopher's Egg (Figure 7.1)

Perform the final processes of the Great Work.

The Egg is also called the Vase of Hermes, for reasons that will be explained in a moment, and it is, of course, hermetically sealed. The final parts of the Work are variously and vaguely described; they are called multiplication and projection. Multiplication is increase in power and/or quantity. Projection is the actual process of transmutation. The powder of projection is the Stone itself. It is enclosed in paper or wax and thrown into mercury or molten lead, which then immediately turns into gold.

Did anyone ever perform the Great Work? Tales of charlatans are many, but there was only one practitioner who according to general opinion ever achieved it. He was Nicholas Flamel, a Parisian public scribe of the fourteenth century, who succeeded after many years, with the aid of an old book and a wise Jew who helped him to read it. At noon on Monday, 17 January, 1382, "thanks to the intercession of the most blessed Virgin Mary," he completed the Work in White in the presence of his wife and helper Pernelle, obtaining about half a pound of purest silver.[4] Continuing, he finally produced a red stone,

> and afterwardes, following always my Book, from word to word, I made projection of the Red stone upon the like quantity of Mercurie, in the presence likewise of Pernelle onely, in the same house, the five-and-twentieth day of Aprill following, the same yeere, about five a clocke in the Evening; which I transmuted truely into almost as much pure Gold, better assuredly than common Golde, more soft, and more plyable. . . . Pernelle . . . understood it as well as I, because she helped mee in my operations, and without doubt, if shee would have enterprised to have done it alone, she had attained to the end and perfection thereof. I had indeed enough when I had once done it, but I found exceeding great pleasure and delight, in seeing and contemplating the Admirable workes of Nature, within the Vessels. (Quoted by Read 1937, p. 66, from Flamel 1624)

Flamel performed three more projections, making (it is said) enough gold to found and endow fourteen hospitals and three chapels and perform

---

[4] In that year January 17 fell on a Friday, but in astrological theory silver belonged to the moon; hence Monday.

FIGURE 7.1. The Philosopher's Egg. From M. Maier, *Atalanta fugiens* (1617).

many other charitable works. He continued a sober and modest style of
life and died in 1417. His gravestone, listing some of his benefactions, is
preserved in the Musée de Cluny.

The news of Flamel's success helped to precipitate an alchemists' gold
rush all over Europe, and it was in the fifteenth and sixteenth centuries
that the fever burned most strongly. Charlatans traveled from town to
town carrying bits of gold that they could pretend to produce out of other
substances, some of them so successfully that they were tortured and
killed for the secrets they were thought to possess. We shall see that even
during the following century people of the fame of Robert Boyle and Isaac
Newton sequestered themselves for weeks at a time, patiently trying to
make gold.

Alchemy as Science

Alchemy was more than a search for gold; it was a science that sought to
illustrate in practice the great intellectual structure of what was already
known, by showing that chemical reactions are governed by the same

principles as everything else. Fundamental to these principles are the four Aristotelian causes. The roles of material, efficient, and formal causes in alchemy are clear enough, but the final cause is just as important; for the alchemist is banking on the Aristotelian premise that everything in the world strives toward perfection, just as the individual human being does, through the power of God. Even base metal struggles upward, and on its behalf the alchemist, by removing obstacles, performs an act of virtue.

Because the principles of Aristotelian science are universal they must often be interpreted as one interprets parables and metaphors and analogies. The analogies are everywhere. The saying of Democritus that man is a little replica of the universe was never forgotten, and the ways of the universe were long explained in terms borrowed from its little replica. For Empedocles, Love and Strife rule the world. For Aristotle, the Unmoved Mover acts upon the world as the beloved acts upon the lover. For Plato the world has a living soul. For everybody in the Middle Ages, the zodiac represented a great curved man, with head at Aries and feet at Pisces (Figure 7.2) with each region of the body assigned, for medical and other purposes, to one of the signs. As time went on the correspondences were further developed until it was hard to look at anything in nature without seeing it as a representation of something else.

The vast system of correspondences enabled the theory of nature to be expressed in many ways, but rarely in simple language. The pictorial images of alchemical processes are good ways to represent them, for images emphasize, better than words ever could, the unity of all principles.

Precisely because of that unity, everything finally makes sense. Of course, God sees and knows all, but more than that, his plan has no gaps in it. "Are not five sparrows sold for two farthings, and not one of them forgotten before God?" (Luke 12:6). This is not just an expression of God's pity; it expresses the wholeness of his plan and the faith of those who, like Alexander Pope, worship a God

> Who sees with equal eye, as God of all
> A hero perish or a sparrow fall,
> Atoms or systems into ruins hurled,
> And now a bubble burst, and now a world.
> *Essay on Man* (1734, 1.87)

We think of a scientific experiment as a compartmentalized act. Just as the efficacy of the Holy Sacraments does not depend on the morals or learning of a consecrated priest, we would not think that the result of an experiment depends on how the scientist lives. For the alchemist it was different. He needed to pay as much attention to the details of his life as to his apparatus and procedures. The typical alchemist was a man, but

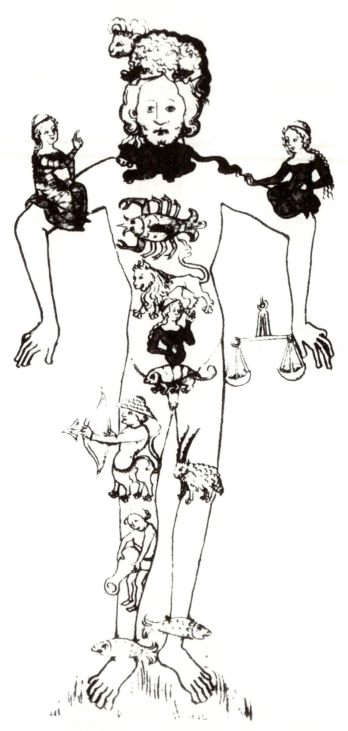

FIGURE 7.2. The zodiacal man, from a twelfth-century manuscript.

some instructions called for him to have a female collaborator with whom he could lawfully be sexually joined but was not. Many of his operations had to be performed fasting and continent, and often he could bitterly blame his failures on the state of his soul. Sometimes he had to go close to the edge of what was allowed. Herbs, stones, and gems gained occult powers from their association with planets and constellations, and these powers could be used, but Saint Thomas warned that if he engraved any letters or words on a gem, or used incantations along with an herb, he might unwittingly enter into a pact with the Devil, to the damnation of his soul. Written or spoken words could not by themselves work magic. There must be an Intelligence to understand them. Space was full of demons and angels, but the practitioner launching a message could not be sure which one was going to receive it. He hoped for an angel, for in huge numbers they traversed the universe performing the tasks that make it obey the will of God. Now they are trivialized into little sexless beings that hang from Christmas trees.

Hermes

The principles of alchemy were venerated not just because they pertained to the magic that people could feel working all around them, but also because they were believed to be mankind's oldest knowledge, imparted directly by God to Adam or to one of the sons of Noah. In about 1460 an agent of Cosimo de' Medici brought back to Florence a Greek manuscript he had found in Macedonia. It contained fourteen treatises purporting to have been written by one Hermes Trismegistus, called Thrice-Greatest Hermes because he was supreme as philosopher, priest, and king. The treatises, known as the *Corpus hermeticum*, were cast in vaguely Egyptian terms, and they aroused intense excitement because their existence was already known and they had long been hoped for. In the *City of God* (VIII.23–24) Augustine condemns the magic they describe for making an image and then inviting a demon to live in it, but in language that makes the reader long for more. Lactantius (1871, 1.6) reports that the treatises contain elements of Christian belief. Both writers stress that they are very old, Lactantius identifying the author with the god Thoth, who "gave laws and letters to the Egyptians" (remember Thoth; he flickers in and out of medieval works under various names: Toc or Toz or Thot the Greek, and we shall meet him again). Modern scholarship finds that the *Corpus hermeticum* dates from A.D. 100–300 and probably originated in the Greek culture of Egypt, with Platonic, Stoic, Jewish, and Persian influences (Yates 1964, Shumaker 1972).

The *Corpus* is very pious in tone, and though not Christian it comes from the same intellectual tradition as much Christian doctrine and is generally in harmony with it. Marsilio Ficino, the Medici family's expert in Greek, was ordered to stop work on his long-awaited translation of Plato and start on the new manuscript.

Ficino translated all of the *Corpus* that was available to him and named it *Pimander*, after the title of its first section. It was received as pre-Christian, but wonderful in its anticipation of Christian thought. From the Egyptian references it was dated at about 2500 B.C., which made it incomparably the oldest wisdom in the world, a thousand years older than anything Greek and correspondingly closer to the divine sources of all knowledge. The *Pimander*'s piety could not be questioned, and Ficino the Platonist became convinced that it was the original source of Plato's teachings. Some of the texts are magical, and they powerfully reinforced prevailing ideas as to how the powers of nature could be influenced. Their age and purity were in contrast to the tattered books of charms and symbols that alchemists exhibited to back up their claims, and at once magic of all kinds gained the attention of scholars and prelates.

Ficino himself became deeply involved. He believed that not only do planets influence our lives but that when they are in some favorable position in the sky their power can be attracted, by suitable magical practices, to aid in some particular purpose. To this end he practiced with the Spirit of the World—one might say the breath of the world—an influence that flows down from the planets to protect us from harm (see Agrippa 1898, ch. 12; Taylor 1951, ch. 2; Walker 1958, chs. 1, 2; Yates 1964, ch. 4). He describes it at the head of this chapter. The goal of natural magic as Ficino practiced it was not to influence external events but to attract this spirit, especially from Jupiter and the sun where it was brightest and warmest and most beneficial, to join with the practitioner's own spirit to make it stronger and better. Ficino clothed himself in a robe of a suitable color and at the right hour, surrounded by the right colors, odors, wines, jewels, metals, arrangements of plants, and much else, played and sang old Orphic hymns in honor of the chosen deity. Sometimes he practiced with demons, so close to the line of damnation that in 1489 he had to defend himself, with the aid of powerful friends, before the pope. What he had actually done is not clear from his writings (Walker 1958, Yates 1964).

In those days every lively mind must have been tempted to dabble in magic. Good and evil spirits were everywhere, their powers manifest in a dozen events each day. Why shouldn't one get in touch with them, perhaps to control them a little? And as for demons, they did not all have horns and tails, and many of them had been worshipped for their beauty and majesty in past ages. Human existence was full of dangers: most people died young, and survivors lived under the whim of princes whose comings and goings, whose wars and combinations they could in no way control. So people appealed to higher powers still, whose domain was a step closer to the source of all goodness and who, if only one knew the right way, might be persuaded to protect them from nature and tyranny. It is not surprising that God-fearing people seeking celestial help and not very sure of the boundary that separated white magic from black sometimes found themselves facing perdition. How could one know and choose when all causes were occult? In the next chapter we shall see how the domain of true science began to grow.

CHAPTER 8

Illumination

Now, as the Creator played, so he also taught Nature, as his image, to play, and to play the very same game that he played for her first.

—JOHANNES KEPLER

LOOK back on the basic scientific problems that the philosophers have been struggling with: What is reality? How is the universe constructed? Why do the heavenly bodies move as they do? What is the fundamental nature of matter? As posed, the first is not really a scientific problem. The rest remain, today as always, at the head of the list. The reason they stay there is their comparative inaccessibility. Neither cosmos nor atom is observable as a whole, and where there is no direct observation progress can only come from patient reconstructions based on what we can observe. Real progress in these directions requires scientific instruments and these, in turn, came from centuries of invention. Before then, natural philosophy needed some things to study that were accessible to simple experiment and simple quantitative description. Also needed was a loosening of the general distrust of sensory information as evidence of the truth. In this chapter we shall see how these needs began to be filled.

Light

In the old writings light and understanding are almost synonymous. When God said "Let there be light" before he created the sun and moon it is as if he did it so that we, the hearers of the tale, could see from the beginning what he was doing. In reading the biblical text and many others it is often hard to distinguish the light of the sun from the light of understanding, the sign that the world around us is accessible to reason and that it is good.

> There was a man, sent from God, whose name was John. The same came for a witness, to bear witness of the Light, that all men through him might believe. He was not that Light, but was sent to bear witness of that Light. That was the true Light, that lighteth every man that cometh into the world. (John 1:6)

These words were not composed as a literary trope. They express a dominant idea of Neoplatonism and we shall see in this chapter that until the Renaissance no clear distinction was made between the light of the eyes by which we see and that of the spirit by which we know. The study of light and vision was more than the study of physical and physiological facts; it was seen as one of the ways in which man could reach toward knowledge that lies beyond the senses.

Obviously, sight is the primary sense. God never said "Let there be sound." Light and sound both bring us messages, sometimes of great subtlety, from the surrounding world. Perhaps they come via the inflowing of some substance, perhaps by some other path, but how is it that the eye forms images in our minds while the ear does not? In some respects sight, telling so much about the shapes of things, is more like touch than hearing; it is as though something reached out from the eye. Aristotle quotes Empedocles as saying that we see by means of fire and that vision is the result of light issuing from the eye as from a lantern, but then he reasonably asks why, if that is the case, we cannot see in the dark. Doubtless the question had already crossed Empedocles' busy mind, and he meant that the eye's light is necessary for sight even if it is not sufficient. Plato is more specific. In the *Timaeus* he is describing how the gods designed and built mankind: "So much of fire as would not burn, but gave a gentle light, they formed into a substance akin to the light of everyday life, and the pure fire which is within us and is related thereto they made to flow through the eyes in a stream smooth and dense" (45b).

This pure fire flows outward and, mixing with the sun's light, affects "the whole stream of vision." It "diffuses the motions of what it touches or what touches it over the whole body, until they reach the soul, causing that perception which we call sight." The whole body is the organ of vision; the eyes only emit the invisible fire. (Recall that in Aristotle's theory we perceive something when its form enters our soul and combines with it.) Vision is an active process, unlike the other senses with which the body merely receives. And what of the external source of light that goes round and round the world, illuminating first one race of mankind and then another? Surely that light also issues from intelligence, our sign that the world is good and that it makes sense.

What is physical light itself? Again Aristotle quotes Empedocles: "the light from the sun arrives first in the intervening space before it comes to the eye or reaches the earth" (*Sense and Sensibilia*, 446a). It is something that travels through space at a finite rate. But Aristotle does not consider that light is a something at all, or that it travels. Rather, it is the actuality of what is transparent.

> Every color has in it the power to set in movement [i.e., to alter] what is actually transparent; that power constitutes its very nature. That is why it is not visible except with the help of light; it is only in light that the color of a thing is seen. (*On the Soul*, 418b)

Thus light is not color; it is what enables color to be seen. It belongs to the medium—air, glass, water—through which we look. "Light is the presence of fire or something resembling fire in what is transparent. . . . The following makes the necessity of a medium clear. If what has color is placed in immediate contact with the eye, it cannot be seen. Color sets in motion what is transparent, e.g. the air, and that, extending continuously from the object [to] the organ, sets the latter in movement" (*On the Soul*, 418b, 419a).

Since nothing travels from object to eye the change is instantaneous, or so it was generally understood, and that is why in 1676 Ole Roemer's demonstration that light travels at a finite rate drove one of the last nails into the coffin of Aristotle's physics.

It is comparatively easy to experiment with light. Mirrors of polished bronze were made very early, and it must soon have been noticed that when light is reflected from a plane surface the incident and reflected rays make equal angles. A ray of light is also refracted, i.e., bent, when it passes from one transparent medium into another; this is the principle of a lens. Many years ago Austin Henry Layard found a lens, ground from rock crystal and dating from the eighth century B.C., in the ruins of the royal palace at Nimrud, near Babylon. The lens is convex on one side and flat on the other, slightly oval in contour and 4 cm across, with a focal length of 11 cm, making it suitable for use as a magnifier (Layard 1856, p. 167). The Greeks knew that a clear, spherical bottle full of water could be used as a burning glass, and people must have observed from the beginning of time how a stick thrust into water appears to be broken at the surface. There were plenty of phenomena to study.

Except for Aristotle's remarks on the nature of light and his unsuccessful attempt to explain the rainbow as a result of reflection from water droplets (*Meteorology*, III), the first theoretical discussion of light is in Euclid's *Optics* (1945), where various principles of perspective are derived geometrically. Some of the results are quite simple, such as the apparent drawing together of parallel lines stretching away from us or: "When the eye approaches a sphere, the part seen will be less, but will seem to be more." Others, depending on deeper theorems of geometry, are more recondite.

Hero of Alexandria, in his *Catoptrics* (*katoptron* means "mirror"), took Euclid's work much further. He discussed the formation of images when light reflects from several plane surfaces. He considered reflection from curved surfaces and calculated the apparent sizes of the images formed, and he also formulated the basic principle in clairvoyant language that anticipated the way in which modern physicists like to express the laws of nature. He showed that the laws of reflection can be derived from the single postulate that the path taken by a ray of light that originates at the eye [Figure 8.1(a)] , reflects from a plane surface, and strikes an object is the one that is the shortest, i.e., if light moves at finite speed, the one that requires the least time (Cohen and Drabkin 1948). Note A shows a modern proof. This attempt to reduce physical laws to principles of economy may seem to echo Aristotelian attempts to relate them to the most perfect geometrical figures, but there is nothing wrong with that part of Aristotle's program—provided, of course, that one gets the principles right. To be able to derive quantitative laws from general qualitative considerations gives a feeling that one is getting to the heart of things.

Ptolemy of Alexandria also wrote an *Optics* (1885), in which he reports (Cohen and Drabkin 1948) the results of careful measurements of the refraction of a beam of light at a water surface. As he also does in the *Almagest* (1984), rather than give us the results of his measurements, Ptolemy gives numbers apparently calculated from a rule derived from them. In modern notation it amounts to

$$r = ai - bi^2,$$

where i is the angle of incidence (Figure 8.1b), r is the angle of refraction, and a and b are numbers determined empirically. When the angle is small, as it usually is with a lens, this is a good approximation, although Ptolemy would have been better off using i^3 instead of i^2. The correct law of refraction was not found until the beginning of the seventeenth century, by Thomas Hariot and Willibrord Snell,

$$\sin i = n \sin r,$$

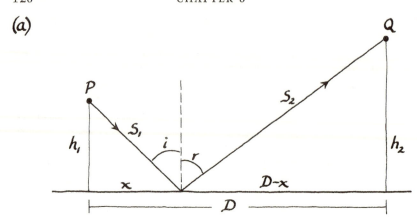

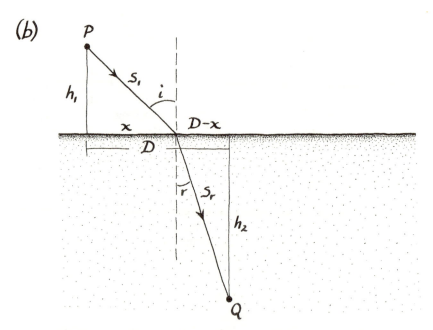

FIGURE 8.1. (a) Reflection, and (b) refraction of light at a plane surface.

in which n is the relative index of refraction of the two media. Long afterward it turned out that Hero's principle of least time also applies to refraction and provides a derivation of this law. A proof was laboriously constructed in 1661 by the French mathematician Pierre Fermat, and a simple version is in Note B.

There are three steps to Ptolemy's formula: to guess that there is a

mathematical regularity; to make the measurements; and to find a formula that fits the data. This work took place in the world of the senses.
Ptolemy was not looking at philosophic truth and so his data were fitted
with the simplest formula he could find. It was otherwise with the planets,
for the theory of their motion had to follow necessarily from certain
hypotheses of circularity. As Gregory of Nyssa (ca. 324–94) wrote later,
the spheres of Earth, Water, and Air constitute the sensible world while
the realm of Ether, where the planets move, is the intelligible world (1960,
76d).

None of these writers has anything to say about how we see, and in
fact their results are largely independent of what light really is. The atomists had a theory: light consists of atoms that have no properties but mass,
size, and motion. They are the bearers of secondary properties such as
color but have none themselves; secondary properties are created in our
minds as a result of their impact on the atoms of our eyes. How, then, are
images formed? This is a major problem, and Lucretius devotes much of
Book IV of *De rerum natura* to it. To make a long story short, atoms that
reside near the surface of a thing are always being thrown off coherently,
and when moving toward us they retain the form of the surface as the
cast-off skin of a snake retains the snake's form. The air is full of these
simulacra, as they are called, flying in every direction. Apparently the idea
is older than Lucretius, for Aristotle seems to say that Democritus
thought these flying images are what we see in dreams (*On Divination in
Sleep*, 464a).

In the Middle Ages these images were explained in Aristotelian terms.
The Latin term for them was *species* and they were regarded as immaterial representations of the object's form. In vision they propagate through
the intervening medium, finally causing the form to be reproduced in the
observer's soul. Until the Renaissance the history of optics is more important for what it says about the development of scientific thinking than for
anything it tells us about the nature of light.

In the eleventh century Ibn al-Haitham (Alhazen) experimented with
mirrors and glass and found out how images are formed. Just as the atomists had said all along, vision results from light going into the eye, not
coming out of it. By representing the rays of light that carry the species of
a distant object as converging on the pupil of the eye, he solved a problem
that had puzzled many: when we look at a mountain, how can its *simu-
lacrum* go through that little hole? One problem he did not solve correctly: if the lens of the eye were allowed to form an image on the retina
it would be an inverted image and we would see everything upside down;
therefore he thought the image is actually formed at the back surface of
the lens, before it becomes inverted. These misconceptions show that even
as some Aristotelian theories were modified, the idea that sense percep-

tions directly transmit their messages to the soul was still strong. The works of Alhazen were the main source of information on optics for European scientists in the later Middle Ages. But as long as light was studied only through more and more competent considerations involving mirrors and refracting surfaces, its nature and its role in the created world remained obscure.

The First Bodily Form

In the beginning, God created the heaven and the earth. And the earth was without form and void, and darkness was upon the face of the deep.

Suppose that "form" is Aristotelian form, and that "without form" means the earth had no properties at all.

And God said, Let there be light.

Saint Gregory of Nyssa, a "father of fathers" of the Eastern Orthodox Church, believed that this light was inherent in the formless matter from which it burst out at God's command, filling all the sensible world (Gregory 1960, 76d). At this moment all the works of creation existed in potentiality, to be actualized in their proper and necessary succession during the next six days. Gregory's vision explains the Bible's mysterious account in philosophic terms of form and substance, and it sat in people's imaginations for more than a thousand years as the picture of what actually happened at the moment of creation. Light became more than the vehicle of sight; it was a causal agent and a central part of the rational scheme.

Robert Grosseteste put light at the center of his metaphysics of nature. We have seen that in the scholastic system form is knowable, but substance—true existence—is not. It therefore becomes important to consider the forms by which a thing is known and to determine which of them brings us truest knowledge of the underlying reality. Here is the beginning of Grosseteste's treatise *On Light* (ca. 1225), and it echoes Gregory.

The first bodily form, which some call corporeity, I judge to be light. For light of itself diffuses itself in every direction, so that a sphere of light as great as you please is engendered instantaneously from a point of light, unless something opaque stands in the way. But corporeity is that upon which of necessity there follows the extension of matter into three dimensions, although nevertheless each of them, namely corporeity and matter, is a substance which in itself is simple and has no dimension at all. . . . Further, men of good sense judge that the first bodily form is more worthy than all the later forms and of a more excellent and noble essence . . . than all bodily things; and it is more like the forms which stand separate—and they are the intelligences—than all bodies are. Therefore light is the first bodily form. (Grosseteste 1939)

Obviously the author is trying to convey a vision that cannot readily be spoken, in which the light of divine understanding as Plato and Saint John knew it becomes joined with the light that comes from the sun. Later in this chapter, Galileo, who tried to ignore the vocabulary of scholasticism in his later writings, will express the same vision once more.[1]

Light was the first form. Suddenly the substance of earth acquired the form of three-dimensional ponderable matter through which God's existence could be known to his creatures. Light actualizes matter and takes part in every new form it adopts. In other words, everything that influences something else acts with the power of light. Ordinary physics is the study of such influences, and therefore the science of optics stands at the center of physics. But optics, according to Grosseteste, is a mathematical science that fully explains the phenomena of light. Therefore, and finally, the path is open for a mathematical physics, a physics in which necessary quantitative conclusions are drawn.

The idea of dimension as the first form of corporeity originates in some undeveloped remarks by Aristotle on the nature of space and dimension (*Physics*, 209a) and was seriously taken up by Averroes the Commentator in the twelfth century. In his *Treatise Concerning the Substance of the Celestial Sphere* (Hyman and Walsh 1967), Averroes explains that matter at the most fundamental level, prime matter, has no form of its own; all its essence is potential. Potentially it can receive any form, but the first one must be extension in three dimensions. Now Grosseteste is describing the process by which that occurs.

Grosseteste is the last great medieval doctor in this book. After him science slowly turns toward more modern ways of thinking. He was one of the builders of an intellectual structure that I consider as beautiful and useless as music or abstract sculpture, but who can say how much more primitive our thinking would be if it had not existed. The structure was never completed; there was intense disagreement over many of the details and procedures, and what was produced may finally have been of less value than the devotion and fervor of the builders. Its creation absorbed much of the intellectual effort of Europe and Islam for two thousand years. Much of it is quite uninspired, as much sculpture and music is, but occasionally it rose to greatness in the clarity and unity of its conceptions

---

[1] For more on these ideas, see Grosseteste (1939); Duhem (1913, vol. 5, pp. 341–58); Crombie (1953, ch. 5).

and in its synthesizing power, a hymn in praise of reason and order and God. It is pointless to pick out this or that factual error of medieval science as a subject for derision. Facts came into this science but they were not central to it. What was central was a vision of the universe that perhaps some moderns still respond to, though the terms in which they think about it will be very different. Robert Grosseteste expresses it in an example:

> Consider the smallest and most insignificant object in the universe, a speck of dust. Its shape is the most perfect known in nature: a sphere.... Consider the human mind meditating on a speck of dust. It presents a mirror of the Trinity in the memory, intelligence and uniting love within the human mind: an image of God who remembers everything, understands everything, and loves without beginning, variability, or end. (Southern 1986, p. 216)[2]

Spectacles

The next great development relating to optics arose not in the sciences at all but in the arts, at the edge of magic, in the work of some Italian experimenter—possibly a Dominican monk from Pisa named Alessandro della Spina. Whoever it was, in about 1285, invented spectacles for far-sightedness. The effect of this invention was a social revolution among the middle-aged as hairdressers and tailors and literary men were able to take up again the work they had had to put down. Earlier a few scientists such as Alhazen and Roger Bacon had noted how lenses could magnify and produce inverted images on a screen, but since there was no mathematical theory these gadgets did not yet belong to optics, and no one had previously studied how they could improve sight.

The early lenses must have been very small, since they were known as *lente*, lentils. They were household objects, not mentioned in learned books, and for three whole centuries even scholars who were entirely dependent on them for their work paid them no more attention than they paid the stools they sat on (Ronchi 1970). When at last lenses were noticed they were given a dignified Latin name, *specillum*, and people began to think about them. In his *Natural Magick*, Giambattista della Porta writes briefly of lenses and mentions spectacles "which are most necessary for the use of man's life; whereof no man hath assigned the effects, nor yet the reasons of them" (1658, XVII.10). Later spectacle-makers developed concave lenses for near-sightedness; by their use telescopes were made and modern astronomy began.

---

[2] Southern translates *atomus* as "speck of dust" to remind us that Robert's idea of atoms was probably much closer to Isidore than to Lucretius.

The Decline of Reason

Except perhaps for a few of the humbler Christians, every thinker so far mentioned has had a vision of a universe that exemplifies a rational order, and has assumed that the task of philosophers was to lay bare as much of it as was possible for humans to understand. The aim was to go a little way, perhaps only a very little way, toward seeing the rational order as God does. God thinks; he does not need to perform experiments to know the truth, and so the philosophers of the Classical world and the Middle Ages sought to create a science that consisted of necessary conclusions drawn ultimately from premises that could not be otherwise—Aristotle's most brilliant effort to show how this can be done is in the opening chapters of *On the Heavens*, in which he declares the principles of order and beauty that determine how the universe really is. From this point of view experiment is no way to find out the truth, and even mathematical reasoning, while exact, remains only descriptive and incapable of answering the question "Why?" We have seen the beginning of a challenge to this dogmatic point of view in Grosseteste's claims for mathematics; now we shall watch as the rebellion against unassisted reason begins to gather momentum.

At Oxford, Roger Bacon (ca. 1214–ca. 1294) was Grosseteste's most brilliant student. In 1233 Bacon went to study in Paris and was appalled at the conditions he found there—professors lecturing on Aristotle from translations so poor that nobody could understand them, but nevertheless refusing to learn Greek, and science taught by syllogisms that proved nothing. He returned to Oxford in 1250 and joined the Franciscan order. Seven years later he was back in Paris, under interdiction and something close to house arrest for not keeping various unorthodox views to himself. He seems to have been one of those people who say exactly what they think. It helps to understand his career in Paris if he is imagined speaking (Latin of course) in the loud and irate British tones that can still occasionally be heard there. During this time Pope Clement IV, overruling Bacon's Franciscan superiors, ordered him to write down his thoughts about the sciences, which he did in three books called *Opus majus*, *Opus minus*, and *Opus tertium*: the Greater Work (1928), the Lesser Work (actually a summary of the greater Work), and the Third Work. A few years later, under Pope Nicholas IV, came a second condemnation, and fourteen years of prison with little chance to write. Bacon died shortly after his release.

Opus majus calls for the reform of universities. There must be serious study of languages, mathematics, and experimental science; Bacon calls the last one the leader of all the sciences, provocatively close to theology's proudly maintained position as their queen. He begins his argument for studying mathematics with the claim that every science needs it and proves this claim in several ways, by authority and by reason. Some of the proofs by reason are worth noting:

- Change always involves some augmentation or diminution, and these must be considered quantitatively.
- The comprehension of mathematical truths is innate in us (this goes back to an argument in Plato's *Meno*, which Bacon has heard of but not read).
- Of all parts of philosophy mathematics is the oldest, and therefore it should be studied first.
- It is not beyond the intellectual grasp of anyone. "For the people at large and those wholly illiterate know how to draw figures and compute and sing, all of which are mathematical operations."
- The clergy, even the most ignorant, are able to grasp mathematical truths, even though they are unable to attain the other sciences.
- Every doubt gives place to certainty and every error is cleared away if a subject can be reduced to mathematical proof.
- Therefore if other sciences are to be rendered free from error they must be founded on mathematical principles (1928, IV.2).

The mathematics Bacon has in mind is mostly Euclidean geometry, and he intends to use it not for computation but (somehow) to construct irrefutable proofs. Euclid's *Elements* is not only concerned with geometry; it also discusses the properties of numbers and proves some theorems about numbers and proportions. It is unfortunate, however, that the proofs are organized as syllogisms, because that form of argument is ill adapted to telling someone how to compute, and Euclid says nothing at all about computation. The Western world was incapable of any but the simplest arithmetic until Leonardo da Pisa, known also as Fibonacci, published his *Liber abaci* (which is not concerned with the abacus) in 1202. This extraordinary book taught the Arabic numerals and showed how to do arithmetic with them, as well as greatly advancing the study of algebra. But even fifty years later in Bacon's time, the new techniques had not spread widely.

Bacon is just as eloquent, and perhaps more persuasive, on the importance of experiment in science:

> I now wish to unfold the principles of experimental science, since without experience nothing can be sufficiently known. For there are two modes of acquiring knowledge, namely, by reasoning and experience. Reasoning draws a conclusion and makes us grant the conclusion, but does not make

the conclusion certain, nor does it remove doubt so that the mind may rest on the intuition of truth, unless the mind discovers it by the path of experience. ... For if a man who has never seen fire should prove by adequate reasoning that fire injures things and destroys them, his mind would not be satisfied thereby, nor would he avoid fire, until he placed his hand or some combustible material in the fire, so that he might prove by experience that which reasoning taught. (1928, VI.I)

As an example, Bacon mentions the notion, first reported by Pliny (1938, XXXVII.15) but apparently by Bacon's time demanded by reason, that a diamond cannot be broken except with the use of goat's blood. He tells us that goat's blood doesn't actually work and that professional diamond cutters do very well without it.

Bacon was in most ways a scientific as well as a theological conservative and the fabric of reasoning in his scientific works is scholastic to the core, but nevertheless he makes the radical claim that experience, as well as authority and reason, can lead to true knowledge. Experience need not be of the world of sense. It can also occur in the mind, as revelation, and all knowledge finally results from the direct union of the knower with the world. Religion is inspired by the experience of revelation, but there is another reason why revelation is necessary: through sensory experience we know only particular cases, and therefore knowledge of general truths must originate elsewhere.

Sitting in confinement, Roger Bacon had little chance to perform experiments, but the door he was banging on had already begun to open, for some of the greatest figures of the church were thinking the same way about experience, though they expressed themselves more moderately. Both Albertus Magnus and Thomas Aquinas were painfully aware that in most centers of learning the philosophy they loved had degenerated into an unproductive grinding of wheels, and they recognized that as long as there was nothing to prevent the selection of hypotheses on whimsical and arbitrary grounds Aristotle's fine logical mill would produce an uneven mixture of meal and chaff. Neither saint was adept at mathematics or agreed with Grosseteste and Bacon on its importance for natural science, but both emphasized the importance of knowledge derived from experience. Such knowledge could never conflict with knowledge from reason or revelation, for all truth issues ultimately from a single divine source. Experience, which necessarily pertains to single instances, and logical argument, which does not deal with them, illuminate truth from different sides. Albert, especially, wrote treatises on all branches of natural science and was probably the greatest biologist between Aristotle and the moderns. As long as that spirit of reconciliation prevailed, what we call science could do no more than accumulate data. There had to be confrontations and victories before it could begin to erect conceptual structures of its own.

Motion

Calls for a more mathematical physics were answered with a number of geometrical analyses of phenomena, especially optical, but for some time little was said about motion. Partly this was because motion was such a general concept, including every kind of change—what could one say about it except to relate it to its causes? But of course everyone realized that local motion was conceptually simpler and more amenable to measurement than the others and attention became focused on it, even if only as an example of the more general concept. How then could local motion be characterized mathematically?

The first problem was to characterize it at all. Begin with rest, absence of motion. Then comes what we call uniform motion, usually called equal motion. As early as 310 B.C. Autolycus of Pitane defines it: "A point is said to be moved with equal movement when it traverses equal and similar quantities in equal times" (Claggett 1959, p. 164). In the Middle Ages definitions like this often had to be rediscovered, and Thomas Bradwardine, writing about 1330, describes some of the efforts made to get ideas straight. For example: some people, he writes, think that local motion varies as the volume of the moving object, but if that were true then a whole object would arrive before its two halves (Claggett 1959, p. 220). Bradwardine, who was briefly Archbishop of Canterbury, belonged to a group of scholars at Merton College, Oxford, who worked on mathematics and on questions relating to motion. They arrived easily at the same definition of equal motion as the one Autolycus had given, but they never went further and expressed a velocity as, say, 10 miles per hour. They felt no more need for that than Euclid would have felt for specifying the size of one of his triangles. For them it was enough to say, for example, that if two objects move at velocities V_1 and V_2 during equal time intervals, then the distances D they travel would be in the ratio

$$D_1/D_2 = V_1/V_2.$$

The next problem was to define the velocity of a nonuniform motion, and this was done by another of the Merton College group, William Heytesbury, who wrote essentially what we now tell students of elementary calculus:

> a nonuniform or instantaneous velocity is not measured by the distance traversed, but by the distance that *would* be traversed by such a point, if it were

moved uniformly over such or such a period of time at that degree of velocity with which it is moved in that assigned instant. (Claggett 1959, p. 256)

In other words, freeze the instantaneous velocity and its magnitude is then defined as the value at which it is frozen.

So much, for the moment, for local motion. Motion in the more general sense was defined as the alteration of the form of the thing moved, and to represent this philosophers developed a concept called the *latitude of forms*. The latitude of a form is some measure of its intensity. Consider the surface of a wall with a lamp near one end of it. At each point on the wall, the intensity of illumination is one of the wall's accidental qualities and therefore one of its forms. Since the intensity is not constant over the surface, one said that it is *difform*, and since it diminishes continuously as one looks further away from the lamp, one said that it is *uniformly difform*. The idea is very naturally illustrated by drawing a graph. In the fifteenth century, with motion considered as an accident, scholars began to speak of the latitude of motion, which in the case of local motion is what we call speed or velocity.

Figure 8.2(a) represents the constant latitude of a form that does not change. Say that, for example, it is a velocity. Then the distance covered depends on both the magnitude of the velocity and the duration of the movement—on the area of the enclosed rectangle. If the motion is not local but something harder to characterize, like the ripening of a peach, the same considerations apply: the faster it ripens and the longer it ripens, the riper it gets.

Consider now the case of local motion that is uniformly difform. In such motion, the speed of the motion changes by equal amounts in equal times. Figure 8.2(b) shows the case of uniform deceleration from motion to rest. The Merton College mathematicians argued, in a roundabout way that is not really quantitative, that the total motion is again given by the area under the line. They found the area by the construction of Figure 8.2(c): evidently the total change is the same as would have occurred had the entire process taken place at a constant rate equal to the momentary rate at the midpoint of the time interval. This is the celebrated Merton rule, proved again and more neatly by Nicolas of Oresme in the following decades (Claggett 1959, p. 347; J. E. Murdoch and E. D. Sylla, in Lindberg 1978).

They are making progress toward a mathematical physics but two things are still lacking: no one has calculated the number that gives a velocity, and no one has said anything about the real world. Not until the late seventeenth century did anyone get around to mentioning the numerical value of a velocity in a scientific treatise. As to the real world, one might be curious as to whether the Merton rule actually applies to anything that happens there. In the real world a freely falling object

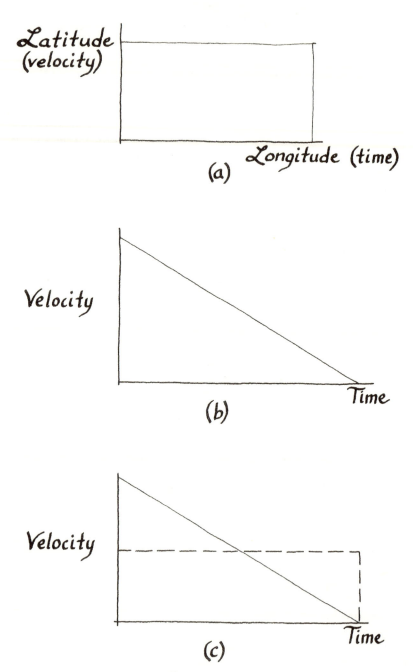

FIGURE 8.2. To illustrate the latitude of forms.

moves faster and faster. Is this perhaps an example of the uniformly accelerated motion contemplated in the Merton rule? Apparently the question of motion in general was much more interesting than the special case of local motion, for it was about three hundred years before Galileo answered "yes" to that question. Also, in the real world things sometimes move violently for a while under the influence of some cause other than the natural tendency toward the earth's center. What is the nature of the motion resulting from a given cause?

Nowadays physicists and mathematicians distinguish clearly between two branches of the science of motion, kinematics and dynamics. Kinematics formulates mathematical descriptions of motion, regardless of cause. Dynamics then picks out, from all the kinematically describable motions that might occur, the one that does occur. Dynamics therefore concerns itself with causes. The distinction was not quite so clear in the Middle Ages, but one can perceive it in the writings of the Merton school.

The starting point for all discussions of dynamics was the huge and portentous error made by Aristotle at the beginning of Book VII of the *Physics*:

Everything that is in motion must be moved by something.

That is, unless some sort of mover ("motor") is acting, there is no motion. Furthermore, everything that moves is moved by *contact* (*Physics*, 202a); there is no action at a distance. Thus the heavenly spheres must continually be kept going by a chain of influences originating in the Prime Mover, and if a thrown ball moves forward for a while before it falls to the ground, that must be because it is kept going by motions set up in the air.

In Book VII of the *Physics* Aristotle comes as close as he ever does to formulating a law of nature, but in that age of great mathematicians he flourished a very short sword and he never went beyond simple proportions. Assume without proof that a heavy object resists being moved in proportion to its bulk, and measure this resistance by a quantity R. Then a force (*dynamis*) equal to F produces a velocity V given by

$$V \propto F/R.$$

In this relation V is the velocity as we know it. *Dynamis* denotes effort in some way, but Aristotle does not say how it is defined nor does he say anything about R. In fact, this relation is not to be read as a quantitative formula at all, but only as a suggestion of the relation between force, resistance, and velocity when we push something.[3]

[3] Similarly, in our own time, when Claude Lévi-Strauss (1969) writes that jaguar = anteater^{-1} or rainbow = Milky Way^{-1}, he is not telling us to do arithmetic.

If we throw a stone into the air, R is its weight, which must be overcome before it can rise; if we drag a heavy box across the floor, R is the frictional force. The truth of the law was not widely questioned before the time of Galileo, but two exceptions ought to be mentioned. One of the most acute scientific commentators of the late classical period was John Philoponus, or John the Lover of Work, who toiled in Alexandria around A.D. 530. In his *Commentary on Aristotle's Physics* (Cohen and Drabkin 1948), Philoponus doubts whether a flying arrow can possibly be kept going by movements of the same air that also resists its motion, and points out an apparent absurdity in Aristotle's formula. The less air there is to keep an arrow going the smaller R is, and so the faster it will move. In the extreme case of motion in a vacuum, with no air to push the arrow, it would be moved infinitely fast, but by what? (Aristotle had already used the approach toward infinite velocity as an argument against the possibility of a void, but without pushing the argument further.) This might, for example, be the case with the motions of the celestial spheres, for the matter around them (if ether is matter) should be very tenuous, but it seems absurd to apply it to an arrow. To avoid such doubtful arguments Philoponus suggests

> it is necessary to assume that some incorporeal motive force is imparted by the projector to the projectile, and that the air set in motion contributes either nothing at all or else very little to this motion of the projectile (Cohen and Drabkin 1948, p. 223).

This "incorporeal motive force" which keeps the object going later gained the name of *impetus*; it corresponds to what we call momentum but we do not consider it as a force.

It is worth noting that Philoponus lived at the same time as Cassiodorus. Geographically, Alexandria is not far from the southern tip of Italy where Cassiodorus, compiling his encyclopedia, hoped to preserve for future generations some shards of knowledge that had once been alive. But the cultural distance between the Romans and the Greeks was greater than the geographical distance, for in Alexandria a few people were still thinking creatively, and if any encyclopedias were compiled there they have not come down to us.

In the Middle Ages some writers were skeptical of the Aristotelian theory, and perhaps the most successful counterproposal was made by John Buridan, who in about 1350 was rector of the University of Paris. The works of Philoponus had been discovered and translated before 1300 and Buridan carried his idea forward by specifying the nature of impetus: the impetus I of a moving body arises jointly from the quantity of matter it contains, M, and the velocity of its motion, V. In symbols,

$$I \propto MV.$$

I think Buridan really meant that there are two numbers and that they should be multiplied. When things have been properly defined this is what we mean by momentum today, but in Buridan's time there was nothing to do with the idea. A formula like this represented a quantitative relation only in a very restricted sense, for the question of how to generate actual numbers to put into it was far in the future.

Nevertheless, we can see in such speculations the dawn of a theory of dynamics. The formulas are not specially important. What is important is that concepts are beginning to be clarified: impetus is introduced and irrelevant notions like the role of air in projectile motion are beginning to fade; but it was not until the seventeenth century that the conceptual basis of physics, which is the intuitive picture in a physicist's mind of how the world works, what is relevant and what is not, became clear enough and correct enough to support a coherent dynamical theory.

We see in this chapter the Aristotelian concept of form reaching the end of its usefulness. I have mentioned how it unified some ideas that we keep distinct: sensory perception and the great complex of ideas belonging to light—vision, intensity of illumination, corporeity, physical cause—and finally, local motion and the beginning of techniques to deal with it quantitatively. With the departure of unifying concepts people began to think of separate problems in separate ways. This permitted the rapid development of certain fields of study while it subtracted, perhaps, from the feeling that one was approaching at least some distant view of God's plan for the world. We shall see as we go on that the hope of a limited modern version of such a view has not completely faded, though in modern terms it is expressed by the acronym GUT, standing for Grand Unified Theory. But more of that later.

CHAPTER 9

The Spheres Are Broken

The die is cast 'and I write this book. Whether it will be read by my contemporaries or by posterity is not important. If God himself has waited six thousand years for someone to contemplate his works, my book can wait for a hundred.

—JOHANNES KEPLER

THE last chapter told that although in the late Middle Ages the palace of truth was still to be approached only by reason, it was becoming more and more necessary for reason to be supported by experience. The relation between experience of the specific case and deduction of the general proposition was changing, and people were looking at nature with eyes no longer blinded by truths that must not be doubted. The turning point, the event from which the modern age in science is dated, turned out to be the introduction and final acceptance of arguments in favor of setting the sun at the world's center instead of the earth. One cannot speak of this as a discovery, for there are no milestones in the universe by which to judge rest and motion, but it turns out that the new view is much more fruitful scientifically than the old one, and of course it radically affects our own view of man's place in the universe to be told that we live on one of several planets of a certain star.

At first the new astronomy differed from the old largely in the way it described things verbally, but a century later it turned out that the new description could be expressed mathematically in a far neater and more precise way than the old. There were no spheres or circles in this description; the classical account of how it all works no longer applied, and the question became imperative: If it is no longer true that the planets move as they do because they are mounted on ethereal shells controlled by a Prime Mover, what new explanation can be found? Thus were born planetary theories in which the planets move under the control of forces exerted by the sun, and there began a search for the mathematical principles that explain why the structure and motion of the solar system shows various numerical regularities. The story told in this chapter shows how very hard it is to deduce fundamental principles from observed facts and it may give us increased respect for Aristotle, whose theory explained so much, so convincingly, for so long.

Copernicus

The third Nicholas, or Nicolaus, Copernicus (1473–1543), or Koppernigk, and there are many other spellings, was born in Thorn, or Torun, on the Vistula in what was then the kingdom of Prussia, now Poland. Though early dedicated to the church he managed to spend eight years in Italian sunlight living the life of a student before he returned in 1506 to the chilly marshes of Ermland, or Varmia, a semi-independent Prussian bishopric, where (without having become a priest) he had been appointed as a canon of the cathedral of Frauenburg, or Frombork. During his early years there his duties were light and he devoted himself to astronomy.[1]

With the Greek he had learned in Italy Copernicus had access to a great body of Classical texts. Concerning the moving earth he quotes Aristarchus and an expositor of Heraclides, and there is no doubt that he was familiar with the ideas of the relativity of motion that had already been expressed by Oresme and Cusanus, though he does not quote them. We have little information about his studies in Italy and do not know when he seized on the idea that occupied him for the rest of his life. He was reluctant to publish anything short of a definitive treatise, which he knew would take an immense amount of work, but in about 1512, at the urging of friends, he circulated a manuscript called the *Commentariolus*, or Brief Outline. It listed the hypotheses of his system: that the earth is a planet and that all the planets revolve on circles and epicycles around the sun, that the apparent daily motion of the stars results from the earth's rotation, and that the regressions of the planetary orbits are easily understood (Figure 9.1) as consequences of the earth's motion. The whole system of six planets plus the moon required 48 circles, each planet being given a

---

[1] In order to counteract the notion projected by Arthur Koestler in a fanciful book called *The Sleepwalkers* (1959), of a timid little man who did not really know what he was doing, it should be mentioned that Copernicus was an active and influential ecclesiastical statesman of wide-ranging intelligence and that he was not timid at all.

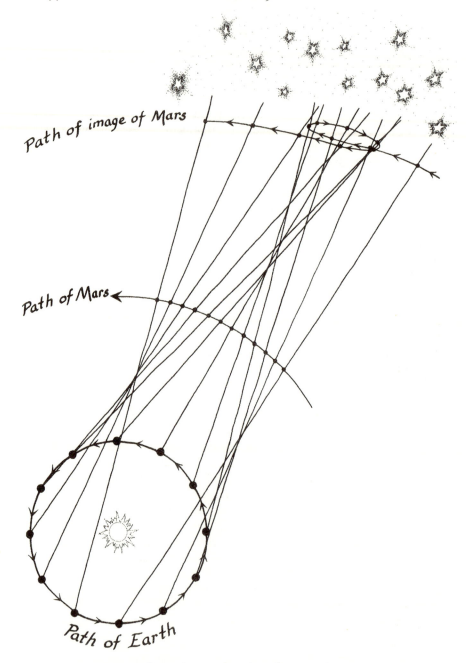

Path of image of Mars

Path of Mars

Path of Earth

FIGURE 9.1. Copernican explanation of retrograde motion.

large one for its orbit around the sun and a number of small epicycles to account for slight departures from uniform circular motion.

The manuscript was copied and many people read it. In 1532 a map of the world was published in Basel (Figure 9.2), in which the discerning eye can see an angel busily turning a crank at the north pole, and there is another at the south pole. A year later Johann Widmanstad, secretary to Pope Clement VII, explained the Copernican theory to His Holiness in the Vatican gardens and was thanked and rewarded. In his diminishing moments of leisure Copernicus worked on a detailed account of his theory but it was still incomplete when in 1539 an enthusiastic young man named Georg Joachim Rheticus arrived in Frauenburg and pushed him toward publication. Copernicus was by now sixty-six years old. He finished his manuscript in some haste and then allowed Rheticus to start work on a fair copy of the manuscript that was finally titled *De revolutionibus orbium celestium* (On the Revolutions of the Celestial Orbs). The term "celestial orbs" refers not to the planets but to the circles that carry them around, for Copernican astronomy was still Ptolemaic in every way but one: it put the sun instead of the earth at the center of the diagram. Of this long book, the first few pages explain the heliocentric model and are all the modern reader ordinarily has time for. The rest is mathematical astronomy.

To prepare his argument, Copernicus illustrates his point that absolute motion is undetectable by quoting not Oresme but Virgil: "Forth from the harbor we sail; the city and harbor move backward" (Aeneid, III.72). But the earth was not to be set moving with a quotation from Virgil, for it was anchored not only by theology but by science. In Aristotelian philosophy there is great emphasis on absolute motion. Stones fall with respect to the earth—that is, they fall absolutely—down is down, and if the natural motion of a celestial body carries it in a circle, that motion is defined in the same absolute sense. The sun, being above the moon, is made of ether and moves in a circle or combination of circles. Aristotelian physics forbids us to suggest with Virgil or Oresme or Cusanus that absolute motion is undefined, though of course there is nothing to prevent one from adopting a moving point of reference for the sake of argument or calculation.

On the subject of natural motion Copernicus contradicts Aristotelian physics. He claims that according to Ptolemy a rotating earth would fly apart (though I have not found this anywhere in Ptolemy); then he argues (1978, 1.8) that if the earth rotates it is a natural motion and not a violent one, and natural motion does not disrupt. Rotation is "the motion appropriate to its form," but the argument would have been incomprehensible to an Aristotelian, for motion of the earth as a whole could be neither violent nor natural. It could not be violent because violent motion must

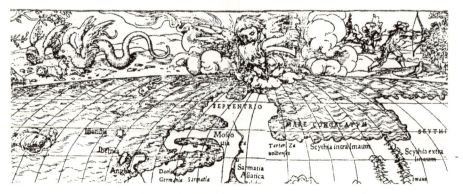

FIGURE 9.2. An angel turns the world. From Grynaeus, *Novus orbis* (1532).

be kept going by force from outside, and we find no such force acting on the earth. It could not be natural because the natural motion of the earth's materials of Earth and Water is down, not in circles. Copernicus argues that circular motion never needs to be kept going and dismisses Aristotle's classification of motion toward the center, away from the center, and around the center as a logical exercise. He needed to adopt this radical tone if he was to meet the most elementary objections, but it did not endear him to traditionalists.

The remaining five books of *De revolutionibus* describe geometrical models for each of the planets and tell how to compute tables. These constructions represent an enormous advance in technique and simplicity over anything previously available—even though, being based on circles and epicycles, they were no more accurate. The working astronomer of those days was involved in producing almanacs for agriculture and casting horoscopes for births, christenings, marriages, medical treatments, and the erection of buildings. He had no more inclination to ponder the conceptual foundations of the system than a modern automotive mechanic has to think about the thermodynamics of combustion.[2]

The book was printed in Nürnberg in 1543 and Rheticus moved there to oversee the publication, but he had to leave before it was complete and asked a friend named Andreas Osiander, well known as a Lutheran theologian and preacher, to take care of the final details. He did so, and one of his acts was to write a preface and insert it, without telling Copernicus, at the very front of the book, even before the dedication to Pope Paul III. Nürnberg is a long way from Frauenburg, and Copernicus, in failing

---

[2] Professor Owen Gingerich tells me he has examined nearly every extant copy of *De revolutionibus* and that while the pages of the technical sections of most copies are darkened with the grime of many fingers, those of Book I are mostly as white as snow.

health, had no control over what was happening. In the preface Osiander writes

> There have already been widespread reports about the novel hypotheses of this work, which declares that the earth moves whereas the sun is at rest in the center of the universe. Hence certain scholars, I have no doubt, are deeply offended and believe that the liberal arts, which were established long ago on a sound basis, should not be thrown into confusion. (Copernicus 1978, p. xvi)

Osiander declares that the astronomer's sole object is to save the phenomena and that for this purpose he may use any mathematical model he chooses. Copernicus puts the sun at the center and uses epicycles. Nobody believes that epicycles are physically real; hence nobody need believe that the sun is really at the center.

Osiander's motives have been much debated. It is unlikely that he could follow the book's astronomical reasoning. He must have found the line of argument in Book 1 inconsistent and philosophically naive; also he hoped to quiet the clamor that could be expected from the learned establishment. Why then is his act condemned as a terrible betrayal? Because in Book 1 Copernicus says explicitly that the ancients were wrong in putting the earth at the center and he gives clear physical arguments to show why they were wrong—for example, the immense sphere of the heavens, rotating once a day at colossal speed, would suffer a huge centrifugal force; yet outside it, according to universal (i.e., Aristotelian) opinion there is "no body, no space, no void, absolutely nothing" to contain it. The heliocentric model is not just a mathematical construction.

As to the reality of the orbs that carry the planets, Copernicus is silent. They occur in the title of his book, but he nowhere says what they are. Since a planet or the center of an epicycle is mounted on the equator of a turning orb, it really makes no difference whether the orb is considered to be a whole sphere or merely the circle of its equator, and in the *De revolutionibus* Copernicus seems to use the terms *orbis, sphaera,* and *circulus* almost interchangeably. If he had had anything to say about the matter he would have said it.

Early in 1543 Copernicus, then seventy years old, suffered a stroke and lingered half-conscious. The first copy of *De revolutionibus* was brought to him and put into his hands. Nobody knows whether he opened it and saw the preface. A few hours later he was gone.

Since the time of Ptolemy there had been astronomers but little making of astronomy. Mostly there had been arguments over undecidable points that had already been decided a thousand times, and readjustments of numbers to bring the tables up to date. For the professionals of his time,

TABLE 9.1
Average Orbital Radii of the Planets

Planet	Copernicus	Modern
Mercury	0.376	0.387
Venus	0.719	0.723
Earth	1.000	1.000
Mars	1.520	1.524
Jupiter	5.219	5.203
Saturn	9.174	9.539

Copernicus was just another such astronomer, and the value of his work was judged by the accuracy and convenience of the tables his method produced. An error of one degree in the computed position of a planet at the moment of a birth (or, equivalently, of four minutes in the time at which the birth was reported to have occurred) could somehow (I don't know how) produce an error of a year in the predicted time of a crucial or dangerous period in the life ahead. Accuracy was all, and the existing tables were not very accurate.

Nevertheless, Rheticus and a few others, pushed perhaps by a nascent Prussian nationalism, recognized the incomparable scientific talents of Copernicus. And, perhaps I should add, his fortitude. Possessed of an idea whose time had come[3] and sentenced to life in a cold provincial backwater, Copernicus set out to change the minds of the professionals, the only people he might reasonably expect to convert. "Astronomy is written for astronomers," he wrote, and after the twenty-five introductory pages he got down to business and presented in the following three hundred a comprehensive treatise on positional astronomy. He showed that with his hypotheses he could calculate good tables. Mostly he used other people's observational data, but with crude equipment he himself had measured a few stellar positions and observed a few eclipses, since an eclipse gives an accurate fix of the relative positions of sun and moon. Many years were to elapse before anyone measured the absolute scale of the solar system, but he was for the first time able to establish a relative scale. Table 9.1 gives his values for the radii of the principal orbital circles of the planets, on which their epicycles ran (rounded off to three decimal places), together with modern values of the average orbital radii, taking the earth's value as unity.

A few years after Copernicus died his cause was helped by another German, Erasmus Reinhold, who published a revised and enlarged version of his tables, the *Prutenic Tables*, which remained standard for sev-

[3] For a discussion that rounds out this cliché see Gingerich (1975a).

enty-five years. By the end of another seventy-five years the epicycles had been swept away, the heliocentric theory was almost universally accepted among astronomers, and *De revolutionibus* was firmly in place in the church's index of forbidden books.

Truth Is Beauty

Johannes Kepler was born in 1571 in the south German town of Weil der Stadt. After an exceptionally bitter childhood in which, however, his talents won him an education, he graduated from the University of Tübingen and shortly afterward, aged twenty-three, found a job teaching mathematics and astronomy at Graz, a small city in Austria. One of his duties was to publish an annual astrological forecast. When he correctly predicted a cold winter and an invasion by the Turks his position became secure, and in 1596 he was able to write and publish his first book, bearing the modest title of *Mysterium cosmographicum* (The Secret of the Universe, 1981; Kepler 1937, vol. 1). In it he constructs a version of the Copernican system that looks like Plato's theory of matter on a cosmic scale. He has found that if he properly nests the five Platonic solids he can roughly account for the orbital radii given by Copernicus. Starting with the sphere of Saturn he places inside it the largest cube that would fit. Inside that he places the largest sphere that would fit: this is (roughly) the sphere of Jupiter, and so on through the other solids and planets. The calculations were a good exercise in geometry. If one follows his figures the results are not impressive—but the world of experience never follows the ideal very closely. The fit is improved if, instead of the mathematical spheres of the mean orbital radii, one assumes spherical shells thick enough to hold the Copernican epicycles. Of course, people who thought the heavenly spheres were made of crystallized ether could hardly think the same of Kepler's polyhedra, for they would get in the way of the celestial motions. Kepler tells us that they are real and rigid but penetrable, whatever that means. Like Aristotle and Ptolemy he sees no reason why their physics should be that of ordinary matter. Thus at the outset of his career we can see Kepler's supreme ambition: to explain the why of the world in terms of structural and harmonic principles that God might have used. Nevertheless, he was pretty much alone in thinking he had made a stupendous discovery.

Kepler did not remain long at Graz. He next increased his knowledge of astronomy and the quantity of good planetary data available to him by moving to Prague as assistant to Tycho Brahe, the supreme observational astronomer. After Tycho's death eighteen months later he stayed on as Imperial Mathematicus at the court of Rudolph II, and it was during the following years that he solved the riddle of the planetary orbits.

He focused on the planet Mars; let us follow him. Tycho's data, taken mostly on a Danish island using great open-air instruments but no telescopes, rarely had errors even as large as a minute of arc (the moon's face is about 30 minutes in diameter). The observations were accurate and were made every night when weather permitted. Consider the intellectual problem. Given that we ride around the sun on a planet whose orbit is unknown, observing the position in the sky (not the distance!) of a planet whose orbit is likewise unknown: Find the two orbits.

At first sight this seems impossible, yet the orbits have a property that opens the way: they repeat themselves periodically. Choose a date on the calendar on which, in several different years, Mars was observed. Effectively the earth is standing still, and therefore even without knowing the earth's motion one can follow Mars around in its orbit. Further, Kepler knew (from Copernicus) that the Martian year is 687 days long. Thus, by collecting observations of Mars taken 687 days apart he could do the same thing again, effectively looking from a moving earth at a stationary Mars or, by reversing the directions of the sightings, from a stationary Mars at a moving earth.

Kepler tells of his work and his conclusions in his *Astronomia nova* (1609; Kepler 1937, vol. 3). It is a work unique in this world, a mathematical memoir. It tells of years of effort, first in one direction, then another. Hypothesis after hypothesis is made and discarded. Gradually a few results become clear: the sun lies exactly in the plane of each planetary orbit, a fact Copernicus had not noticed. Then a result of stunning simplicity: Mars, in its orbit, moves faster when it is nearer the sun, in such a way that the line from the sun to Mars sweeps out equal areas in equal times (Figure 9.3). This conclusion was first suggested by quite erroneous notions, to be explained later, concerning the reasons why the planets move as they do, but it was confirmed with data taken when Mars was at the apsides of its orbit, the points where it is nearest and furthest from the sun. Kepler could not believe that such a beautiful law would hold only at two points of the orbit of one planet and concluded, correctly as it turned out, that it applies over the entire orbit of every solar planet. It is now known as Kepler's second law of planetary motion—second because it is logically second, though it was first in time. We shall see in Chapter 10 how he was led to this law by essentially medieval theories of causation.

There remained the problem of finding the shape of the orbits. Kepler chose to study Mars, and his conclusion, that it is an ellipse, was arrived at in a way that is so extraordinary, so simple, so unparalleled in history,

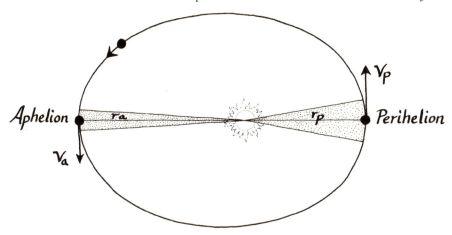

FIGURE 9.3. Kepler's second law: the radius to the planet sweeps out equal areas in equal times.

that I will present it here, not as Kepler found it, but as he might have done if everything had gone smoothly.

First, a remark about ellipses. To construct an ellipse, put two pins into your paper and make a loop of thread that fits loosely around them. Slip a pencil into the loop, pull it outward until the thread is tight, and draw a closed curve by moving the pencil all the way around the pins. Figure 9.4 shows the result, with a few labels attached. The places of the pins are called the foci of the ellipse; the distances a and b are the semimajor and semiminor axes. The closer the foci are to one another, the more the ellipse looks like a circle. The distance f, which measures the separation, is related to a by

$$f = ea ,$$

and this defines the quantity e, called the eccentricity, which varies between 0 and 1. If it is 0, the ellipse is a circle.

Now I must derive a property of ellipses that one does not notice at once. Let the pencil be at P. The loop of thread passes around the point marked O; its length is $2a + 2f$. Now let the pencil be at Q. The length is $2r + 2f$. Since the thread does not stretch, these quantities are equal and

$$r = a.$$

From Tycho's observations, after many struggles, Kepler had established that the orbit of Mars is some kind of squashed circle that looks like

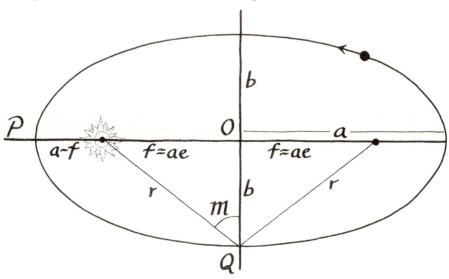

FIGURE 9.4. Anatomy of an ellipse. (Actual planetary orbits are nearly circular.)

Figure 9.4, though he did not yet know it was an ellipse. He knew that the ratio of the greatest to the smallest diameter is

$$a/b = 1.00432.$$

He also knew that the angle M is $5°18'$. The ratio r/b is called the secant of M, and one day he happened to look up its value in his trigonometry tables. It is 1.00429. He remembered "as if waking from a dream" having seen the previous number (1609, p. 267). He knew the data were not perfectly accurate; were the numbers trying to tell him that the two quantities are really the same? If so, what follows? If $r/b = a/b$, then $r = a$, which, as just shown, is a property of an ellipse. You, Reader, see the conclusion, but Kepler did not, since in those days ellipses were not in the astronomer's toolkit. The chase took him up and down many more hills; finally he discovered that the relation $r = a$ is true of all ellipses and guessed the truth, which he could then verify by further calculations from the data. When generalized to all the planets it is known as Kepler's first law: the paths of the planets are ellipses having the sun at one focus. Note that here, as with the law of areas, Kepler guessed the whole truth about the solar system from two points of a single orbit.

The phrase "as if waking from a dream" tells what Kepler thought about scientific discovery. Later, in the *Harmonices mundi* (1619, v) he wrote "to understand is to compare what is externally perceived with one's own ideas, and to judge that they agree. Proclus expressed this very beautifully with the word 'awakening,' as from sleep" (1937, vol. 6, p. 226).[4]

It is possible to create elliptical orbits by combining circular motions that take place at nonuniform rates, but if the spheres do not revolve uniformly there is no point in having spheres at all. Kepler destroyed them, and by the end of the century intellectuals impatient with traditional philosophy had swept them up and thrown them away. It helped that the circle was no longer regarded outside the universities as the most beautiful geometrical form. The circular arches of Roman and medieval bridges were being supplanted by the dynamic spring of elliptical arches, and Galileo's preference for circles, to be discussed below, must be taken as an example of conservative taste.

Celestial Harmonies

Nine years later Kepler made his third great discovery. The first and second laws refer to the planets individually, but the third law joins them all. It has always been known that the outer planets take longer to complete their orbits—this is why they were considered to be further away. We have seen in Table 9.1 Copernicus's values for the mean orbital radii. Table 9.2 lists Kepler's slightly better values, the orbital periods, the cubes of the radii, the squares of the periods, and, to show how nearly circular the orbits are, modern values of the orbital eccentricities *e*.

Columns 5 and 6 of Table 9.2 yield Kepler's third law, first published in the *Harmonices mundi* (1937, vol. 6, p. 302). In modern terms, "measured in terms of the values for the earth, the cubes of the mean orbital radii of the planets equal the squares of their periods." It is also called the harmonic law, since it involves not a physical mechanism but a relation between the numbers of the solar system and thus belongs to the Pythagorean tradition of harmony as law. The table shows that in this law the moon is not to be considered a planet. It circles the earth, whereas the planets circle the sun.

In order to explain the law of areas Kepler developed a dynamical theory that explained how planets move around the sun. Though the theory is false, it contains an innovation that perhaps makes it worthy to be called the beginning of modern physics: the reason why planets move as they do is assumed to lie not in the planets' inherent properties (e.g.,

---

[4] Wolfgang Pauli's study, "The influence of archetypal ideas on the scientific theories of Kepler" (Jung and Pauli 1955), collects Kepler's writings on the origins of scientific ideas and adds Pauli's own reflections.

TABLE 9.2
Planetary Data Compared

1 Planet	2 Radius R(AU)[a]	3 Period P(days)	4 Period P(years)	5 R^3	6 P^2 (years²)	7 Eccen- tricity, e
Mercury	0.388	88	0.241	0.0584	0.0580	0.206
Venus	0.724	225	0.616	0.3795	0.3795	0.0068
Earth	1.000	365.25	1.000	1.0000	1.0000	0.017
Mars	1.524	687	1.881	3.540	3.538	0.093
Jupiter	5.200	4333	11.863	140.61	140.73	0.048
Saturn	9.510	10759	29.456	860.08	867.69	0.051
Moon	0.00257	27.32	0.0748	1.7×10^{-8}	0.0056	variable

Reproduced, with additions, from Gingerich (1975b).
[a] AU stands for astronomical unit, a unit of distance equal to the mean radius of the earth's orbit.

he did not say that the natural motion of a planetary body is in an ellipse or that it is carried on a moving belt), but in a force exerted by the sun. This is the first attempt in history to formulate and apply quantitative mathematical laws relating motion to applied force, and it is discussed briefly in the next chapter.

Kepler made other discoveries connecting the different planetary orbits that were even more purely Pythagorean in spirit. According to the area law a planet moves more quickly when it is at perihelion, the point of the orbit when it is closest to the sun, than when it is at aphelion, the most distant point. He computed the angular velocities of the planets, as viewed from the sun, at these two points of their orbits, expressing them as the number of minutes of arc traversed during one day. Table 9.3 gives the results for the three outer planets (the others are similar), and shows how close the ratios of these velocities, in the same orbit and between different orbits, are to some of the ratios that occur in Pythagorean harmony. On these harmonic principles Kepler goes on to write in musical notation the music of the planetary orbits, no longer music of the spheres.

The third law, the other principle of celestial harmony, is a cornerstone of modern celestial mechanics, but the subtler harmonies just mentioned have died away. The reason is that the third law holds for all planetary systems, however they may have started, whereas the harmonies just exhibited are peculiar to our solar system in our own epoch. The orbital elements are slowly changing and we happen to catch them at an opportune moment. Why do they come out so perfectly? For two reasons: first, they don't come out perfectly; you have to fudge, and if you fudge a little it is not hard to fit randomly chosen ratios with simple fractions. Second,

TABLE 9.3
Planetary Angular Velocities and Their Ratios

Planet	Velocity at aphelion		Velocity at perihelion	
Saturn	(a)	1'46"	(b)	2'15"
Jupiter	(c)	4'30"	(d)	5'30"
Mars	(e)	26'14"	(f)	38'1"

Ratios

$a/b = 3.93/5 \approx 4/5$ (major third)
$c/d = 4.91/6 \approx 5/6$ (minor third)
$e/f = 2.07/3 \approx 2/3$ (fifth)
$a/d = 0.96/3 \approx 1/3$ (octave plus fifth)
$b/c = 1.00/2 \approx 1/2$ (octave)
$d/e = 5.03/24 \approx 5/24$
$c/e = 2.07/3 \approx 2/3$ (fifth)

Source: Kepler, *Harmonices mundi* V.4 (Kepler 1937, vol. 6, p. 302).

Kepler looked at several other sets of orbital parameters before he settled on this one. If one throws away data one doesn't like and then massages the rest one can prove almost anything. So today we have forgotten these harmonies but for Kepler they were his supreme discovery, pointing to the world's formal cause; for they revealed more clearly than any other evidence that the Creator had designed the world according to principles of mathematical beauty. The first and second laws told how; the harmonies told why. Ironically, the solar system does embody harmonies of exactly the kind that Kepler was looking for (King-Heale 1975), but he never noticed them and they are seen today not as the result of a free choice made at the beginning of things but as relations that naturally come into being in a system of satellites that runs for a long enough time undisturbed.

For Kepler what, finally, was the universe? Toward the end of his career he wrote an *Epitome of Copernican Astronomy* (1937, vol. 7; bks. IV and V, translated in Ptolemy, Copernicus, and Kepler 1952) intended for the general reader, in which he explains not only how the universe is arranged but why it is arranged that way. The explanation takes many pages; I mention only two topics. First, the form of the universe. It consists of three parts that symbolize the Holy Trinity: at the center the spherical sun, at the outside, concentric with it, the sphere of fixed stars, and in the intervening space the planetary system. Figure 9.5 shows the dimensions Kepler derives for the system (IV.4), all conjectured from numerological

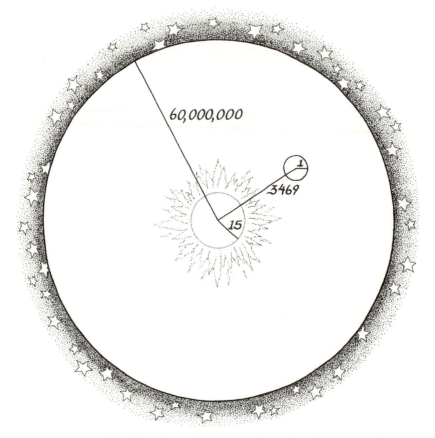

FIGURE 9.5. The dimensions of Kepler's universe (not drawn to scale).

arguments since Cassini's measurement of the distance to the sun was 50 years in the future.

So Kepler did not break all the spheres. Aside from the planets he kept two, the sun and the outer one with stars mounted on it. He seems to have had no interest in stars. The medieval universe was small enough so that they could shine with the sun's light but Kepler put them much too far away for that, and he must have thought they were self-luminous. Why did he retain the outer sphere? First, I think, because the ether that fills the universe must have a boundary to contain it. Second, the idea of a universe of concentric spheres was almost two thousand years old and had survived many vicissitudes. It must have had a powerful grip on the imagination of mankind. But the important reason, not separate from the other two, is that for Kepler a sphere represented the three persons of the Trinity; for it combines the three separate dimensions of space into the only figure in which they cannot be distinguished: who can say what point of a sphere ought, intrinsically, to be called the top?

Worn out by his exertions, by religious persecution for his liberal Protestant views, and by poverty, Kepler died in his sixtieth year while journeying to the Emperor's court at Regensburg to beg for enormous arrears of salary earned and promised long before. His talents have been compared to Mozart's: the fertility of ideas, speed of execution (he once wrote a book in a week), and, in the slow decline of a worsening personal situation, the same indefatigable energy. It is remarkable that his three laws are the *only* three exact and general mathematical laws of planetary motion, applying not only to this but to all similar planetary systems. And he contributed a further revolutionary idea: that the planets move in their orbits not because it is their natural motion, not because they are guided by divine intelligences (which he thought for some time they were), but because the sun exerts a force that causes them to move as they do. That was one of his best ideas.

Can a life and mind like Kepler's be understood? Of course not, but in the *City of God* Augustine makes a remark that Kepler, who with eyes weakened by smallpox had read so much, probably knew: "Man has no reason to philosophize except to attain happiness" (XIX.1.3).

The Triumph of Pythagoras

It would be wrong to think of Kepler as a solitary eccentric in pursuit of numerical fantasies, for it was during his time that Pythagoras's reputation rose to its summit. In the Renaissance he was revered as the first thinker who ever had the modesty and wisdom to call himself not *sophos*, a wise man, but *philosophos*, a lover of wisdom. He was a law-giver, an expert in magic, the first monotheist (and hence proto-Christian). He founded each of the four disciplines that made up the *quadrivium*, the arrangement of all secular knowledge as set down by Boethius, which formed the curriculum of medieval studies: arithmetic (numbers in themselves); music (numbers in motion); geometry (mathematical form); and astronomy (mathematical form in motion). Plato, the passion of the Renaissance, was said to be ninth in the line of his intellectual descendants, and the numerical harmonies explicit in the *Timaeus* were thought to be implied though unexpressed in Plato's other works. Pythagoras went robed in white with a gold crown on his head; his golden thigh betokened divinity, and to him was attributed a comprehensive vision of man and nature, of ethics and science, united in the harmony of form and number that affected Renaissance thought to an extent that is only now beginning to be appreciated.

During the Renaissance the doctrines of numerical ordering were nowhere more widely articulated than in works on architecture. Once manuscript-scouts had found a few copies of the *De architectura* of Vitruvius, Italian architects went hurrying about Rome to study the proportions of the classical buildings that remained and verify the numerical principles they had read in Vitruvius. On the whole, the enterprise was not very successful. Like Pythagorean harmony, classical architecture tended to base its proportions on the ratios of small integers, but they were not constrained by such prescriptions. Their numbers were not integers crudely imposed on the design, for columns swelled and tapered and were spaced more closely at the ends of a platform than in the middle, and who therefore could say afterward just what the numbers were?

In the same way, the Renaissance humanists thought it was more important to plan the Pythagorean proportions into a work than to have them obvious when it was done. From Pythagoras they learned that the tones of music can be expressed as numbers measured in space and from Vitruvius that harmonious relations in music have their counterparts in harmonies of visual proportion. From Plato they learned that God ever geometrizes and that his Ideas are shapes and numbers. First was the universe, constructed on the mathematical principles that Galileo and Kepler would later try to grasp; then came Man, the microcosm. Every aspect of human existence reflected relations in the great and largely unknown universe. Mind, thoughts, body, the structure of society—they all imaged the mind and thoughts that govern the universe and the God-given ordering of angels, planets, and the other hierarchies of nature.

The numerical principles of harmony appeared in literature as well as in the visual arts. An example is Spenser's *Faerie Queene*, the second part of which appeared in 1596, the same year as Kepler's *Mysterium cosmographicum*. Modern criticism (Fowler 1964, Heninger 1974) shows that Spenser's vision of the world as expressed in numerical structure and imagery was as pervasively Pythagorean as Kepler's. The entire structure of *The Faerie Queene* is numerological: the numbers of characters and their names and personalities, the numbers of episodes and their arrangement, the planetary symbolism all show the poet's determination to express in his work the principles that order the world.

In music the Pythagorean tradition as transmitted by Boethius was particularly durable. Although Boethius seems to have been studied mostly in the universities (Carpenter 1958) his indirect influence among practical musicians remained, and more than a century after Kepler and Spenser, when both astronomy and poetry had gone on to new preoccupations, musicians still felt the pull of the old numbers. I have mentioned that in the *Timaeus* (35ff) Plato relates the structure and motions of the world to certain numerical ratios. David Humphreys (1983) has analyzed compositions of J. S. Bach for their use of the numbers of the *Timaeus*, "a work which Bach undoubtedly knew well and which forms the best exposition of the numerical world-order Bach intended to depict." The details of Bach's music express a conviction that the whole world is governed by Pythagorean harmony. Bach seems to have taken this over and incorporated it into the tonal and dynamic structure of at least certain compositions. The analysis is so complicated that I regret I cannot summarize it here, for it is no more possible to detect Bach's numbers by hearing the music than it is to find the Pythagorean correspondences between form and number by looking at the sky.

A Message from the Stars

As soon as it was published, Kepler sent a copy of his *Mysterium cosmographicum*, with its nonsense about nested Pythagorean solids, to the eminent Florentine mathematician Galileo Galilei (1564–1642). The addressee avoided saying what he thought of it by a device often used by people who receive books that do not appeal to them: he replied at once, pretending that he wanted to express his thanks even before examining the gift. He said he was glad to see that it was based on the Copernican theory, which he had accepted many years ago, but that since the immortal Copernicus was still an object of ridicule among fools he had not dared to say anything in public. The fools, then and later, were not high churchmen but university professors for whom the teaching of philosophy still amounted, by definition, to expounding Aristotle. From manuscript notes prepared for his students at Pavia and later published under the name *Treatise on the Sphere* (1968, vol. 2, p. 202) we can see that a decade after his letter to Kepler he was still publicly expounding the Ptolemaic system in traditional terms: the stellar sphere, Earth and Water and their natural motions, the immobility of the earth (from the fact that birds and clouds are not left behind). He does not go into mechanical details but a casual reference to "the sphere and its circles" suggests that he has no new thoughts.

About 1604, someone in Italy made a telescope by mounting two lenses in a metal tube, but scientists had not yet turned their attention to lenses and the invention passed them by. Four years later the telescope was invented again in Holland and in 1609, simultaneously with Kepler's publication of the elliptical orbits, one reached Italy. At once Galileo started grinding lenses, mounting them in increasingly long tubes, and looking upward. He discovered the four principal moons of Jupiter, revolving around the planet so quickly that their positions changed from night to night. Then he saw that the Milky Way was composed of countless stars, an observation that suggested, though it did not require, that some stars may be very much more distant than others. Simultaneously with other observers he studied the mountains on the moon and tried to measure their altitudes. He observed sunspots and he discovered two new planetary phenomena: that the image of Saturn has a complicated shape involving loops of some kind, and that Venus shows phases like the moon's. Galileo announced these discoveries in a book called *Sidereus nuncius* (A Messenger [or Message] from the Stars); it was his first book, and he was forty-six years old. Though still cautious about advocating the Copernican model, he suggested that the moons circling Jupiter might resemble planets circling the sun, thus upsetting some of the more conservative savants to the point that they refused to look through the telescope at all or else declared that Galileo was using the instrument to produce optical illusions.

We should not think of the images Galileo saw as being like those we see through a telescope of moderate power today. Modern lenses are made by advanced techniques, using optically homogeneous glass. If the density of the glass varies from point to point, if it has bits of crystal in it, or if the lens is not perfectly shaped, the image is fuzzy and distorted and the same star may even be seen in two places at once. Also the first instruments were very hard to use. Their mountings were primitive and their fields of view so narrow that an object swam out of sight almost as soon as it was located. It was easy for conservative people to question Galileo's good sense in basing positive arguments on such pale and fuzzy images.

Take a single example—sunspots. The sun, orbiting above the lunar sphere, was supposed to be composed of pure ether and immune to alteration; yet the telescope showed dark smudges on its face, in no visible order or pattern, which altered from week to week. It was natural to attribute these observations to conditions of the earth's atmosphere or to some artifact of the telescope. Illustrations from the period show the dilemma. Figure 9.6 reproduces the frontispiece of a book on sunspots by Christoph Scheiner, s.j., a bitter opponent of Galileo (Scheiner 1626). The sun (*sol*) lights the lower part. At the right, a telescope projects an image of sunspots on a sheet of paper. This is labeled *sensus*, the evidence of the senses. To the left is a book labeled *auctoritas profana*, secular authority, its little lantern supplementing the sun's light. This book contains the kind of knowledge the senses provide, and no more. Above these images the source of illumination is divine (ihs). *Ratio*, reason, perceives by this light that the order of the spots is regular, and the same light shines on the book of *auctoritas sacra*, where real truth is written.

By 1610 Galileo was in an excellent position to make his views known. He was highly regarded in Rome and on good terms with Pope Paul V, and he had just been appointed Court Mathematician to Cosimo de'

FIGURE 9.6. According to the senses sunspots are disordered; according to reason they are neatly arranged. From C. Scheiner, *Rosa ursina, sive sol* (1626–1630), courtesy of the Linda Hall Library, Kansas City, and Dr. William Ashworth.

Medici, Grand Duke of Tuscany. It seems to have been the conservatives' unwillingness to accept his telescopic discoveries, more than his long-standing conviction in favor of Copernicus, that moved him to take the offensive. The very fact of his new appointment showed that the *auctoritas profana* that he personified was extending its scope, and that the *auctoritas sacra* was accordingly diminishing. By his brilliant and abrasive polemical writings Galileo hastened what was already in progress and forced into polar opposition the representatives of sacred and profane authority who without him might very well have resolved their differences peacefully.

The story of Galileo's confrontations with the Inquisition and his eventual defeat and punishment has been told often and well (e.g., by de Santillana 1955, Langford 1971) and it is not relevant to this account. What is relevant is that to the end of his life Galileo, the great scientist and public figure, the originator of a quantitative science of mechanics, the center of the scientific culture of his day, believed in the Copernican system as Copernicus himself had given it—eccentrics, epicycles, and all. In his *Narrative and Demonstrations Concerning Sunspots* (1611) he tells exactly what he thinks on the subject:

> I do not deny the circular motions around the earth or another center (slightly) different from the earth's, and even less other circular motions entirely separate from the earth. . . . I am most certain that there are circular motions described by eccentrics and epicycles, but that to execute them nature really uses the farrago of spheres and orbs the astronomers talk about—this, I think, is unnecessary to believe, since they are adopted in order to facilitate astronomical computations. My own opinion lies between that of those astronomers who assume not only that the planets move eccentrically but that there are eccentric orbs and spheres that actually guide them, and those philosophers who equally deny both the orbits and any motions around centers other than the earth's. (1968, vol. 5, p. 102)

I understand this to mean that the planets follow the Copernican epicyclic paths without needing any mechanical apparatus to guide them. Galileo had seen some of Kepler's books and disapproved of their undisciplined volubility, but even if he failed to plow through them, it is known that he was familiar with the first and second laws, if not the third. Galileo's work was in mathematics and physics. He built telescopes with which he made a number of important astronomical discoveries but he was not much involved in positional astronomy, and so perhaps he never appreciated Kepler's achievement in describing planetary motion with very simple and straightforward mathematics, far more accurately than could be done using the machinery of Ptolemy and Copernicus.

As he thought about the natural world, Galileo followed the fourteenth-century tradition in being more interested in relations than in numbers. Unlike Kepler he had little idea of causal laws, and for him the mathematization of physics amounted to a description, in clear and exact mathematical language, of observed natural phenomena. In those days there were no numerical data in physics to match in accuracy those of astronomy. Kepler needed to get quantitative agreement between theory and data, while both he and Galileo hoped to describe, in the transparent language of mathematics, how things really are. Geometry was almost exclusively the study of what could be constructed and proved using two tools, a compass and a straight-edge. There were, though, a few departures from this rule. The theory of conic sections—ellipses, parabolas, and hyperbolas—had been worked out in detail by Euclid and Apollonius, but an ellipse cannot be constructed by compass and straight-edge; you need two pins and a piece of string, and the need for those plebian materials served to isolate conic sections from the great tradition. When God drew a circle on the face of the deep (Proverbs 8:27) he used a compass, or so everyone assumed (Figure 9.7); he certainly never used two pins and a piece of string.[5]

So finally Galileo was trapped into the inconsistency of declaring that one of the Copernican hypotheses, the centrality of the sun, must be taken as literal truth, while the eccentrics and epicycles were mathematical but not physical facts. His failure to profit from Kepler's explanation of planetary motion weakened his arguments and may have contributed to his defeat, and later it drew a wry comment from Einstein: "That this decisive step left no trace in Galileo's life work grotesquely illustrates the fact that creative minds are often not receptive" (Galileo 1967, p. xvi). I have already quoted Gaston Bachelard's remark about scientific explanation: "It does not consist in replacing what is concrete but confused with what is theoretical and simple, but rather in replacing confusion with an intelligible complexity." For Galileo it was Copernicus rather than Kepler who offered the intelligible complexity.

Sense and Matter

In 1577 a comet appeared and Tycho Brahe observed it carefully. Since nothing above the sphere of the moon was ever supposed to change, comets, like meteors, were considered to be phenomena in the upper atmosphere, and Aristotle (*Meteorology*, 344a) had explained them plausibly. Above the earth's atmosphere, ether fills the sphere of the universe. This sphere rotates once a day and the nightly motion of the planets,

[5] For more on the reasons for Galileo's stubbornness, see Panofsky (1954).

FIGURE 9.7. The Creation. French manuscript illumination, thirteenth century, Vienna.

which are visible concentrations of ether, shows that the ether turns with it. But the earth's atmosphere does not turn, and so there must be turbulence and friction in the boundary layer between the sphere of fire, attached to the earth, and the turning ether. Comets and meteors were phenomena produced by the heat of this friction. Tycho, in observing the comet, looked for signs of parallax that could have enabled him to establish the height of the layer, but he found none. He concluded that the comet was far above the atmosphere, and in fact far above the moon, a novelty in a region where novelties should not occur. Aristotelian physics is a theory that explains very well our qualitative observations of the world, but here was a quantitative test that it appeared to fail.

The theory, and the entire mode of argument on which it rested, was already under attack from many sides. Copernicus had written that the earth is not at the center of the spheres but this could not be demonstrated by astronomical measurements and so his challenge was on a doctrinal plane. The challenge issued by Tycho's measurement was different. Either Tycho was wrong or the carefully shaped traditional theory of nature was wrong, and the discomfort produced by this situation boiled over into controversy when in the autumn of 1618 three conspicuous comets were seen in quick succession. Much of what was written was irrelevant to the central issue, but it roused the elderly Galileo to write a book replying to criticisms of some of his opinions and ridiculing the science of his opponents.

Galileo was probably the best Italian prose stylist of his time and his sarcasm is merciless as *The Assayer* cheerfully tears the opponents to pieces. It cannot have helped his popularity in official circles a few years later, when his troubles began. In other places *The Assayer* is a very serious book, for Galileo took the occasion of a scientific squabble (in which he was far from being an innocent victim) in order to set out his idea of what science ought to be. Since the Copernican theory, which he had been warned not to defend, was not involved, he could express himself freely,

> I seem to detect in Sarsi [one of the disputants] the belief that to philosophize one must always lean on the opinion of some celebrated author, as if our own minds were condemned to remain childless unless married to the discourse of someone else. Perhaps he thinks that philosophy is a work of fiction written by some man, like the *Iliad* or *Orlando Furioso*, in which the least important thing is whether what is written in them is true. Signor Sarsi, that is not how it is. Philosophy is written in that great book—I mean the universe—that forever stands open before our eyes, but you cannot read it until you have first learned to understand the language and recognize the symbols in which it is written. It is written in the language of mathematics and its symbols are triangles, circles, and other geometrical figures without which

one does not understand a word, without which one wanders through a dark labyrinth in vain. (1968, vol. 6, p. 232)

Galileo's boundless memory, which was rumored to contain all of Latin poetry, may also have held some bits of Robert Grosseteste, for the tone of this passage is remarkably similar to one I have already quoted from him. Each writer expressed his vision of a time when geometrical principles, like those of optics, would be used to build a mathematical philosophy of nature. Except for some elementary principles of statics, Kepler's planetary laws, and Galileo's own law governing the motion of falling bodies (Chapter 11) there were not yet more than hints of what such a theory could be, but Galileo's certainty that this was the path to follow is shown in his comparison of human knowledge to God's. The human mind grasps so few mathematical propositions that, compared with God's knowledge, their quantity is nothing. But with regard to those it does understand, proclaims Galileo's spokesman Salviati in the *Dialogue on the Great World Systems* (Galileo 1968, vol. 7, p. 128; 1953, p. 114), "I believe its understanding equals the divine understanding in objective certainty, for it understands the necessity of things, and nothing can be more certain than that." In other words, to the small extent that we are able to understand the world mathematically, we are understanding its necessary truths as God does. We can imagine the voice in which Simplicio replies "That seems to me a very bold remark," and in fact it might have been a fatal remark had not Galileo's dialogue form allowed him to include Simplicio's protest.

Cardinal Maffeo Barberini had been a friend and admirer of Galileo, but as Pope Urban VIII there were words he could not overlook. Every reader of the *Dialogue* knew very well that Salviati represented Galileo as he set his companions straight on one question after another, and that Simplicio (the name is enough) represented the Pope. It was not just Galileo's Copernicanism that got him into trouble with the Vatican.

The following year, as the rope began to tighten around Galileo's neck, Francesco Niccolini, who was the Tuscan ambassador to Rome, described an interview with the Pope in which he tried to explain that the solar system Galileo described was not an impossibility. He said that Galileo specially emphasized God's freedom in the matter by writing that, since he could have made the universe in infinitely many ways, one of them could have been the Copernican way. But the Pope was not to be deflected from the main point. "Red in the face, he replied to me that the blessed God is not to be necessitated, and I, seeing him so enraged, did not wish to dispute him in matters of which I know little." Niccolini changed the subject (Galileo 1968, vol. 15, pp. 67–68). Galileo's written words illustrate his lack of finesse; the Pope's outburst shows that he saw where the real threat lay, and that Galileo's unorthodox astronomical views were not the main reason behind his approaching arrest "on vehement suspicion of heresy." The Pope could not permit the necessity expressed in that "language of mathematics" (see also Wisan 1986).

One of the many questions at issue in the exchanges of sarcasm prompted by the arrival of the comets was how the light and heat of comets could be produced by friction in the upper atmosphere, and this led Galileo in *The Assayer* to consider the nature of light and heat, and of sensory information generally. Light, even if its nature was unknown, was studied objectively in the science of optics, but heat was known only as a sensation. In the Aristotelian tradition sensation is a message which, transmitted by some medium, causes the sensitive part of the soul (*On the Soul*, 431a) to participate in the substantial form of the thing perceived. The substantial form was present in both places at once. Sensations of heat and cold bring us news of the relative balance of the qualities of heat and cold in the four elements that compose a thing. But it is absolutely impossible to bring such perceived qualities under the dominion of mathematics in order to understand heat as God does, and so Galileo was led to reflect on the nature of sense data in general and what their role is in a scientific account of the world. He probably knew the words of Democritus, "Sweet exists by convention, bitter by convention, color by convention; atoms and void exist in reality"; at any rate, his thoughts led him back to the distinction between primary and secondary qualities and to question the relation between the perceiver and the perceived. The reality underlying our sensations, he argued, must be the reality of a few primary qualities: size, shape, weight, position, and velocity. This is the reality of the measurable properties of atoms. And at this point in the discussion, violating his custom of not speculating about subjects out of range of observation, he becomes an atomist.

First, he must try to establish that there is a wide difference between primary and secondary qualities: "I have been thinking that these tastes, odors, colors and so on, for the subject in whom they reside, are only words pertaining to the sensorium" (1968, vol. 6, p. 348; the sensorium is that part of the organism in which sensory data were thought to be collected and interpreted). Galileo then gives an example of a sensory experience that clearly originates in the sensorium: tickling. The sensation can be very intense; yet obviously it originates in us and not in some special property of the hand or the feather that is tickling us. Next he suggests that some substances give off tiny particles which, striking parts of our bodies that are far more sensitive than the skin with which we feel the sense of touch, produce sensations of taste and smell according as the particles differ in shape, velocity, and number. Sounds, also, can be explained, "not as the result of some special sonorous or transonorous property," but as the result of vibrations of the air that surrounds us. "I think that if ears, tongues, and noses were taken away, then shapes, numbers, and motions would remain but not odors, tastes, or sounds, which, except as applied to living creatures, I take to be nothing but words" (p. 336).

Galileo next considers heat, which he takes to be the sensation produced in our skin by the arrival of certain very fine particles such as originate in fire:

> But apart from the shape, number, motion, penetration, and touch of its particles I do not think fire has any other quality that is heat. . . . If we rub two hard objects together, either we grind their constituents into very small flying particles or we open an exit for particles already inside, and when these moving particles penetrate our bodies and move through them, our *anima sensitiva* [the sensitive part of the soul] feels the touch of these particles, pleasantly or painfully, as the sensation we call heat. And perhaps, if the rubbing stops at producing very small particles, they move only for a short time and produce warmth, but if the process continues until finally it liberates atoms, the smallest particles of all, then light bursts forth and expands instantaneously through space by virtue of its—I do not know whether I should say subtlety, rarity, immateriality, or some other property different from these that has no name. (1968, vol. 6, p. 351)

So, following his thought to the end, Galileo arrives, like Grosseteste and other thinkers before him, at light that bursts forth from the innermost recesses of matter. For him, though, it is physical light and no longer the divine intelligence that planned it all.

Recent research (Redondi 1987) has shown that in the eyes of the Inquisition, Galileo's atomism may have been his most serious offense. The church's doctrine of transubstantiation asserts that whenever Mass is celebrated, the substance of the consecrated bread and wine is miraculously and entirely changed into substance of the body and blood of Christ, while its accidents remain those of bread and wine. This doctrine, fixed after long and bitter controversy, depends on the Aristotelian disjunction between substance and accidents, and it is undermined by the claim that accidents are only impressions produced in the mind by the action of substance.

From the church's standpoint, Galileo's Copernicanism was the least of his offenses, but it was the one for which he could be tried and silenced before the public without exposing to daylight the more dangerous consequences of his ideas. The trial was held in Rome in 1633, and Galileo spent his remaining years near Florence under house arrest. His last book, the *Two New Sciences*, appeared in the Protestant city of Leyden in 1638; he died four years later.

Kepler and Galileo created a great new wave of science. They taught that observation and mathematical theory are the two paths that lead to scientific truth, but their lives and works show the iron grip of ideas in which they were brought up. Kepler tried to deal quantitatively with the abstract and difficult idea of force. In the next chapter we shall see how this idea becomes central when people try to explain how things actually work.

CHAPTER 10

Influences

On the contrary, the elliptical shape of the planetary orbits and
the laws of the movements by which such a figure is traced,
suggest more the nature of balance and material necessity than
conception and determination by a mind.

— JOHANNES KEPLER

IN the last chapter we saw the beginning of a mechanical and mathematical style of thinking that prevailed over one that dealt in innate qualities.
The new philosophy emphasized material and efficient causes, while
formal and final causes, whose operations no one could claim to observe
directly, were less spoken of though not forgotten.

In the Aristotelian philosophy, natural motion was the rule and violent
motion the exception. The cause of an object's natural motion was usually supposed to be material and to reside in the object itself. But arguments concerning natural motion assume that the directions of space are
fixed to the motionless center of the earth: up is up and down is down
and circular refers to motion around the earth. With the overthrow of
that system the question of motion was once more an open one: if motion
is no longer explained by Aristotelian hypotheses then how can it be
explained? We shall see in Chapter 11 how Newton answered this question. It will turn out that some of the hardest conceptual problems in his
physics center on the nature of force, a concept that from the beginning
has been very hard to define. In this chapter force begins to be thought
about without being clearly defined—that is why I used the name "influ-

ences." First come astrological influences. Then we shall see William Gilbert, a brilliant and skeptical scientist, struggling to understand the nature of magnetic force and finally, in defeat, falling back on the old Aristotelian categories. Next Kepler attributes planetary motion to a force exerted by the sun, but here again Aristotle mixes things up. Isaac Newton gets it right, but his first steps are taken so cautiously that the word "force" needs hardly to be used at all. Gradually the ideas are clarified, and in the *Principia* force makes a somewhat ambiguous appearance.

During all this time, however, concepts of force were central in the design of structures and machinery, and we shall see how they were specifically applied to the understanding of simple machines. The chapter goes on to discuss some questions of scientific method raised by these theories, and ends with a brief account of the how atomism revived and was cleansed of the philosophical objections that had accompanied it since antiquity and how ideas concerning interatomic forces began to give some coherence to theories of matter.

The Power of Planets

I have already mentioned how influences rained down from the heavenly spheres to control human activities, or at least to affect them. The rain continued throughout the Middle Ages and Renaissance, and of course it still falls over much of the globe. Jacob Burckhardt (1944) tells us "Pope Paul III never held a consistory until the star-gazers had fixed the hour," and even today the invitation to a wedding in India usually says that the ceremony will start at some particular odd minute. In Italy astrology had eminent opponents, among them the humanist Giovanni Pico della Mirandola. That there are influences they did not deny, but they condemned the casting of horoscopes at birth, and astrological forecasting generally, on several grounds: in deciding to undertake a venture now we may know what the planetary influences will be in the critical season but we cannot know what the exact circumstances will be at that time and so we cannot predict how those influences will act. Further, we know that predicted calamities often fail to occur. Nothing is automatic, and so anyone who lets the stars control his plans is prevented from using his own good judgment when he needs it most. And as for horoscopes, they may say something about character but they have no predictive value at all, since most of a child's characteristics are determined at various stages before and after birth and the precise moment of birth has no special significance. And over all the discussions reigned the undoubted fact that if a human being is not free to choose at each instant between good and evil acts, the

balance between salvation and damnation will be tipped not by free will
but by the stars. Thus the church found it necessary, again and again, to
repeat its condemnation of the proposition "That our will is subject to
the power of the heavenly bodies," as Tempier expressed it in the condemnations of 1277.

A century after Pico, Johannes Kepler, whose duties at the Imperial
court included the casting of horoscopes, expresses his faith and his puz-
zlement:

> How does the face of the sky affect the character of a man at the moment of
> his birth? It affects the human being as long as he lives in no other way than
> the knots which the peasant haphazardly puts around the pumpkin. They do
> not make the pumpkin grow but decide its shape. So does the sky; it does not
> give the human being morals, happiness, children, fortune and wife, but it
> shapes everything in which the human being is engaged. (Letter to Herwart
> von Hohenburg, 1599, in Baumgardt 1951, p. 51)

Kepler goes on to say he cannot understand how it is that the planets
can influence one's life at the moment of birth but not afterward. A
decade later, he knows, and he writes a little book called *Tertius interveniens*. The rest of it is mostly in German; the title can be translated as "The
intervening third party; that is, a warning to several theologians, doctors,
and philosophers, especially Dr. Philipp Feselius, that as they rightly cast
out the stargazers' superstitions they must not throw out the baby with
the bath and, in so doing, unwittingly do harm to their profession" (1937,
vol. 4, p. 145). The little human soul is a point, but *potentially* it is a
sphere in the image of God, and at the moment of birth its spherical
potentiality is affected by light rays from the planets. It is the geometrical
configuration of these rays (and not at all their relation to the distant
patterns of the zodiac!) that determine the influences (see Pauli, in Jung
and Pauli 1955, pp. 180ff). Kepler believed with Grosseteste and Ficino
that light was an agency by which an object can act on another object
distant from itself, but clearly there were other such agencies.

Action at a Distance

Magnetite, the magnetic oxide of iron, is found in association with surface deposits of iron; possibly one of the original sites was called Mag-
nesia, but at any rate it was known early that a piece of magnetite will
cause a piece of iron to move by some influence that seems to act through
empty space and even through a thin sheet of wood or bronze placed in
between. According to Aristotle "Thales said the stone has a soul because
it moves iron" (*On the Soul*, 405a). I have quoted Thales earlier as saying

that everything is alive and full of gods. Instead of "alive," it would have been a better translation to say that everything has a soul. Soul was the name for the source of life and action and consciousness in a human or animal and by extension, in a stone as well. The magnet is able to exert its peculiar influence because soul is anchored in that particular kind of stone.

To say that a magnet has a soul is first of all to say that it has an unusual ability to affect certain other objects, but of course, in a cosmos imagined as full of life, it meant more than that. In the *Timaeus*, God "put intelligence in soul, and soul in body, that he might be the creator of a work which was by nature fairest and best. On this wise, using the language of probability, we may say that the world came into being—a living creature truly endowed with soul and intelligence by the providence of God" (30b).

Later Plato writes that a lover of nature ought first to explore what manifests intelligence, and only afterward investigate "those things which, being moved by others, are compelled to move others," i.e., things that act like machines. In Aristotelian terms, to give only an efficient cause is not enough to explain anything, for cause is bound up with intelligence and purpose. Even Kepler, striving to explain the motions of planets no longer driven by celestial spheres, considered for a while that they might be guided by angelic intelligences, and when Grosseteste wrote that everything that causes change in something else is acting with the power of light, it is by associating light with intelligence and through intelligence with action that we can make sense of his words.

It is not easy to imagine the struggles of a Renaissance scientist trying to decide how one event causes another. We have been taught that there are different modes of causation that have nothing to do with each other, and we do not stop to puzzle whether the process by which we catch a disease is related to the process by which the position of the moon affects our luck or a magnet lifts a nail. Aristotelian doctrine admitted only a few different kinds of causes and so brought a spurious unity to the explanations of matters like these. Until a new doctrine grew up, a weak understanding of causal relations outside the purely mechanical sphere is the reason why the intellectual giants of the seventeenth century continued to believe that astronomy, alchemy, and the processes of life all involved similar actions. Until the end of the century and perhaps beyond, cardinals and courtiers, even a pope, went through picturesque rituals with lighted candles to ward off the evil effects of eclipses and hostile planets (Walker 1958, p. 207), while witches continued to stick pins into waxen images of intended victims, trusting that the principle "like affects like" would lead to their demise.

I have mentioned Aristotle's declaration that every motion is caused by direct contact with something else (*Physics*, 202a); nevertheless, there were so many notions of action at a distance in Kepler's time that it would be tedious to describe them. In his book *De magnete* (1600, 1958) William Gilbert (1540–1603), physician to Queen Elizabeth I, copies pages of explanations of magnetism and points out that the only way to test their truth is by experiment, which he has done. The principle "like affects like" does not explain why iron is attracted by a lodestone, which is a black mineral; the principle "like attracts similar" does not explain why a lodestone does not attract metals other than iron, or why amber, which does not resemble lodestone at all, can be made to attract it if the amber is first rubbed with a cloth. Gilbert has no use for explanations like one he quotes from Guilelmus Puteanus, which attributes the attraction of a magnet to "its substantial form as if it were a prime mover, producing motion by itself, as if by its own most powerful nature and natural temperament or instrumentality through which its substantial form becomes effective in its operations" (Gilbert 1600, p. 63), and he dismisses such talk with winged words: "Those explanations of magnetic effects which in philosophic schools are derived from the four elements and the four primary qualities [hotness and coldness, wetness and dryness]—we leave them to the roaches and the worms." In his writing Gilbert is developing a maxim that by the end of the century will be almost universal: pay attention only to those explanations whose truth or falsity can be tested by experiment.

Gilbert, far from the main scientific centers, studied magnetic actions in the small and in the large. On the laboratory scale he verified that every bar magnet has ends of opposite polarity, and by comparing measurements of the earth's magnetism in different places with measurements made in the neighborhood of a small spherical lodestone he called a *terella* (Figure 10.1) he arrived at the astounding (but essentially correct) conclusion that they are the same, provided the polarity of the lodestone is oriented roughly, but only roughly, north–south. Gilbert is hard on his predecessors. Cardano says that a wound made by a magnetized needle is painless—did he try it? Fracastoro says a lodestone attracts silver, Scaliger says a diamond attracts iron, Mattioli says that garlic cuts off magnetic action—why didn't any of them try it?

Gilbert has a theory of other kinds of interaction: electric, gravitational, and the interaction that transfers infectious disease. Their material causes are *effluvia*, vapors that flow from one body to another. These are physical vapors; air, for example, is the effluvium of earth that causes gravity. This means that there is no gravity above the sphere of air, and Gilbert ascribes the moon's motion around the earth to the earth's magnetism. But precisely because he had studied magnetic phenomena so

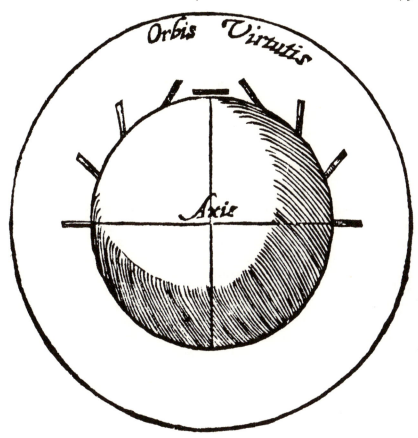

FIGURE 10.1. The magnetic field at the earth's surface. From Gilbert's *De magnete* (1600).

carefully he was obliged to conclude that there is no magnetic effluvium, since it cannot explain why the magnetic force acts undiminished through screens of dense material without magnetizing them. Magnetism can have no material cause. But the ancients have taught us that the planets have souls to guide their motions. By fixing the earth at the center of things they denied it a soul, but now we know better. The earth and the other planets are guided by the sun's magnetic force and kept going (for otherwise they would stop) by the energy of its light. Thus every magnetic particle partakes of the soul that guides the earth in its orbit (Gilbert 1600, v.12) and is the formal cause of magnetism. With Thales, Queen Elizabeth's doctor finally concludes that a magnet is alive and has a soul.

Kepler's Dynamics

Having derived a wonderfully simple and accurate representation of planetary motions, Kepler tried hard to make a theory as to why they should
be that way. What he made was wrong from the start and was eradicated
by Newton's *Principia*, but as this was the first time in history that anyone
had ever put together a quantitative theory to account for the motion of
anything, it has a certain interest.

Kepler's fatal error was to regard planetary motion as violent motion,
so that it had to be kept going. He began by supposing that if the planets
were not continually swept along their orbits by a force from the sun they
would stand still as the stars do. This led him to propose (*Astronomia
nova*, ch. 57)[1] that the sun rotates on its axis and that a *species*, analogous
to the *species* responsible for the phenomena of vision, issues from it like
the spokes of a turning wheel to drive the planets along. Almost immediately afterward, the first telescopic observations of the sun revealed sunspots, and when these were followed from day to day they showed that
the sun does indeed rotate. This unfortunate discovery served to embed
Kepler firmly in an erroneous argument. How does the driving force
depend on the distance? Let us start from the second law and apply it at
the extreme points of the orbit (Figure 10.2). Consider a short period of
time, t, during which the planet, at velocity v, moves a distance vt. Since
the two shaded figures are very nearly triangles we may say that their
equal areas are given by

$$\tfrac{1}{2}v_a r_a t = \tfrac{1}{2}v_b r_b t, \qquad \text{whence} \qquad v_a r_a = v_b r_b.$$

But, according to Aristotle's law of motion, interpreted literally as an
algebraic formula, the velocity is proportional to the driving force, and
so we guess that the force obeys

$$Fr = \text{const} \qquad \text{or} \qquad F \propto 1/r.$$

Kepler was not entirely happy about this, for he knew that the intensity
of light varies inversely as the square of the distance. He mentions in letters that he would have preferred the inverse square for his driving force,
but the simple inverse was what he got.

There is another force. In elliptical motion a planet moves nearer and

[1] See also Kepler (1937, vol. 9, p. 301); Ptolemy, Copernicus, and Kepler (1952, p. 898);
Dreyer (1906, ch. 15).

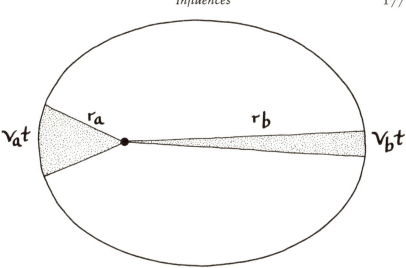

FIGURE 10.2. Velocities at the apsides of a planetary orbit.

further from the sun; Kepler concluded that there must also be a radial force that alternates between attraction and repulsion. Knowing from Gilbert's work that the earth acts as a great magnet he supposed that the sun is a magnet also, but of the special kind that today we call a monopole, a single north or south pole. Such a thing is not known in nature. A simple bar magnet has a north and a south pole, but if you try to isolate a pole by breaking the magnet in two, you are left with two bar magnets of the same kind as before.[2] If a magnet revolves around the solar monopole without turning (here I simplify Kepler a bit), it will be alternately attracted and repelled depending on which pole is nearer the sun (Figure 10.3), and the orbit will no longer be circular.

Nine years after these speculations Kepler discovered the harmonic law. A quick calculation must have shown him that whereas in his theory the area law requires $F \propto 1/r$, the harmonic law requires $F \propto 1/\sqrt{r}$, and therefore the whole thing is wrong. He seems not to have written any more on the subject.

Independently of his planetary theory Kepler entertained some ideas that fall much nearer the ultimately successful Newtonian theory of universal gravitation. In a letter of 1605 (Jammer 1957, p. 82) he imagines the earth set motionless somewhere in space and then a larger earth brought up and placed near it. He suggests that "the first would become a heavy body in relation to the second and would be attracted by the latter

[2] In recent years experimental physicists have tried to find elementary particles that are monopoles but have not yet succeeded.

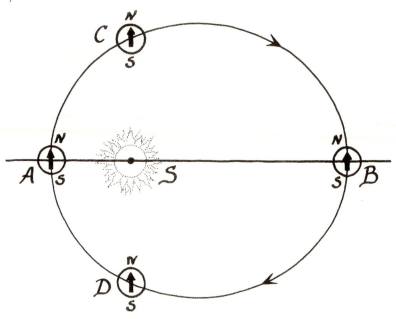

FIGURE 10.3. Kepler's explanation of elliptical orbits. When the magnetic planet is at A or B it experiences no force toward or away from the sun. When it is at C the force is away from the sun; at D the force is toward the sun.

just as a stone is attracted by the earth." The terminology reflects the delicate balance in his thought: to say that the earth becomes a heavy body says that it acquires the quality that causes it to fall, and yet "attracted" shows that he has gone beyond the classical explanation and is thinking about a force.

The letter is evidence of a shift of ideas that was taking place at this time, away from the philosophical tradition that rooted all explanations in modes of being, in which every human act and material change was determined by inherent qualities (think of the doctrine of original sin). For Democritus the modes of being are atoms, and even though their variety was infinite, so few properties were ascribed to them that their explanatory power, even in the dextrous hands of Lucretius, was very small. Aristotle and his successors defined the qualities more generally and therefore less precisely, so that they provided vague causal explanations of an immense variety of events in nature and mankind. The scientists we have been studying were trained in this mode of argument, but they had begun to realize that innate occult qualities are exactly what we cannot know, and that therefore the explanations at the heart of science were being phrased in such a way that their truth could never be established. Kepler has taken only a small step. He suggests that the stone falls because it is attracted by the earth. The earth exerts the force; we can feel it acting on our bodies. But what causes the force? The cause was hidden. It was occult.

The earliest suggestion I have encountered that the sun does not need to act on a planet to keep it going but only to prevent it from wandering off into space comes from the astronomer Jeremiah Horrocks (1619–41). He attended Cambridge University until he was sixteen but was essentially self-taught since Cambridge had no professor of mathematics in his time. He had no money, and so his only option was to take orders (though he was below canonical age) and presently he was installed as curate in a remote village in Lancashire. There he undertook to revise Kepler's Rudolphine astronomical tables so as to improve their accuracy and, by watching through his little telescope one of Venus's rare passages across the sun's disc, was able to provide a crucial and very exact observational point. He was the first to show that the moon travels around the earth in an elliptical orbit and he seems to have anticipated Newton in thinking of it as continually falling towards the earth. Most of his work has disappeared, but a note among his surviving papers (1672) suggests "a magnetic influence of the earth which attracts all heavy bodies to itself, as the magnet attracts a piece of iron." He died at twenty-one. Had Horrocks lived, the young Newton would probably have found him at Cambridge when he arrived.

The Law of Gravity

Isaac Newton (1642–1727) was eighteen when he arrived in Cambridge from his native Lincolnshire. For four years he was, as we would say, on Financial Aid, earning his keep at Trinity College by doing menial jobs. It seems that Isaac Barrow was his tutor, and if so he was lucky, since Barrow was one of the leading lights at Cambridge. Formerly professor of Greek, he was now professor of mathematics and what passed in those days for physics, and he must have staggered under the mass of paper that flowed in from the young Newton.[3] Barrow "would frequently say that truly he himself knew something of mathematics, still he reckon'd himself but a child in comparison with his pupil Newton" (Stukeley 1936).

In 1665 plague broke out in Cambridge and continued intermittently

---

[3] The collected mathematical works (Newton 1967) devote 1500 printed pages, including commentary and, in some cases, translation from Latin, to the mathematics that Newton produced between 1664 and 1671 and thought worth keeping.

for two years. Newton spent most of this time at home in Lincolnshire, working hard. He produced mathematics that ended in his version of differential calculus, which he called the theory of fluxions. Possibly in Cambridge he had already discovered that white light is a mixture of light of all colors; at home he made more experiments. Finally, at about this time he formulated some ideas about gravitation. He returned to Cambridge in 1667. Two years later Barrow, wishing to devote himself to religion, resigned his chair in favor of Newton, who at twenty-seven became Lucasian Professor of mathematics. The next twenty years were occupied with immense labors in mathematics, dynamics, astronomy, alchemy, chemistry, theology, and Biblical chronology, apparently all at the same time. The first three found their culmination in the publication of the *Philosophiae naturalis principia mathematica* (Mathematical Principles of Natural Philosophy) in 1687. Figure 10.4 shows Newton at the time he was writing the *Principia*. The mind that speaks to us through the gaunt face and staring eyes has toiled for years, day and night, to construct the single, unified philosophical system in which all those lines of thought would converge to show mankind whatever it can possibly understand of the true nature of God.

We shall discuss the work in dynamics later; for the moment let us go back to the beginning and see how Newton discovered the law of gravity and how, in his mind, it served to explain the phenomena. To do this I must refer to two discoveries of Galileo's that Newton knew at this time: that a moving body does not need a force to keep it going and that a freely falling body increases its speed at a constant rate; i.e., it falls with constant acceleration. Both will be discussed in the next chapter. The following reconstruction of his thoughts in 1666 agrees with his meager published reminiscences of the period and is perhaps not too far from the truth.

"As he sat alone in a garden, he fell into a speculation on the power of gravity." (Pemberton 1728, preface). Let us say that he focused his thoughts on an apple hanging from the bough of a tree—there is a tradition to this effect. The twig breaks; the apple falls toward the earth. Now consider the motion of the moon. It too must be attracted toward the earth, for if it did not it would move off into space and vanish in the distance. Can it be that the moon falls toward the earth on account of its weight, just as the apple does? If so, then perhaps the planets are kept in their orbits in the same way, by the attraction of the sun.

Note that Newton is concerned only with the force that keeps the moon and planets from wandering away. He has already absorbed via Descartes the Galilean principle that the planets need no force to keep them going. He considers that the moon moves as a thrown ball does, moving forward and at the same time falling toward the earth. A revolution about

FIGURE 10.4. Isaac Newton at the time he finished the *Principia*. Reproduced by courtesy of the Portsmouth Estates.

a central body consists in combining these two motions so that the distance to the central body changes little (this is illustrated in Figure C in the Notes). There is no reason to think that the force of attraction is the same at all distances from the gravitating body. Let us try to find out how it depends on the distance.

I will continue the discussion in modern notation and modern terms, as one explains it to a student. The outcome will be a formula for the dependence of the gravitational attraction on distance. Newton must have gone through an argument amounting to the same thing. Then I will show Newton's reasoning in the early days as he struggled with the elementary theory of planetary motion.

We start with the formula that tells how much force is needed to keep an object falling toward a central point. If the object's velocity is v, the force toward an attracting center that keeps it moving at a constant distance r from the center is given by the proportion

$$F \propto v^2/r$$

(the orbits of the planets are not circles but ellipses; still, except for Mercury's, the ellipses are close to circles). Most likely Newton had worked out the proportion for himself, and Note C suggests how he may have done it.

Ordinarily one does not talk of the speed of a planet in its orbit, but rather of its period T, the time it takes for a complete circuit. The relation with v is simple,

$$v = 2\pi r/T,$$

or, as a proportion,

$$v \propto r/T.$$

If the force that keeps the planets in their orbits is equal to the gravitational force, we have

$$F \propto \left(\frac{r}{T}\right)^2/r = r/T^2.$$

But Kepler's harmonic law tells how T varies with r:

$$T^2 \propto r^3.$$

Therefore

$$F \propto r/r^3 = 1/r^2.$$

This is the desired result: only if the force of gravity varies inversely as the square of the distance from the gravitating center will Kepler's law hold true.

My justification for the foregoing reconstruction is some early notes (Herivel 1965) and a remark in a letter to Edmund Halley written twenty years later, "But for the duplicate proportion [i.e., the square] I can affirm tht I gathered it from Keplers Theorem about 20 yeares ago" (*Correspondence*, 1959, 14 July 1686).

Expressed in the algebraic notation the foregoing argument can be followed by a schoolchild. For Newton it was much more complicated, since people in his day had not learned to write formulas. I know of no place in Newton's published works, his correspondence, or his private mathematical papers where he writes out in the algebraic notations of the time (which were not very different from ours) the fundamental laws attributed to him. I think this reflects the continuing authority of the ancients, especially Euclid. You will remember that although a number like 5/8 was considered as a quantity whose numerical magnitude is exactly known, a number like $\sqrt{2}$ was not, and it was represented only as the length of a certain line; this probably explains why Euclid ordinarily represents all numbers, except occasionally integers, as the lengths of lines. If a number is a length, the product of two numbers is an area, but what is a number like the quantity T? Is it a length and also a time? And what is r/T^2? Before simple algebraic notation could be used in physics a new generation had to grow up for whom Euclid was no more than a respected classic. That happened a few years later.

The conclusion drawn from Kepler's law suggests trying to understand the motion of the moon, but now there is a complication. The orbital radii of the planets are so vast compared with their sizes and that of the sun that all these bodies may reasonably be considered as gravitating points. The only force of gravity about which we know anything directly is the force at the surface of the earth. It is produced by the forces of attraction exerted by every atom of the earth's composition. Summing all these forces while taking account of their varying magnitudes and directions is not an easy problem, and Newton did not finally solve it until almost twenty years later when he was writing the *Principia* (see Book 1, Proposition 76). He found that if the forces vary as the inverse square (and in no other interesting case), the force of attraction toward a spherically symmetric mass, which the earth very nearly is, is the same as if all the mass were concentrated at the center. Newton seems to have assumed that this holds, exactly or approximately, in his early thoughts about the moon's motion.

The lunar calculation went something like this. All falling objects have the same gravitational acceleration toward the earth. As Ptolemy and other authors had shown, the moon is distant from the earth by about

60 earth radii. If the attraction varies inversely as the square of the distance from the earth's center, the moon's downward acceleration should be about $(1/60)^2$ the acceleration of a falling object near the earth's surface.

The moon's motion can now be calculated by an argument much the same as the one in modern style that I have just used to derive the law of gravity, but it would not show how Newton proceeded. Here is a calculation that follows notes on the back of a torn piece of legal parchment found among Newton's papers that can be dated at 1665 or 1666, possibly to the time spent in Lincolnshire (Herivel 1965, p. 183). Although the entire argument depends on the comparison of accelerations, Newton does not use that word here or anywhere in the *Principia*. The derivation will be easier to understand if I mention acceleration but it will not appear in the formulas.

Newton's proof begins by assuming a strange theorem he must have derived elsewhere. Suppose that a body moves uniformly in a circle of radius r, always accelerating toward the center, and that a second body is given a constant acceleration in a straight line equal in magnitude to the centripetal acceleration of the first. Then, during the time when the first body travels a distance equal to r along its curved path, the second one travels a distance $r/2$ along its straight path.

This verbal statement is equivalent to the formula v^2/r for the acceleration of a particle that moves in a circle. It holds regardless of how fast the first body moves on its circular path. Note C gives a proof that Newton might have constructed.

The argument is now quite simple, for the theorem allows us to compare the moon's motion in its curved path with the fall of a stone near the earth, whose rate we can measure. Note D gives the algebra; the result is that if we assume 25,000 miles for the earth's circumference, a figure Newton may have used, the moon's period comes out as 29.3 days, which is very near the correct value of about 29.5. Newton did not publish these calculations for another twenty years, probably because he did not know how much of the error, if any, to assign to his simplifying assumptions that the earth's mass may be taken as concentrated at its center, that the moon moves around the earth in a circle, and that the earth is stationary in space, whereas in reality it orbits the sun.

Both the simple calculations given above depend on the assumption of circular motion, and so they can only be approximate. It was essential to show that the inverse-square law follows from elliptical motion just as surely as it follows from circular motion, and this seems to have taken a long time (see the Introduction to vol. 6 of Newton 1967). At length, in 1679, Robert Hooke (1635–1703), a brilliant savant connected with the Royal Society, suggested to Newton that one can represent the planet's motion in terms of two components, one along the orbit and the other toward the central body. Then Newton was able to complete the proof, which appears in Book 1 of the *Principia* as Proposition 11. Finally, he thought his way to the principle of universal gravitation: if the earth attracts the apple and the moon, then the apple and the moon must attract the earth. The sun attracts the planets. Now the crucial question: does the moon attract the apple? Does every object in the universe attract every other? Newton's guess that this is the case is known as the principle of universal gravitation. Having guessed it he was ready to write the *Principia*.

Other Views of Gravity

The law of gravity accounts for the phenomena of the solar system but to seventeenth-century minds it was a very incomplete statement, for traditional doctrine held that a knowledge of physics is a knowledge of causes. What was the cause of this force? I have already mentioned Gilbert's theory of effluvia, which attributes the fall of an object to a sort of breath of air flowing in and out of the earth. Galileo adopts Aristotle's explanation while avoiding his terminology:

> All parts of the earth cooperate to form the whole. It follows that they have equal tendencies to come together from every direction so as to unite in the closest possible way, which is spherically. Why must we not then believe that the sun and moon and the other celestial bodies are also round in shape by exactly such a concordant instinct and natural cooperation of their component parts? (1968, vol. 7, p. 58)

These remarkable words occur in the *Dialogue on the Great World Systems*, written by Galileo at the end of his career. They are uttered by his spokesman Salviati, and they are of course tuned to his companions' Aristotelian ways of thinking, but see how naturally they come out. It seems that a stone feels almost a moral obligation to return to the earth from which it has been snatched, but that if you put it close to the moon it would want to fall to the moon. Writing thirty-two years after the publication of Gilbert's book, Galileo seems to be telling us, in conventional language, his understanding of gravity. He never speculated in print how the force is actually exerted, and in the *Two New Sciences* he specifically declines to do so.

In 1644 René Descartes (1596–1650) proposed a mechanism to account for planetary motion. For him space and matter are the same thing, extension. There is no matter that does not occupy space; there is no space unoccupied by matter. Around every star forms a vortex of ethereal matter in which planets are carried around like twigs in a stream. In the *Principia* Newton takes pains to show that this mechanism would produce planetary motions in violation of Kepler's laws; nevertheless, forty years after the *Principia* appeared the Cartesian theory was still being taught in Cambridge University. The only concession to modernity was that successive editions of the English translation of Rohault's *Traité de Physique* (1671) had accumulated a mass of footnotes, inserted by the translator, that refuted statements made in the text and taught New-tonian physics instead (Rohault 1729).

Viewed against the conceptual richness of Gilbert's effluvia, Galileo's piece of matter that wants to form a sphere, and Descartes' gigantic vor-tices, Newton's simple assertion of a recipe giving the magnitude of grav-itational attraction seemed conceptually barren. Admittedly, it enabled him to do calculations that could be observationally verified and that nobody else could do, but to European philosophers it was not natural philosophy at all if it did not explain causes. There was a long and bitter controversy, and we must pause to take note of it.

Mechanical Philosophy

At the very beginning of modern physical science, when Galileo and Des-cartes had so few successes to their credit, they wrote with unbounded confidence about the method they used. I have already quoted Galileo that the book of nature is written in mathematical symbols, and similar statements abound in the writings of Descartes. At the beginning, on the beach of an unexplored ocean of truth, they experienced a sense of secu-rity that has perhaps never again been felt, for they saw the ocean as finite and mankind able to explore it. We have found through the years, and more than ever recently, that the ocean grows larger every day, but for Galileo and his successors the world was made of atoms, planets circled around the sun, and light and sound were signals that moved through space from source to eye and ear. The general nature of the physical world was clear to them; they were beginning a science with vast but finite pos-sibilities.

The details were yet to be filled in. They were filled in by the only pos-sible means, guesswork, which is the normal procedure of science. A sci-entist, guided by observed facts, frames hypotheses that imply that certain other statements also are facts. If experiment shows them to be facts, the hypotheses hold up; otherwise they are dismissed. The explanatory struc-ture of science depends on these hypotheses, and the sword of Damocles hangs over every one. There is no certainty in scientific theory.

Allowing for individual differences, the correctness of the trial-and-error approach would have been upheld by many scientists in the seventeenth century and it is pretty much the way people think today. Descartes did not agree, but his theories, deductively arrived at, were dealt with very harshly by experiment. "Of the mechanical truths which were easily attainable in the beginning of the seventeenth century," wrote William Whewell (1859, vol. 2, p. 52), "Galileo took hold of as many and Descartes of as few as was well possible for a man of genius."

Isaac Barrow held a low opinion of the "scientific method." In a widely used textbook he wrote that in science as then practiced, "for the dispatch of every Question or the Explication of a Phaenomenon, a new and distinct Hypotheses is invented. From whence it happens that in what is called and accounted the same Science are found Hypotheses without number" (Barrow 1734, p. 54, quoted by Kargon 1966, p. 120). The science Barrow sought was not to be cobbled together out of lucky guesswork. Rather, it was to be built with the certainty of mathematics. Barrow's science differed from Aristotle's: where Aristotle based his arguments on certain universal principles Barrow proposed to base his on experiment, and where Aristotle used logic to draw necessary conclusions, Barrow intended to use mathematics. Mathematics needs axioms. They must be "exceedingly familiar, indubitable, and few in number." And the physical hypotheses assumed must be very few and "undeniable by any sane mind" (Barrow 1860, p. 66, quoted by Burtt 1954, p. 153).

Newton adopted Barrow's program as his own. He was willing enough to discuss hypotheses with his friends, or even to introduce them into his writing, if it was understood that the truth of what he was saying was independent of any hypotheses he might make as to why it was true, for the truth consisted of experimental fact and mathematical law. European skeptics complained that Newton's utterances were designed to cloak the fact that he did not know what was going on. To say that the stone falls because it gravitates toward the earth was to explain it in terms of natural motion and occult causes. They recalled the scene in Molière's play *Le Malade imaginaire* (1673) in which a bachelor candidate in medicine is examined by doctors in barbarous pseudo-Latin:

BACHELOR

Mihi a docto doctore	The learned doctor asks me
Domandatur causam et rationem quare	The cause and reason why
Opium facit dormire?	Opium produces sleep?
A quoi respondeo	To which I answer
Quia est in eo	Because it possesses
Virtus dormitiva	A dormitive power
Cujus est natura	Whose nature it is
Sensus assoupire.	To bring drowsiness to the senses.

The chorus judges this to be a spectacularly good answer.

In fact Newton did not think the matter ended with the bare statement that the force exists. There must be some physical explanation, and in his mind the first hypothesis one ought to make is that it is transmitted by some medium that fills all space. In an often-quoted letter to the theologian Richard Bentley, Newton wrote, in about 1691,

> That gravity should be innate, inherent and essential to Matter, so that one Body may act upon another at a Distance thro' a *Vacuum*, without the Mediation of any thing else, by and through which their Action and Force may be conveyed from one to another, is to me so great an Absurdity that I believe no Man who has in philosophical Matters a competent Faculty of thinking, can ever fall into it. Gravity must be caused by an Agent acting constantly according to certain Laws; but whether this Agent be material or immaterial, I have left to the Consideration of my Readers. (Newton 1958, p. 302; Bentley 1838, p. 212.)

Newton's imagination teemed with hypotheses, and the ether was one that remained with him. In a long letter to Henry Oldenburg, secretary of the Royal Society, who for years acted as a sort of scientific postal service, he suggests several ethers—one that is set into wavelike motion by the passage of light, another that transmits electric actions, another for gravity, "not of the maine body of flegmatic aether, but of something very thinly & subtly diffused through it, perhaps of an unctuous or Gummy, tenacious & Springy nature, and bearing much the same relation to aether, wch the vital aereall Spirit requisite for the conservation of flame and vital motions, does to Air" (Newton 1959, vol. 1, p. 395).[4] The "vital aereall spirit" will be discussed below.

Newton's writings make his position perfectly clear. In the magnificent General Scholium that ends the *Principia* he writes

> Hitherto we have explained the phenomena of the heavens and of our sea by the power of gravity, but have not yet assigned the cause of this power. This is certain, that it must proceed from a cause that operates to the very center of the sun and planets, without suffering the least diminution of its force, . . . and in receding from the sun decreases accurately as the inverse square of the distances as far as the orbit of Saturn. . . . But hitherto I have not been able to discover the causes of these phenomena, and I frame no hypotheses; for whatever is not deduced from the phenomena is to be called an hypothesis; and hypotheses, whether metaphysical or physical, whether of occult qualities or mechanical, have no place in experimental philosophy. In this philosophy particular propositions are inferred from the phenomena, and afterwards rendered general by induction. Thus it was that the impenetrability,

---

[4] I have omitted a parenthetical remark that Newton later withdrew.

the mobility, and the impulsive force of bodies, and the laws of motion and of gravitation, were discovered. And to us it is enough that gravity does really exist, and act according to the laws which we have explained, and abundantly serves to account for all the motions of the celestial bodies, and of our sea. (1934, p. 546)

Hypotheses are propositions that cannot be deduced from experiment. The words "I frame no hypotheses" mean "Think all the crazy thoughts you want to but don't build them into the theory." Yet Newton was mistaken in requiring theories to be based on propositions "inferred from the phenomena, and afterwards rendered general by induction," for that process cannot lead to necessary conclusions. Every observation has theories built into it; every theory has hypotheses built into it, and they all may be wrong. The proof? Newton's own theory of gravity, which he himself puts forward as an example, assumes that gravity is a force and that the space in which the calculations are carried out is the flat space of Euclidean geometry. Einstein's theory is better than Newton's, in that it agrees better with refined observations. In it gravity is not a force but a manifestation of curvature in the underlying spacetime. I have quoted the complaint of Averroes that none of the celestial models he knew was derived from observation by necessary arguments. They still are not, and they can never be. As the medieval logicians had understood perfectly well, one can rigorously argue from the general to the specific but not from the specific to the general. It is always possible to have more than one theory based on a given set of observational facts.

Newton was interested in a third class of propositions. At the end of successive editions of the *Opticks* appear a number of "Queries," propositions that were yet to be proved or disproved by experiment but could be. By the fourth edition (1730) there are seventy pages of them. Here we find (Query 21) Newton's conjecture as to why there is gravity: the ether outside a massive body such as the sun or a planet is denser at greater distances. Thus an object near the body experiences a larger density on the further side, and is forced inward as a result. The same ether transmits waves with velocity even greater than that of the particles of light, causing the particles to vibrate to and fro as they move and thus giving rise to the wavelike phenomena that are observed (Queries 17, 21, and 29). Newton's mind generated hypotheses as copiously as any mind ever has, but he at least tried very hard to be careful how he used them.

This account of Newton's work has hardly touched the idea of force. The word is not defined in the *Principia*, and when it does occur it is used in different ways. But engineers interested in machines and structures had need of it, and we pause to see how it entered their calculations

The Marvel Is No Marvel

Simon Stevin (1548–1620) was born in Bruges. He was essentially self-taught, beginning as a bookkeeper and military engineer, but he entered university at thirty-five and emerged a scientist. He was one of the first European advocates of the decimal system, arguing for decimal coinage and decimal systems of weights and measures, and was one of the few university-educated people of his age who worked in applied science. In 1605 Stevin published *Wisconstige Gedachtenissen* ("Scientific Thoughts"), in which he says he will be concerned with what might be called pure machinery, machinery in which the parts do not stick or bind but move without friction—magic machinery that is governed not by practicality but by the principles of pure science.

One of Stevin's problems was to find the force that must be exerted to keep a ball that rests on an inclined surface from moving down the surface (Figure 10.5). Figure 10.6 shows his solution as illustrated on the title page of his book. The motto *Wonder en is gheen wonder* says that the marvel is no marvel, and is directed against some of the things he had been taught at the university. He had probably read the following passage from Aristotle's *Mechanics*. (Although that book is certainly spurious it was considered canonical in those days.) The author marvels that a lever can multiply force and then explains:

> The original cause of all such phenomena is the circle. It is quite natural that this should be so; for there is nothing strange in a lesser marvel being caused by a greater marvel, and it is a very great marvel that contraries should be present together, and the circle is made up of contraries. For to begin with, it is formed by motion and rest, things which by nature are opposed to one another. (*Mechanics*, 847b)

And so on. Here is the gist of Stevin's idea. First of all, assume an idealized chain of little balls that moves without friction. If, as in the picture, the triangle's longer leg is twice the shorter one, the chain will have twice as many weights on the long side as on the short one, but they must both pull downhill with equal force, for otherwise the whole chain would start to revolve around the triangle, one way or the other, and "this motion would have no end, which is absurd." He concludes that if the vertical distance between the top and bottom of the plane is kept constant, the force required to keep a single ball from moving varies inversely as the length of the plane. A simple and entirely correct derivation.

But what made Stevin so sure that the perpetually revolving loop of chain is absurd? Of course, in Aristotelian physics both natural and vio-

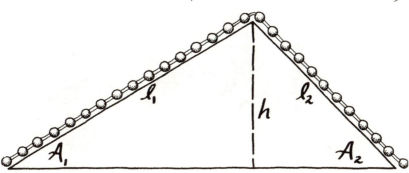

FIGURE 10.5. Illustrating Stevin's argument.

lent motion come to an end. The only motion that continues forever is that of bodies above the sphere of the moon, and Stevin's chain is not there. But that is academic talk, and in spite of it, the invention of perpetual-motion machines was a special hobby of Medieval and Renaissance engineers. They designed water wheels to pump the water back up again while at the same time turning a millstone; arrangements of magnets that would cause a wheel to turn; wheels with movable balls or hammers (Figure 10.7) attached to them so that there were always more weights pulling downward on one side than needed to be lifted on the other; a self-blowing windmill, and so on. Perpetual motion was an old subject by that time, and Stevin and almost everyone else knew it didn't happen.

I give this elementary argument because it illustrates a principle that was becoming clear in Stevin's time, that except for trivialities one cannot generate a scientific fact by reason alone, however one may dress it up with mathematics. There is nothing self-contradictory about a world in which perpetual motion occurs. It violates no principle of logic or mathematics; it just is not the world we live in. Stevin's argument illustrates how a physical conclusion must stand on physical assumptions, and it pointed the way for those who sought to create a quantitative science.[5]

The instructed reader will have scented the conservation of energy in Stevin's proof. The general concept of energy was three centuries in the future, but a rule derivable from the energy principle was already used in Stevin's time. Consider the law of the lever, well known to everyone since Archimedes and perhaps before, Figure 10.8. At equilibrium,

$$F_1 l_1 = F_2 l_2.$$

[5] For a more detailed discussion of Stevin's very original reasoning see Mach (1942, pp. 32ff).

FIGURE 10.6. "The marvel is no marvel." From Simon Stevin's *Wisconstige Gedachtnissen* (1605).

Imagine now that the lever is given a little tilt around the fulcrum. The distances *h* through which the ends move will be proportional to the lengths *l*, so that this can also be written

$$F_1 h_1 = F_2 h_2$$

Since early in the seventeenth century the product of the force on an object multiplied by the distance it moves the object has been called the

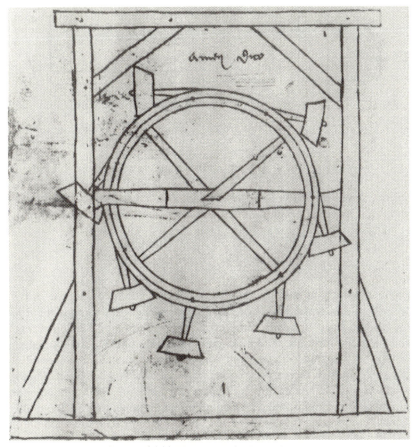

FIGURE 10.7. A medieval perpetual-motion machine, from the notebook of Villard de Honecourt, ca. 1250. Someone has written on it "Amen dico" (I say Amen).

work done by the force. The last equation states that the work put into lifting a weight a little way with an idealized lever equals the work gotten out of it, and is an expression of what is termed the principle of virtual work.[6] This is the idea at the bottom of Stevin's proof concerning the chain, and he later used it to derive the law of the lever. Since in modern terms work is a form of energy, we are saying that the energy that goes into an idealized frictionless machine equals the energy that comes out of it. This principle has been used in the design of machinery ever since.

So much for statics, the theory of forces in things that do not move. The next step was to study motion and that turned out to require an

_____
[6] For the dawn of the idea of virtual work in the Middle Ages, see Claggett (1959, ch. 2).

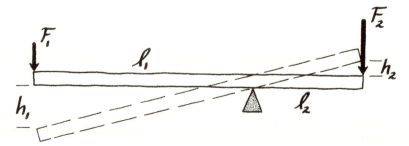

FIGURE 10.8. To illustrate virtual work.

entirely new way of thinking, but first let us note the revival of atomism
and some speculations about the forces atoms exert on one another.

Atoms in the Seventeenth Century

I have shown how, though atomism was never accepted doctrine in the
schools, the essential ideas got past the barriers erected against them in
the Middle Ages. In addition there was the text of Lucretius, which survived in a few dusty manuscripts, one of which the papal secretary Poggio
Bracciolini found in Germany and brought to Italy in 1414. It was printed
in 1473 and often thereafter, and as people learned to read classical Latin
it was very widely read. Probably more than anyone else Lucretius was
responsible for the flowering of atomic speculation in the early seventeenth century, since not only did he express the central idea in vivid Latin
poetry but he also showed in many examples how to think atomically.

   In reading the works of this period one notices that some writers refer
to atoms and others to particles or corpuscles. Most writers seem to have
understood that the word "atom" should refer to something truly elementary, uncuttable, moving in empty space, but once past the era of the
four classical elements they were properly hesitant to claim that any of
the chemical substances they knew were actually elements. It was better
to speak of particles, or corpuscles, understanding that there was no
reason to think that even the particles Lucretius talked about had to be
atoms, for air and water and sound might be compounds instead. In
modern terms, the distinction people were making was the distinction
between atoms and molecules, but those ideas were not developed
enough for anyone to make it clearly.

   The most original and for many years the most influential atomic
theory of the scientific renaissance was that of René Descartes, who
worked out an elaborate theory of particles without any empty spaces

between them (Descartes 1983, III), but here as elsewhere, the philosopher's imagination overcame whatever tendency he may have felt to perform the relevant experiments; his ideas on the structure of matter produced an immense literature but had little effect on later scientific developments and I shall not describe them. A more conventional Lucretian view was expressed by Pierre Gassendi (1592–1665), a French priest (his family name was Gassend), thoroughly orthodox in his religious beliefs. He published a book *Observations on the Tenth Book of Diogenes Laertius* (1649), and later several others, in which he removed the fatalistic philosophy that affronted theologians and explained, clearly and with no use of scholastic vocabulary, the physical theory of Democritus, Epicurus, and Lucretius. He explained that there is no reason to insist that atoms are uncreated, eternal, or infinitely numerous, or that their motions are self-generated (i.e., natural motions), or that they move at random. That the world originated (as Lucretius claimed) in a fortuitous grouping of atoms Gassendi says is refuted if one merely looks around: evidences of design are everywhere. Once the irrelevant philosophical accretions have been thrown away we are free to believe that in the beginning, God created atoms with

> their own weights, sizes and shapes in all their inconceivable variety, and likewise the impetus appropriate to each for motion, effort, and change, as for being loosened, rising up, leaping forward, striking, repelling, bouncing back, seizing, enfolding, holding and retaining, etc., as He saw fit to give them for their assigned ends and purposes. (Gassendi, *Physica*, trans. in Sambursky 1974, p. 251)

Note the word "impetus," which shows that Gassendi is not worrying about what kind of force keeps an atom jumping around from moment to moment: once God starts it, its own impetus is enough. Gassendi's reasonable and nondogmatic language endeared him to a generation of intellectuals who wanted nothing more to do with scholasticism. In England during the Commonwealth the atomic philosophy was tainted with materialism and the term "corpuscularian infidel" was more than once thrown at those who professed it, but after the Restoration a group of Royalists who had taken refuge in France and associated with Gassendi and other French scientists returned to England. There, in the tolerant atmosphere of the Stuart monarchy, atoms took root and flourished (Charleton 1654, 1966; Kargon 1966).

Robert Boyle (1626–91) was one of the rare people of his time whose leisure and means allowed him to devote himself to scientific research, and his labor enriched chemistry and physics wherever it touched them. *The Sceptical Chymist* (1661) refutes chemical theories based on the four elements or on the alchemical trio of mercury, sulfur, and salt—not on doctrinal grounds but on their inability to account for the results of experiment. He is sure that none of those seven substances are elements (actually, mercury and sulfur are), and though he believes that elements exist and even suspects that there may be many different kinds, he is not sure that any are known. He is also not sure that elements retain their identity, giving as an example that the wife of the French naturalist Rondelet "kept a fish in a Glass of water without any other Food for three years, in which space it continually augmented, till at last it could not come out of the Place where it was put in, and at length was too big for the glass itself, though that were of large capacity" (1661, p. 405). Boyle concludes that "Earth it self may be produc'd out of water," though since he does not consider either earth or water as elements this does not really contradict the hypothesis that atoms are imperishable. He believed that gold could be made and hoped to do it; not for any commonplace reason, since he was already rich enough, but to fortify the principles of his chemistry. His enthusiasm led him to press successfully for the repeal of an English law forbidding the practice of alchemy. Thus it was that half a century after Ben Jonson's *The Alchemist* (1610) spoke for most educated people in satirizing the chicanery of the artists and the superstition of their patrons, the study of alchemy was once again taken up by such luminaries as Robert Boyle, John Locke, and Isaac Newton.

Newton and the Unseen World

The last of the seventeenth-century atomists I shall mention is Newton, whose other labors were periodically interrupted by weeks in the laboratory. Apparently he went through the standard chemical experiments and found in them nothing to satisfy his quest for the truth that ties everything together, and so he went on to the broader and more speculative domains of alchemy.

Newton's ideas of the structure of matter descended from Gassendi. He starts with atoms.

> All these things being considered, it seems probable to me, that God in the Beginning form'd Matter in solid, massy, hard, impenetrable, movable Particles, of such Sizes and Figures, and with such other Properties, and in such Proportion to Space, as most conduced to the End for which he form'd them; and that these primitive Particles being Solids, are incomparably harder than any porous Bodies compounded of them; even so very hard, as never to wear or break in pieces; no ordinary Power being able to divide what God himself made One in the first Creation. (*Opticks*, Query 31)

In Newton's chemical theory these particles are made of a substance that was generally called *prima materia*, the first matter, and it seems that they were not all alike. They are bound by strong forces into clusters having the character of the elementary chemical substances; Newton calls these clusters particles of the first composition, and they correspond roughly to our idea of an atom. Particles of the second composition are further aggregations that correspond roughly to molecules, and the sequence goes further (Thackray 1970, ch. 2; Dobbs 1975, ch. 6). The process of disrupting these clusters is called putrefaction, and in a published fragment "On the Nature of Acids," Newton suggests that it can occur even at the first level: "If Gold could once be brought to ferment and putrefie, it might be turn'd into any other Body whatsoever. And so of Tin, or any other Bodies; as common nourishment is turn'd into the Bodies of Animals and Vegetables."[7] Obviously if gold can be taken apart it can be put together, but Newton did not consider that the processes involved in transmutations at this level, any more than those that change grass into beef, are those of chemistry. Believing that one could not synthesize a cutlet by ordinary chemical procedures, he went deeper.

For thirty years Newton seems to have kept in quiet, almost clandestine contact with a group of British alchemists (Westfall 1980, ch. 8). At his death, his library contained over a hundred books on alchemy, and the surviving alchemical writings in his own hand amount to more than a million words, mostly copied out of works he did not own. He took these books very seriously, read and compared them as no one else ever had, indexed them, and tried to translate their loose symbolic terminologies into exact and uniform language. We must pause to speculate on what he hoped to accomplish.

Every atomist, from the earliest days until our own, has been obliged to face the problem of how a human being, considered as a machine made of atoms, can have a free will. We have seen the solution offered by Epicurus and Lucretius. Descartes brought the question to a head, arguing, as we shall see in the next chapter, that the realms of mind and matter are almost entirely unconnected. Yet if that is true, how can mind control body, or God exert any influence on the material world, which surely he made and populated by his direct action? For Newton the forms of living creatures could not possibly have arisen, as Lucretius taught, by the fortuitous clumping of atoms. "All that diversity of natural things which we find suited to different times and places could arise from nothing but the ideas and will of a Being necessarily existing" (*Principia*, Book III, General Scholium).

---

[7] In Harris (1710, vol. 2, repr. in Newton 1958). Other versions appear in Newton (1959, vol. 3, pp. 205–14).

But how does will act on matter? Newton despised Descartes as an atheist, and yet in his own increasingly materialistic view of nature the question remained. How do God and spirit act? The answer lies beyond the domains of physics and chemistry, and beginning about 1669 Newton seems to have pursued it into the domain of alchemy. He never published his ideas on this deep question, and the private papers that survive express developing thoughts rather than final conclusions. In an unpublished alchemical manuscript known as "Of Nature's Obvious Laws and Processes in Vegetation," which can be dated to about 1674 (Burndy MS 16),[8] Newton distinguishes between the processes of the living world and those of ordinary chemistry: "Nature's actions are either vegetable ... or purely mechanical." Like gravity, mechanical forces originate in the properties of matter, but the "vegetable spirit," working throughout all nature, is not a mechanical linkage; it is the principle whereby nature propagates and sustains herself by means of "the seeds or seminal virtues of things, those are her only agents, her fire, her soule, her life."[9] Perhaps this is the action that connects, on the most fundamental level, those "impenetrable, movable particles," but at any rate Newton is not expressing a new idea, for the vegetable spirit corresponds, in the material world, to the spirit of the world invoked by the old magicians (Agrippa 1898). Marsilio Ficino describes it in the words that begin Chapter 7 and sought to draw it into human affairs by charms and incantations.

Newton's words recall the language of the books of Aristotle that he had been made to read in his first years at Cambridge: the nutritive, or vegetative soul, present in all life, is the most fundamental of the four Aristotelian divisions of the soul (*On the Soul*, 432a). "The acts in which it manifests itself are reproduction and the use of food, [it] is the cause of the living body as the original source of local movement" (415a, b). And what is it in actuality? If it involves a material substance there must be very little of it. A dead creature does not weigh appreciably less than the same creature alive, and yet something has gone. Newton repeats, possibly for the last time, the answer of so many who asked the same question before him: "This spirit perhaps is the body of light because both have a prodigious active principle, both are perpetual workers." "The body of light"—the first corporeal form. Whatever this spirit may be, alchemy, for Newton, is the branch of philosophy that studies it, and the successful transmuter of metals will pass beyond chemical interactions and enter the realm of the Divine.

Newton tried for "the work in common Gold" and got as far as pre-

---

[8] See Westfall (1980, ch. 8), Dobbs (1982). The spelling in these hasty notes has been modernized.

[9] In Latin, *vegetare* means to animate.

paring philosophic mercury. Often his notes represent him as rediscovering the knowledge of the past. In the remote age of Hermes Trismegistus, under an uncorrupted monotheism, knowledge had risen to heights that perhaps in his own day, under his own guidance, in an enlightened culture rid of the false philosophies of Plato, Aristotle, and trinitarian theology, it might once more attain (Manuel 1962, 1974, Dobbs 1975).

Newton's studies of the writings and opinions of the church fathers were based on thorough study of the texts. Few scholars have ever known them as well as he did. But as his imagination reached back toward the beginning of things the evidence, of course, became slimmer, and for the earliest events he was reduced to trying by means of minute textual comparisons to fit the stories told in the classical myths with those told in the Bible, considering the latter as absolutely true though subject to interpretation. His conclusions (Manuel 1963) are hardly relevant to our story, but unless one appreciates their dreamlike quality, his labors in alchemy can hardly be understood. Let him tell us of the reign of King Misphragmuthosis, who shortly after 1125 B.C. drove the Shepherd Kings from Egypt and reestablished the monarchy (Newton 1728, p. 10). The new state deified its own heroes,

> whence their Gods *Ammon* and *Rhea*, or *Uranus* and *Titea*; *Osiris* and *Isis*; *Orus* and *Bubaste*; and their Secretary *Thoth*; and Generals *Hercules* and *Pan*; and Admiral *Japetus*, *Neptune*, or *Typhon*; were all of them *Thebans*, and flourished after the expulsion of the Shepherds. Homer places *Thebes* in *Ethiopia*. (Newton 1728, p. 202)

So Hercules and Pan began as Egyptian generals, and note the secretary Thoth. As Lactantius has already told us and everyone knew from the Hermetic texts, he is Hermes Trismegistus himself, the transmitter of the oldest arts. Newton's private notes reveal intensive study of the Hermetic texts but the *Chronology* does not mention them. Possibly this was in deference to learned opinion, since in 1614, a generation before Newton was born, Isaac Casaubon had shown to the satisfaction of everyone except lovers of alchemy that the Hermetic treatises were of comparatively recent date (Yates 1969, Shumaker 1972). But to Newton it was only the text that was recent; the ideas still belonged to the oldest and purest tradition.

Perhaps Newton's alchemical ideas should not be taken seriously, but let us reduce them to a few propositions. The variety of particles required to explain the variety of substances in the world is surely very great. At the level of the first composition things become much simpler and there is only first matter in various combinations. Newton clearly believed that the actions at this level of composition are of a different character from those at the higher levels, for the alchemical techniques by which he sought to study them were different from the operations of ordinary chemistry. As to how the actions were actually exerted, through a material medium or through actions at a distance, Newton's ideas changed (Dobbs 1982) but perhaps that was not so important as the fact that they were there.

Readers of the *Principia* may be startled by the final paragraph of that immense work:

> And now we might add something concerning a most subtle spirit which pervades and lies hidden in all gross bodies; by the force and action of which spirit the particles of bodies attract one another at near distances and cohere, if contiguous; and electric bodies operate to greater distances; . . . and all sensation is excited, and the members of animal bodies move at the command of the will, . . . But these are things that cannot be explained in few words. (Newton 1934, p. 547)

Newton is writing of the vegetable spirit, the most fundamental action in nature, the one he has sought all his life. Ordinarily one does not introduce a new topic at the end of a long book, but for Newton, overseeing the preparation of a new edition in the last year or so of his life and looking back on his achievements in science and mathematics and history and theology, it was the great unsolved problem, perhaps the last seal to be broken before mankind could begin to see the light of universal truth.

To end this short account of Newton's alchemy let me compress modern nuclear physics into two sentences: *The individuality of atoms is determined by the structure of their nuclei which are composed of particles of a single basic kind called nucleons bound together by strong forces. When these bonds are broken it is possible to restructure nuclei so as to transmute one element into another (for example to make gold), but the techniques used go beyond those of ordinary chemistry.* Before we dismiss Newton's alchemical speculations, might we not admit the possibility of some scientific insight at the bottom of them?

Except for a fairly general agreement that there are chemical elements and that each corresponds to a special kind of atom, atomic theory at the death of Newton had advanced little beyond Lucretius. Chemists still could not say how chemical reactions actually take place or how the properties of compounds are related to their atomic structure, and there remained the stubborn question, what is the first matter that these "solid, massy, hard, impenetrable, movable particles" are made of? In 1758 the Serbian Jesuit Roger Boscovich published *A Theory of Natural Philosophy Reduced to a Single Law of Natural Forces* (1922) that did away with first matter altogether. The theory is quite complicated and time has washed away most of its hypotheses except for a single idea, that the only reason to think of particles (he does not speak of atoms) as extended in space is that solid matter is almost incompressible. That is, particles exert a strong repulsive force when one tries to push them together, just as (in solids) they exert a strong attractive force when one tries to pull them apart. The conventional theory explained these two forces differently, the first as an effect of finite size and the second as an effect of action at a distance. The particle according to Boscovich is not a thing at all but only a center of force. All its actions are actions at a distance, repulsive at short distance and attractive at long ones. In between, things got complicated, but we need not bother. The basic idea survives today. The particles that physicists regard as elementary—perhaps things like electrons and quarks, perhaps conjectured objects more elementary still—are generally represented in mathematical theory as points, and the phenomena of nature are explained by their interactions or, in the language with which this chapter began, their influences. As to how these influences are exerted, we shall see in Chapter 17 that there are valiant attempts to explain them all by a single kind of argument.

CHAPTER II

They Move According to Number

I can teach you the words but not the truths, which are things.
—GALILEO

EVERYONE has an idea of what time is and knows exactly how to measure it with a clock, but to capture it in a form suitable for physical theory took many years. The theory of static forces as applied to structures and machines, and that of optics as applied to mirrors and lenses, were well understood in the early years of the seventeenth century, but they were of little help to those who sought to understand time-dependent phenomena. Where do you start when you want to explain why something happens? What questions do you ask? What do you assume? The difficulty of these questions will probably be better appreciated by readers innocent of physics than by those trained in it.

Centuries of effort had gone into the attempt to understand phenomena of motion in the light of Aristotle's axiom that force produces velocity. Kepler thought he had succeeded, but his theory conflicts with his own harmonic law. Galileo made an immense contribution; in his *Two New Sciences* (1638, 1914), which was widely read and which will be discussed a little further on, he described experiments tending to show that a ball once started rolling on a level surface would keep on indefinitely in violation of Aristotle's law, which said that without a force to keep it going it would stop. Half a century later the assertion served, in generalized form, as Newton's first law of motion. This chapter is mostly about Newton. It reconstructs some of the arguments that led to the law of gravity and describes how when he had found that planets are governed by the same laws as laboratory-sized objects he tried generalizing his principles further into a theory of the universe. These principles consist essentially of mathematical propositions together with instructions for applying them in the physical world. They tell very accurately how things happen, but Newton's critics found that the most important questions remained unanswered because mathematical propositions, by their nature, have nothing to say about why anything happens. After a digression on Descartes' vision of a mechanical world the chapter ends with an epistolary war between Newton and Leibniz that shows how widely two men of genius could differ as to what makes good science.

Laws of Motion

It was harder to find the laws of motion than the laws of statics because motion is change, and on that subject the descendants of Aristotle had so much to say. Aristotle had started with a definition of time and some statements about it that are hard to object to,

> time is the number of movement in respect of the before and after, and it is continuous since it is an attribute of what is continuous. . . . It is clear, too, that time is not described as fast or slow, but as many or few and as long or short. . . . Further there is the same time everywhere at once. . . . Not only do we measure the time by the movement but also the movement by the time, since they define each other. (*Physics*, 220)

All this is the most obvious good sense. The remark that time is not described as fast or slow puts to bed any talk about the motion or flow of time except as a metaphor for something that can be said more precisely in other ways, and if the remark were better understood it would have lightened the modern philosophical literature of many tedious pages that assume time to be a kind of motion at the same time that it is a number; we shall return to this matter later in the chapter. Time is simple enough as long as we take it for granted, and that, in the early development of dynamics, is what was done.

As we have seen, ideas and definitions relating to uniform motion and also to motion uniformly accelerated were clearly expressed and their consequences explored during the Middle Ages, but not in the context of natural phenomena, since phenomena were not looked upon as sources of general truths. As time brought changes in this view, one of the first questions to be cleared up was the motion of an object in free fall. Aristotle mentions it several times; for example, in *On the Heavens* (294a): "a little bit of earth, let loose in mid-air, moves and will not stay still, and the more there is of it the faster it moves." In the course of time it was concluded from remarks such as this that Aristotle says an object's rate of fall is proportional to its mass. As I have said earlier, Aristotle seems to have had no concept of a quantitative natural law. His words express nothing more than the conviction, drawn perhaps from watching oak leaves and acorns, that heavy things fall faster than light ones. Besides, almost everyone knows that a brick does not fall several hundred times more rapidly than a dried pea, and John Philoponus seems to have tried an experiment:

> For if you let fall from the same height two weights, one of which is many times as heavy as the other, you will see that the ratio of the times required for the motion does not depend on the ratio of the weights, but that the difference in time is a very small one. (Cohen and Drabkin 1948, p. 220)

But whereas we would say that this is the case only insofar as air resistance can be neglected, for John Philoponus the equality of rates is an effect of air resistance. In a vacuum, he says, the rate of fall would be proportional to the weight.

Simon Stevin also describes testing the Aristotelian belief with an experiment. He dropped two lead balls, one ten times heavier than the other, onto a board about 30 feet below and noted that they made a single sound as they hit (Stevin 1955, vol. 1, p. 511). Later, Galileo repeated the experiment and also argued that the "Aristotelian" principle, taken with the rest of Aristotelian dynamics, does not even make sense. Suppose we take a small stone, which wants to fall slowly, and tie it to a heavier one, which wants to fall faster. The lighter one will exert a retarding force and so the combination will fall more slowly than the heavier one would if it fell alone. But now consider the two stones tied together as a single object. They should fall faster, not more slowly, than the large stone and thus Aristotle's supposed principle leads to a contradiction (Galileo 1914, p. 63).

If all reasonably heavy objects (heavy enough that the resistance of the air is not important) fall at the same increasing rate, how does that rate increase? Galileo experimented long and carefully on this and related questions (Drake 1978) and at one point presented his results in a particularly vivid way:

> The straight-line acceleration of heavy bodies takes place according to the odd numbers, starting from one. Choose whatever equal intervals of time you wish. Then if the moving body, starting from rest, travels a distance of 1 ell, say, during the first interval, it will travel 3 ells during the second, then 5, then 7 and so on, continuing with the successive odd numbers. (1968, vol. 7, p. 248)

How far has the body traveled at the end of N intervals of time? Pythagoras has already answered that question for us with his pebbles—the sum of the first N odd numbers is N^2: "I may say that the distance traversed varies as the square of the time." This is one way to characterize the motion of a freely falling object, but there is another. We have seen that Nicolas Oresme and the scholars of Merton College analyzed a hypothetical motion that is "uniformly difform" (we would call it uniformly accelerated), and decided that in such motion also, the distance traversed increases as the square of the time. Therefore we can state, and Galileo did, that free fall takes place at constant acceleration: a falling body gains velocity at a constant rate.

Finally, Galileo established another fundamental principle. He found by experiment that if a ball is allowed to roll down one incline and up another with friction reduced as far as possible, then, regardless of the slopes of the two inclines, the ball rolls up the second one until it reaches a height that is very nearly equal to the height from which it started. Suppose now that the second plane is made flatter and flatter. The ball will have to travel further and further to regain its initial height. Thus, if the second plane were made smooth and perfectly level, it would go on for a very long way, showing that once started, violent motion can persist indefinitely. If true, this result destroys the central principle of Aristotelian dynamics, that everything that moves is moved by something, and brings into question the rest of the doctrine of cause and change. And if we accept that ordinary matter, once started, tends to keep going until stopped, then there is no further reason to make the planets of an ethereal substance whose motion is neither briefly up nor briefly down but perpetually in a circle. Any substance could be expected to move perpetually in a circle as long as a circular track was laid out for it. For Galileo, as we have seen, it was. The principle that motion of ordinary matter persists unchanged in the absence of externally applied forces is known as the principle of inertia.

These results of Galileo's studies are set forth in two books written for the general public in the form of dialogues. In them two disputants, Salviati and Simplicio, contest to win the mind of Sagredo, an intelligent, educated, and conservative layman. Salviati speaks for Galileo, who was unable to speak for himself because of the prohibitions of his church, and Simplicio expresses the traditional views of the schools. The *Dialogue Concerning the Two Principal World Systems—Ptolemaic and Copernican* (1632, called the *Dialogo)* and the *Discourses and Mathematical Demonstrations Concerning Two New Sciences* (1638, called the *Discorsi*; the sciences are statics and dynamics) are among the best works of scientific popularization ever written. Clear, witty, and psychologically astute, they present their speakers as characters with strengths and weaknesses, and they helped to create a new age of scientific thought with their emphasis on observation, common sense, clear language, and persuasion by reasonable arguments.

Newtonian Dynamics

During Galileo's lifetime the idea of force was still clouded with Aristotelian metaphysics. In statics, force could be assimilated to weight and could therefore be measured and defined, but as the cause of motion it was harder to understand. What, for example, causes the uniform acceleration of a falling stone? Nowadays we say glibly that it is the force of gravity, but that is hard to know directly—how can one detect or measure the force on an object in free fall? The idea of force is expressed in writings of the period by words like *conatus* (denoting effort or striving) that still retain the implication that the cause of motion dwells inside the object. It seems to have become clear to Newton, however, that the force of gravity acts on a falling body exactly as it acts on one held in the hand, remaining constant in magnitude as the body falls. Constant force produces constant acceleration; this must have suggested the possibility that varying force produces varying acceleration, that the two are proportional. Further, the larger the quantity of matter and the more dense it is, the less acceleration will be produced by a given force. Therefore Newton defines the mass m of an object as the product of its volume and its density and supposes that the acceleration varies as

$$a \propto F/m,$$

and by the choice of a suitable unit to measure the force this becomes the celebrated formula

$$F = ma.$$

We have now encountered all three of Newton's laws of motion, and can state them as follows,

I. A body persists at rest or in uniform motion in a straight line unless acted upon by an externally applied force.

II. A force F externally applied to a body of mass m produces an acceleration proportional to F/m in the same direction as the force.

III. When one body exerts a force on another, the forces of action and reaction are equal and opposite.

These words and formulas are by no means direct translations of the laws as they are stated in the *Principia*, for the latter, though clear enough in application, retain traces of the confusions of thought out of which Newton dragged them. For example, in his statement of the second law he uses *vis*, usually translated "force," to designate an impulse such as a hammer delivers; we don't know how much force it exerts but we can determine the effect of the blow. And as I have mentioned, Newton nowhere mentions acceleration; he expresses the second law by saying that *vis* changes momentum. This statement is followed by a corollary that proves the parallelogram law as it applies to a body with unbalanced forces acting on it and here Newton uses *vis* again, but he can hardly be thinking of a body struck simultaneously by two hammers moving in different directions. In a moment we shall look at Newton's proof of Kepler's area law. There he uses *impulsus* to denote the sudden impulse he has previously called *vis*.

The first law goes beyond Galileo's principle of inertia, since Galileo had envisaged a body rolling or sliding along a level surface on earth rather than moving freely in space. It has often been remarked that this law is a trivial consequence of the second one when the force is zero, but there is more to it than that, for if we imagine that the universe defines a scale of distances, the first law, by specifying constant velocity, establishes a scale of time for it also. Without this the second law has no meaning.

The emphasis on externally applied forces is intended to make clear that internally applied forces have no effect on motion. A child who pushes on the dashboard to make the car go faster does not help at all.

The first law was suggested by Galileo (though Newton seems to have derived the idea from reading Descartes), and special cases of the Second Law had been used by others, notably Christiaan Huygens, before Newton gave it general form. The third law is the only one that is Newton's own, and if one considers it skeptically it is perhaps the most surprising of the three. If I push on a heavy box and am finally able to make it slide across the floor, folk wisdom explains that it is because I am able to push on the box harder than it can push back on me. The third law states that the two forces are always equal, whether or not the box moves.

When combined with the law of gravity the second law leads at once to important conclusions. Consider, for example, the motion of an object of mass m dropped just above the surface of the earth. In modern terms the force on the object is given by GMm/R^2, where G is the constant that gives the strength of gravitational forces, M is the mass of the earth, and R is its radius. Setting this into the second law gives

$$ma = GMm/R^2,$$

from which m cancels, leaving $a = GM/R^2$, independent of m. Thus, as had been observed by Philoponus, Stevin, Galileo and others, everything falls at the same rate unless impeded by the resistance of air. In fact, the same cancellation takes place whenever the two laws are used together, so that a grapefruit placed in the same orbit as the moon would move along the same path at the same speed.[1]

Of the multitude of new results in the *Principia*, the ones most often mentioned are the proofs of Kepler's first two laws of planetary motion, that the planets move in ellipses and that the line from the sun to a planet sweeps out equal areas in equal times. Let us look first at the area law, for it turns out to be the more general, being true for any kind of orbital motion provided only that the force governing it is directed straight toward the attracting center.

In Newton's geometrical repertory there was no way to follow the planet along its curved path, and so he represents the path as a sequence of straight-line segments (Fig. 11.1). Instead of a continuously acting force, he imagines that at regular intervals of time the planet receives a sharp impulse, as if with a hammer, directed toward the sun. During the first interval the radius vector sweeps out an area ABS; if the force did not act the planet would continue to c and the area BcS can, with a modicum of thought, be seen equal to ABS. But suppose the impulse toward S arrives when the planet reaches B, sending it to C instead of c. If Cc is parallel to BS, triangles BCS and BcS are equal; thus BCS = ABS, and so on along the path. "Now let the number of those triangles be augmented, and their breadth diminished *in infinitum*"; the broken line approaches the orbital curve, the areas remain equal, and the result is established (1934, p. 40). The result is true, and the argument is plausible but it proves nothing. To show that in the limit the result really follows was beyond Newton's powers, probably beyond his conception of proof. One reason why the *Principia* is not easy reading is that it is sometimes hard to be sure, as step A leads to step B, whether B really follows mathematically from A or whether it merely follows it on the page.

The consequences of the foregoing argument went far beyond the derivation of Kepler's law of areas, for in dealing with motion the mathematicians of those days did not use time as a variable quantity, and until well into the eighteenth century every planetary movement was calculated not in terms of time but of the even increment of area swept out by the radius to the moving planet. Note E shows how easy it is to derive the

---

[1] That is not quite true. The moon is massive enough to produce a measurable effect on the earth's motion, which the grapefruit is not, and this in turn slightly affects the orbit of the moon.

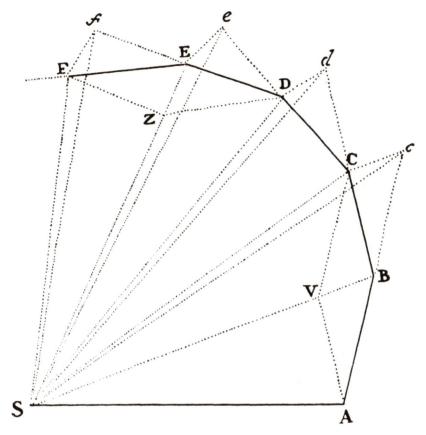

FIGURE 11.1. To illustrate Newton's proof of Kepler's law of areas. From the *Principia*.

area law from Newton's by simple calculus if only we introduce time as a variable; Note F is an example of the earlier mode.

As for Kepler's first law, the law of ellipses, Newton does not give a direct proof in the *Principia*. He proves the converse (Book 1, Proposition 11), that if a body moves in an ellipse around a center of force located at one of its foci the force varies inversely as the square of the distance, but as Jean Bernoulli (1667–1748) pointed out (Newton 1967, vol. 6, p. 148), and as many others have noted since (Weinstock 1982), he never got around to proving the law itself. For want of a proof by Newton, Appendix F reproduces one given by Bernoulli in 1710. Its significance is evident even to those who do not follow the mathematics in detail. Compare it with Newton's argument for the area law, which is really much easier to prove. Bernoulli needs no diagram, no artifice like that of hammer blows, and the entire discussion is contained in a few lines of very ingenious calculus.

Newton seems to have obtained some of his results using his own version of calculus, but there is no sign of it in his explanation of mechanics. Instead, the derivations follow the lines of the one I have presented, proved by the methods of Euclidean geometry and tedious because they were poorly adapted to the work he was doing. The methods are global and synthetic: as the particle moves in its orbit the magnitude and direction of the force, and hence the particle's acceleration, change as it goes. The problem amounts to constructing an entire ellipse from this specification, and what makes it difficult is that the specification has to do with time while the ellipse is a fixed shape, so that the geometrical construction has to eliminate the time somehow. The solution by calculus does the same thing using the flexible analytic procedures of algebra rather than rigid procedures of geometrical construction. It is much easier.

A second edition of the *Principia* appeared in 1713, twenty-six years after the first. Newton was seventy-one years old and most of the work was done by a younger man, Roger Cotes, with his close cooperation. The second edition is only a corrected and augmented version of the first and, given the circumstances, it could not have been anything else. But it was a disaster for British science, for it sanctified archaic methods already rendered obsolete by the rapid development of calculus on the Continent. A third edition appeared shortly after Newton's death. It would be almost a century before Britain once again had anything to say.

The *Principia* contains more mathematics than physics and is not so much a treatise on mechanics as it is a report on Newton's own work and speculations. It contains nothing about statics, a subject highly developed in his time, and nothing about the motions of elastic objects. On the whole, his interest is mainly in astronomical questions. Book II, largely on the motion of bodies through resisting media, seems to have been motivated by curiosity as to how planets manage to move through the ether, though he arrives at no conclusions, and there are interesting sections on hydrodynamics that were designed to destroy Descartes' explanation of gravity as the effect of ethereal vortices. Perhaps the use of ancient mathematics was intended to add authority to the *Principia*, but on reading it one understands the words of William Whewell,

> The ponderous instrument of synthesis, so effective in his hands, has never since been grasped by one who could use it for such purposes; and we gaze at it with admiring curiosity, as on some gigantic implement of war, which stands idle among the memorials of ancient days, and makes us wonder what manner of man he was who could wield as a weapon of war what we can hardly lift as a burden. (Whewell 1857, vol. 2, p. 128)

What's the Real Reason Why They Move?

I have said earlier that in Galileo's published works on motion he restricts himself to kinematical questions: he explains how things move but not why. There is, however, good evidence that Galileo thought about the causes of motion, tried various hypotheses, and was dissatisfied with all of them. He never seems to have questioned the Aristotelian distinction between natural and violent motion. He could assert that an object with no force to retard it can move naturally a long way without slowing down, but when there is a force, he believed with Aristotle that the fundamental dynamical law connects force with velocity and not, as Newton later showed, with change of velocity. As we have seen in Chapter 10, Galileo was dissatisfied with the Aristotelian formula $V \propto F/R$ relating velocity with force and resistance, but never got beyond the idea that although natural motion can persist without a force, violent motion requires that force be exerted all the time. He deduced from his own law of inertia that the force that starts a violent motion can under the circumstances of the law remain with it indefinitely.

How, then, do we understand the play of forces on a ball that is thrown into the air? For Aristotle, as the ball rises, air rushes into the space it vacates and in doing so pushes the ball higher. In about 1590 Galileo, then in his twenties and lecturing at Pisa, prepared a long series of manuscript notes on mechanics of which the dominating theme is the refutation of Aristotle. He contemptuously dismisses the theory of propulsion by air, arguing that if it were true one could throw a balled up piece of paper further than an iron ball because they both set the air in motion and the paper is obviously easier for air to push (Galileo 1960, p. 77). Later he carefully analyzes the motion of a ball that is thrown upward (p. 89). The thrower imparts a motive force greater than the weight of the ball. As the ball rises the motive force continually diminishes. When it becomes less than the weight the ball starts downward, moving faster as it approaches the ground. Thus he explains the ball's acceleration as it approaches the ground without needing to assume an increase in weight. This is an example of a law of motion Galileo considered briefly at the time,

$$V \propto F - R.$$

Here F is the diminishing motive force that remains in the body after it is thrown and R is the weight that resists the ball's rise (see also Galileo

1914, p. 165). Galileo never doubts that in explaining motion one ought to focus on the relation between force and velocity.

A few years ago John Clement of the University of Massachusetts reported the results of questionnaires and interviews with two groups of university students designed to find out how they explain the same motion (Clement 1980). Most of group 1 had taken physics in high school but not in college, while group 2 had completed the college course in physics required of engineering majors. They were all shown the diagram of Figure 11.2 and told that it represents the trajectory of a coin thrown upward. To the question, "What are the forces on the coin when it is at the point P?" the answer in the spirit of Newton is that (neglecting air resistance) the only force is the coin's weight (this gives it a downward acceleration that keeps on diminishing the upward velocity until, at the top, it becomes a downward velocity). Of Clement's group 1, 12 percent answered correctly, while in group 2 the number of right answers had increased to 28 percent. The rest thought that the coin continues to rise at P because it is propelled by an upward force larger than its weight: "the force from your hand," "the force of the throw," "the original force," "the velocity is pulling it upwards," so that as Clement remarks, they were easily and naturally using ideas the young Galileo had written down four hundred years ago, and which were not new then. Since only 16 percent of the students tested seem to have had their views changed by a university course in physics, the dynamics of the fourteenth century must be deeply embedded in the way our culture understands and talks about motion. Newton knew the same temptation; perhaps for a while he gave in to it, for in one of the drafts of an unpublished study for the *Principia* he defines impressed force as "an action exerted on a body to change its state either of resting still or of moving uniformly straight on," and then he adds "This force consists in that action alone, and does not endure in the body after the action is over" (Newton 1967, vol. 6, p. 95). For further examples of Aristotelian thinking in today's students see Viennot (1979) and McCloskey et al. (1980).

It seems to me that the Aristotelian theory underlying these answers is reinforced by everyday experience, and that if the purpose of science is to explain what happens and to suggest what we should do to attain a given result, the theory is actually more useful (as long as we do not need numbers) than a correct application of Newtonian physics. If I run harder or push with more force on my bicycle the expected and actual result is that a moment later I am once more moving at constant speed but faster: force produces velocity.

The law of action and reaction fares no better. A question to ask the unwary at dinner is, "If I push a spoon across the table with my finger, is the force with which the spoon pushes back on my finger less than or the

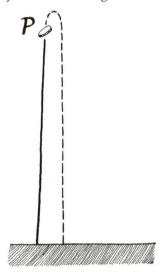

FIGURE 11.2. What is the force on the rising coin?

same as the force with which my finger pushes the spoon?" The New-tonian answer is that it is always the same. Other than those trained in science, how many people, facing their inner convictions with perfect courage, really believe it?

This brief digression suggests several remarks. One is that we can live surrounded by the products of science without thinking scientifically; another is that it must have taken an immense effort for Isaac Newton, using primitive mathematics and terminology, to shake off these natural ways of perceiving and thinking. We know force with our nerves and muscles. We perceive velocity, but how many people actually notice accel-eration as they observe the world, or use the words "velocity" and "accel-eration" with the sense of their exact meanings? A teacher of young stu-dents had better be aware that there may be a gap of many centuries between the blackboard and the first row of seats.

Newton's Universe

The last book of the *Principia* is called "On the System of the World," and it sets out to explain the solar system and the structure of the universe as Newton understood it. Of course this was before there was any con-ception of galactic or extragalactic distances, but little that Newton said would have been different had he known about them. He assumes that one can apply on the scale of interplanetary and interstellar distances the same laws that one learned in the laboratory. It is true that Newton first found the law of gravity through rough calculations of lunar and planetary motions, but the real validation of the concept of mass as he used it was by laboratory experiments with pendulums, for which he claims an accuracy of one part in a thousand (*Principia*, III, Proposition 6). The distance from the sun to Saturn is about 10^{12} times the length of one of these pendulums, and Saturn's mass is about 10^{27} times greater. This was the first time that quantitative principles established in a small room had been extrapolated on so gigantic a scale, and the success of his extrapolations strengthened Newton's often expressed conviction that nature is very conformable to itself.

Newton reaches still further. He uses on the planetary scale the results of laboratory experiments and the assumptions of ordinary Euclidean geometry and they work. Now he considers the entire universe. The space of geometry has no favored points or regions. Every one is like every other; it seems to follow that the universe is infinite, and in his letters to Richard Bentley Newton suggests that it is, arguing that any finite distribution of stars in space would inevitably collapse under its own gravitation.

Extrapolations on such a scale need to be controlled by specific assumptions, and Newton starts by writing them down. Of course they involve hypotheses as to the nature of the universe, but Newton's sensitivity to the word "hypotheses" leads him not to refer to them in his "Rules for Philosophizing":

I. *We are to admit no more causes of natural things than such as are both true and sufficient to explain their appearances.* To this purpose the philosophers say that Nature does nothing in vain, and more is vain when less will serve; for Nature is pleased with simplicity, and affects not the pomp of superfluous causes.

II. *Therefore to the same natural effects we must as far as possible assign the same causes.* As to respiration in a man and in a beast; the descent of stones in Europe and America; the light of our culinary fire and of the sun; the reflection of light in the earth, and in the planets.

The purpose of these two rules was to distinguish Newton's method from that of scientists who invented a new hypothesis to account for every new effect.

III. *The qualities of bodies which admit neither intensification nor remission of degrees and which are found to belong to all bodies within the reach of our experiments are to be esteemed the universal qualities of all bodies whatsoever.*

I can best illustrate the thrust of this rule by examples. Newton's laws of motion and of gravity apply exactly (he claimed) to all bodies that have been studied. It is not a question of degree. There are none to which they apply better than others. Therefore we shall assume that they apply without alteration to all bodies, outside the solar system as well as within it.

IV. *In experimental philosophy we are to look on propositions inferred by general induction from phenomena as accurately or very nearly true, notwithstanding any contrary hypotheses that may be imagined till such time as other phenomena occur, by which they may either be made more accurate, or liable to exceptions.* This rule we follow, in order that the argument of induction may not be evaded by hypotheses.

Do not conclude that Newton imagined some miraculous logical process by which one could attain to general laws by studying individual instances. Rather he is using "induction" in the sense established by Francis Bacon (e.g., *Novum organum*, 1.105): it is the process of trial and error. Try everything; throw out what doesn't work; what remains may be valuable, but without good luck the progress toward truth will at best be asymptotic.

Newton continues the "System of the World" by analyzing planetary and lunar motions with calculations of stupefying originality and power, though sometimes his goals lie so far beyond the reach of any mathematics then known that they are not very successful. Let us pass to the end of the work to see what is the larger setting of the system and how it all began. First, Newton points out that we have in the cometary system a fair example of a system of satellites started at random: the planes of their orbits, unlike those of the planets, lie in every orientation. Also in their long and narrow orbital ellipses they spend the vast majority of their time very far from the sun and are very cold, so that whatever a comet may be, it cannot support life. The six primary planets and ten satellites revolve around the sun in orbits that are nearly perfect circles, nearly in the same orbital plane and all in the same direction.

This most beautiful system of the sun, planets, and comets could only proceed from the counsel and dominion of an intelligent and powerful Being.
. . . and lest the systems of the fixed stars should, by their gravity, fall on each
other, he hath placed those systems at immense distances from one another.
. . . As a blind man has no idea of colors, so we have no idea of the manner

by which the all-wise God perceives and understands all things. . . . We know him only by his most wise and excellent contrivances of things, and final causes; we admire him for his perfections; but we reverence and adore him on account of his dominion; for we adore him as his servants; and a god without dominion, providence, and final causes, is nothing else but Fate and Nature. (Newton 1934, pp. 544–46)

At the creation God placed the stars far apart, but of course there is no mechanical reason why in the course of time their mutual gravitation would not cause them to fall together into some central region. In Query 28 of the *Opticks*, Newton raises this question in the course of a long paragraph on God's role in the creation:

Whence is it that Nature doth nothing in vain; and whence arises all that Order and Beauty that we see in the world? To what end are Comets, and whence is it that Planets move all one and the same way in Orbs concentrick, while Comets move all manner of ways in Orbs very eccentrick; and what hinders the fix'd Stars from falling upon one another?

The answer to each of the questions is to be found in God. Newton seems to have understood that even an infinite distribution of stars would be unstable and would tend to collapse into one or more centers of condensation unless it were continually tended. But there is more. Newton's calculations had convinced him that planetary orbits are unstable, and that the mutual interactions of the planets will, in the long run, disturb them. Obviously this would be a disaster. If, for example, the earth were to move significantly nearer to or further from the sun our climate would be catastrophically affected. Therefore, in Query 31 Newton writes,

For while Comets move in very excentrick Orbs in all manner of Positions, blind fate would never make all the Planets move one and the same way in Orbs concentrick, some inconsiderable Irregularities excepted, which may have risen from the mutual Actions of Comets and Planets upon one another, and which will be apt to increase, until this System wants a Reformation. Such a wonderful Uniformity in the Planetary System must be allowed the Effect of Choice.

This passage suggests two remarks. First, as to its correctness. It has been shown by the difficult methods of modern celestial mechanics that the orbits Newton feared are unstable are actually stable, in the sense that in the long run gravitational interaction tends to bring the planets into orbits that are more rather than less nearly circular and coplanar. Second, it gives a further indication that God's creation cannot function indefinitely without divine readjustment, that God, having once set the universe in motion, does not just sit back and watch it run. He intervenes on the cosmological scale just as, in the preceding chapter, we saw him exerting his own kind of force on the particles of matter.

Newton worshipped the God who had planned the world, made it, and continued to protect it. His goal was to understand something of the plan. We have seen that many before him sensed that there was a plan and that they were part of it, but as Newton and other scientists of his time began to reveal it they were helped by their theology. The God of the scriptures as they understood them must have created a world that is both orderly and lawful. Since he is rational it cannot be otherwise. But God is not trapped by logic. He invented the laws of nature and the world was his free creation; he made it as he chose, and we see him laying his hand on it from time to time. The divine plan can therefore not be uncovered through arguments of necessity any more than it was created that way. Reason and observation must be relied on equally to reveal the machinery of the universe. As Newton's readers pondered this machinery they must have recalled the writings of Descartes.

The Clock and the Clockmaker

I have not written much of Descartes because he did not contribute much to the science we are seeking. Like Aristotle, he studied anatomy and physiology with great care, but he inherited the conviction that the real truth about nature is learned from reason and not from the trial-and-error procedures of experimental science. Leibniz's physics (if one can call it that) springs from the same conviction, the effects of which linger in philosophical writings even in our own day.

Descartes' physical speculations soon disappeared from view. In scientific philosophy, however, he found the door that led from the medieval schoolroom into the open air, and he was the first to pass through it.

It will be convenient to begin this short account with Descartes' division of the world into two distinct categories: *res extensa*, that which is extended, and *res cogitans*, that which thinks. The first refers to material substance that occupies space; the second is not material but comprises pattern, function, dynamism. These two categories "can be understood from this common concept: that they are things which need only the participation of God in order to exist" (*Principles of Philosophy*, 1983, no. 52). The material world includes both inorganic and organic nature and it functions like a machine. Even an animal is considered as a mechanism consisting of particles of matter interacting and moving according to mathematical laws of motion. Life and death are attributes of the physical body. The soul belongs to that which thinks. It is in no sense located in or coextensive with the body, and its continued activity arises from a sort of conservation of the motion of all the particles of the universe. Soul completes the existence of the body; Descartes' famous postulate "*Cogito ergo sum*"—"I think, therefore I am"—points to the necessary connection between what thinks and what is extended. Soul is not alive and therefore it cannot die: this is the true meaning of its immortality.

We cannot appreciate how radical these ideas were unless we compare what he says with the views of then current Aristotelian philosophy according to which knowledge consisted in the soul's participation in the substantial form of a thing. The fact of knowledge united soul and thing; Descartes proclaimed that they were by nature distinct. We are of course involved in the universe, but if we are to think scientifically about it we must act as if we were not, and try to understand its forms and behavior as independent of human participation. This Cartesian dualism, as it is called, is a basic ingredient of the modern point of view.

How does everything work? As long as everything is understood as every thing, we are asking about physical mechanisms, and the answers must invoke only the kinds of cause that Aristotle called efficient. The others sink into insignificance, are mere verbal formulas, as when you might say in the language of final causes that the kettle is boiling because you are going to have a cup of tea. But the doctrine of efficient causes worked only if one did not push one's questions too far. Ask the reason for gravity or magnetism, or how a candle gives light, and mechanical reasoning had nothing to say. Descartes' critics objected not so much that there are mechanisms, but that the realm of nonmechanical actions is so large that it may very well overlap the realm of the *res cogitans* that Descartes hoped so much to keep separate and, apparently, to keep small in comparison with the vast realm of the physical universe. These minor holes in the system did not deter Descartes from claiming, in the French edition of the *Principles of Philosophy* (no. 199), that "there is nothing visible or perceptible in this world that I have not explained."

Of course, the soul does become aware of what is sensed by the bodily organs. It perceives the secondary properties of things, the colors and tastes and smells that cannot be reduced to mathematical description. How does that happen? Descartes' theory of the process goes back to the teachings of Aristotle and Galen.[2]

In Book III of *On the Soul*, Aristotle asserts that when our senses contribute their various bits of information about something, our total impression of it is generated in an organ not directly connected with the

[2] It is found in the *Treatise on Man*, written about 1634, the *Dioptrics* (1637), the *Principles of Philosophy* (1644), and the *Passions of the Soul*, written in 1645. The standard edition is Descartes (1964).

outside world as the sense organs are; he called it common sense and located it in the heart. In the Middle Ages it was believed to be located, along with other faculties like reasoning, imagination, and memory, in one of the ventricles, which are small sacs of fluid within the brain that are now understood to regulate its hydrostatic pressure. Inside and between the ventricles flowed a subtle vapor known as spirit, an accepted item of human physiology since the time of Aristotle (recall what Ficino and Newton thought about spirit). Spirit was created in the body and flowed through it, so extremely volatile that it served as a sort of mean between body and soul and thus could be the medium through which they interact. The psychology of the Middle Ages and Renaissance paid special attention to how it was distributed in the brain and how it moved.

Refining this idea in the *Passions of the Soul*, (no. 31), Descartes identifies the actual organ of common sense as a small structure known as the pineal gland (it is probably a vestigial sense organ and has no known function in man). "The slightest movements on the part of this gland may alter very greatly the course of these spirits, and conversely any change, however slight, taking place in the course of the spirits may do much to change the movements of the gland" (1983, p. 140).

The gland also responds to the soul, changing its position as one thought succeeds another (no. 34). (The reason we cannot think two thoughts at once is that the gland cannot be in two positions at once.) When it moves it agitates the spirit, which in turn leads to our nervous and muscular responses. All of this, even the motion of the gland that initiates the mental and physical acts in which we recognize free will, is a wholly mechanical process. Then is there free will? Descartes thinks there is. If in some situation we had perfect understanding of what was going on, he believes that we would always, predictably, choose what is right, but since that never happens we evaluate the information we have as best we can and act accordingly, and in doing this we are absolutely free (*Principles*, nos. 31–44). I am not sure whether Descartes held that the thought processes of the *res cogitans* take place independently of the material body. If so, I can see nothing logically wrong with his theory.

I have discussed Descartes' mechanization of human thought and action in order to stress the universal nature of his claim that the world is a machine that runs automatically: for him, any idea of a causal connection that cannot be explained in terms of a physical mechanism belongs to magic and superstition.

Under the influence of Descartes the operations of the world began to be seen as open to rational inquiry in a way that had not existed since Epicurus, but there was the central difference in that whereas Epicurus banished the gods like a bad dream, Descartes placed God at the center as Creator and as giver of law, purpose, and meaning. Even so, many people felt terribly alone and undefended in a world that the Divine Watchmaker (echoing Plato's Divine Craftsman) had made and wound up and set and which now ran without his help. Perhaps there were virtues in prayer and faithfulness, but what was the use of them? Some time before 1622 Blaise Pascal wrote among the notes later published as his *Pensées*, "Le silence de ces espaces infinis m'effraie," "the silence of these infinite spaces terrifies me" (no. 206). Only lately these spaces had been filled with legions of angels and other spirits that sweetly cared for the world in its infinite nuance and complexity according to the will of God. Now they were empty. Pascal dismisses the Cartesian cosmology as "useless, dubious, and painful. And if it were true we do not think that all of philosophy would be worth one hour of effort" (no. 79).

Two generations later Bernard de Fontenelle struck a different note in his *Conversations on the Plurality of Worlds* (1686), a playful but solidly based introduction to Copernican astronomy. It is embroidered with talk of endless inhabited worlds and its tone is suitable for the instruction of the very rich:

> Imagine the great sages of the past—the Pythagorases, the Platos, the Aristotles, all the people of whom there is so much talk these days, spending an evening at the Opera. Suppose they see the tragedy in which Phaeton is carried away by the winds, and suppose they do not notice the ropes and do not know about the backstage machinery. One says "Phaeton was carried away by a certain occult power." Another: "Phaeton is composed of certain numbers that cause him to rise." Another: "Phaeton has a certain affinity for the upper part of the theatre and he is uneasy when he is not there." Another: "Phaeton was not made for flying but he would rather fly than leave the upper part of the theatre empty." . . . At last come Descartes and a few other moderns who say "Phaeton rises because he is suspended from ropes and because a weight heavier than him descends." . . .
>
> —From what you say, said the Marquise, philosophy has become very mechanical?
>
> —So much so, I replied, that I fear we shall soon be ashamed of it. They claim that the universe is in large what a watch is in little, and that everything happens by regular movements depending on the arrangement of the parts. Admit the truth. Had you not once a more sublime idea of the universe, and did you not give it more honor than it deserves? (Fontenelle 1966, p. 18)

If you try to escape generalities and illustrate your philosophy with specific examples, this is what you get. His book appeared the year before Newton's *Principia*, and it is against this Cartesian background that we must assess Newton's claim that his cosmic system could not function without the occasional intervention of God.

Britain at War

In 1705 a dispute arose between Newton and Leibniz as to who had first invented the methods of differential calculus. Under the prodding of third parties concerned with national glory the terms of the controversy expanded until Leibniz finally became the voice of all those for whom Newton's physics amounted to a mathematical tour de force that spouted numbers but explained nothing. After various private letters, a public correspondence began (see Alexander 1956) with a letter which Leibniz addressed in November 1715, ostensibly to his intellectual protégée, Caroline, Princess of Wales. In it he attacked at several points, of which I mention only a few that are relevant here.

> Sir Isaac Newton, and his followers, have also a very odd opinion concerning the work of God. According to their doctrine, God Almighty wants to wind up his watch from time to time: otherwise it would cease to move. He had not, it seems, sufficient foresight to make it a perpetual motion. Nay, the machine of God's making, is so imperfect, according to these gentlemen; that he is obliged to clean it now and then by an extraordinary concourse, and even to mend it, as a clockmaker mends his work. (Alexander 1956, p. 11)

Sir Isaac, seventy-two years old and a busy public man, got his friend and collaborator Samuel Clarke to reply. This objection of Leibniz, Clarke wrote, imagines the universe as the work of an ordinary workman.

> But with regard to God, the case is quite different; because he not only composes or puts things together, but is himself the author and continual preserver of their original forces or moving powers; and consequently 'tis not a diminution, but the true glory of his workmanship, that nothing is done without his continual government and inspection. (p. 14)

This is a positive rejection of Descartes' model of the universe: God never intended to make a universe that he could then leave to itself. The points having been made, we need not follow the further bursts of rhetoric but can sample another objection to Newton's system that has already been mentioned. Leibniz writes, " 'Tis also a supernatural thing, that bodies should attract one another at a distance, without any intermediate means; and that a body should move round, without receding in the tangent [i.e., without leaving its orbit], though nothing hinder it from receding. For these effects cannot be explained by the nature of things" (p. 43). Newton is accused not only of bad science, but of sliding back toward the old occult causes, the witchcraft and spells that did their damage at a distance "without any intermediate means," from which mankind was slowly being delivered by the efforts of enlightened minds.

Leibniz died at the end of 1716, but the correspondence[3] did not end then. In 1722 Newton delivered a final blast in an anonymous review of the entire correspondence, which had meanwhile been published as a book (Collins 1713). Newton was eighty, and I quote from the end of his review as a last example of his polemical style. Comparing himself with Leibniz, he writes

> The one teaches that philosophers are to argue from phaenomena and experiments to the causes thereof, and thence to the causes of those causes, and so on till we come to the first cause: the other that all actions of the first cause are miracles, and all the laws imprest on nature by the will of God, are miracles and occult qualities, and therefore not to be considered in philosophy. But must the constant and universal laws of nature, if derived from the power of God, or the action of a cause not yet known to us, be called miracles and occult qualities, that is to say, wonders and absurdities? Must all the arguments for a God taken from the phaenomena of nature be exploded by new hard names? And must experimental philosophy be exploded as miraculous and absurd, because it asserts nothing more than can be proved by experiments, and we cannot yet prove by experiments that all the phaenomena in nature can be solved by mere mechanical causes? Certainly these things deserve to be better considered. (Anon. 1809)

This magnificent defense of scientific method comes from the man who had written that Hercules, otherwise known as Sesostris, was an Egyptian king who set up pillars recording his conquests and "fought the *Africans* with clubs, and thence is painted with a club in his hand" (1728, p. 214); who had attempted the Work in Red and got as far as sophic mercury; who wrote, "In God's house (which is the universe), there are many mansions, and He governs them by agents which can pass through the heavens from one mansion to another" (Brewster 1855, vol. 2, p. 354); and who, in his last recorded scientific conversation, "seemed to doubt whether there were not intelligent beings superior to us, who superintended the revolutions of these heavenly bodies by the direction of the Supreme Being" (More 1934, p. 663).

In Massachusetts the learned Cotton Mather (later elected Fellow of the Royal Society) saw the dark side of that celestial world. In 1689, two years after the *Principia* and three years before the Salem witch trials, Mather wrote, "There is confined unto the *Atmosphere* of our *Air* a vast *Power*, or *Army* of *Evil Spirits*, under the government of a Prince who employs them in a continued Opposition to the designs of GOD" (Mather

---

[3] For more about it see Koyré (1965, ch. 3).

1691, p. 96). Mather is specific (p. 72) as to where these spirits are allowed to go: as high as the sphere of air, a few miles, and no higher. They did not fly around and produce their physical effects by means of the physical forces analyzed in Newton's theory. Occult influences were still in action everywhere.

John Maynard Keynes went to great trouble and expense to preserve some of Newton's papers when they appeared on the auction block. Fresh from the experience of reading them he wrote in his essay *Newton the Man*.

> Newton was not the first of the age of reason. He was the last of the magicians, the last of the Babylonians and Sumerians, the last great mind that looked out on the visible and intellectual world with the same eyes as those who began to build our intellectual inheritance rather less than 10,000 years ago. Isaac Newton, a posthumous child born with no father on Christmas Day, 1642, was the last wonder-child to whom the Magi could do sincere and appropriate homage. (Keynes 1972, vol. 10, p. 363)

Newton died in 1727 and lies in Westminster Abbey. The inscription on his ornate tomb can be translated "Let all mankind rejoice that an ornament to humanity of this kind, and so great an ornament, has existed." He was great for his mathematical inventions, for trying to develop the science of dynamics in many directions starting from a few simple principles, and for the wonderful fertility of the approximate methods with which he struggled forward when he could no longer proceed by rigorous deduction.[4] The educated public saw him differently. I have mentioned that according to Saint Gregory the realms of Earth, Water, and Air are the sensible world while that of Ether is the intelligible world. Newton not only showed how the intelligible world might be understood from first principles; he also erased Gregory's distinction, for he showed that the same principles govern the world below.

---

[4] For a critical review of what was and was not accomplished in the *Principia* see Truesdell (1968, chs. 2 and 3).

CHAPTER 12

Time, Space, and Form

And if the lights of heaven should cease and the potter's wheel
still turn, would there be no time?

—SAINT AUGUSTINE

GEOMETRY is the science that erects mathematical structures in mathematical space and studies their properties, but what has this to do with
physical space? Evidently geometry is in some sense the science of things
in physical space, since the whole subject originated in the solution of
problems in surveying, engineering, and navigation, and it actually
works. If we try to look more closely the exact connection is elusive.
Mathematical space represents physical space as a map represents a coun-
tryside. But what is this physical space that it represents? In the time of
Aristotle physical space was regarded as full and fullness was regarded as
an essential property of space; one did not think of space as an empty
container into which the five elements had been poured. A thrown ball
kept moving because air pushed it from behind, and the entire great apparatus of heaven and earth was controlled by influences transmitted in the
ether.

Think now about Kepler's ellipses. They are not defined or controlled
by any substances, either material or immaterial, the way orbits were in
Greek astronomy. They exist in a new kind of space, empty of all properties except mathematical ones, a background space. Through this space,
along those ellipses, planets move at varying speeds. Therefore the background space defines not only a body's position but also its state of
motion. Motion cannot be defined except by using the idea of time, so

that the background space not only serves to define distances and directions; it also contains a clock. To describe the real world as space that is purely geometrical would be almost useless. We need some kind of space-time. We know mathematical space because it is an intellectual construction. We know physical space because we see and feel the things it contains. But things are things. Is physical space also an intellectual construction?

I make these remarks and raise the final question to show that it is not obvious what to say about so apparently simple a subject as space, and that what we say cannot be disconnected from what we say about time. This chapter traces the emergence of the background space and of a vocabulary suitable for dealing with time. Chapter 14, on relativity, will discuss spacetime. This one ends with some considerations about the equations of dynamics and other physical laws and the sense in which holding these abstractions in our minds encourages us to say we understand the world we experience. The contrast serves to suggest some new aspects of old ideas of form.

The Mathematization of Space

Throughout the *Principia* Newton makes continual use of the theorems and constructions of Euclidean geometry; thus he is assuming that the space of physical phenomena is what we call Euclidean. There is no natural edge to this space: it is infinite. In it two parallel lines never meet, however far they may be extended; the sum of the three interior angles of a triangle is equal to two right angles; and so on. But there is more to it than this, for Newton's axioms assign other properties to space, connected with properties of time, when he says that a body with no force acting on it never changes its motion. To have understood the necessity of this connection and to have made it successfully was a great achievement, but before enlarging on it let us look at the historical background.

Democritus had an idea of space that seems to us natural and uncomplicated. Even if space has no physical existence, it is a logical entity we can reasonably discuss. We can assign it mathematical properties and talk about what it contains. Democritus seems to have taught that space is infinite in extent, populated with an infinity of atoms whose random groupings form an infinity of worlds. (Kirk et al. 1983. ch. 15). But this way of thinking went against the main currents of Greek thought. In Plato's *Timaeus* the forms of material objects, somehow generated from the disembodied Idea, are imposed on a medium called the receptacle (*Timaeus*, 49a), which has been variously explained and which, as I have said earlier, is perhaps space filled with matter awaiting its form. Evidently Plato, like Parmenides, had no use for the idea of emptiness.

Aristotle, of course, is more explicit, but this does not mean he is clear.[1] To avoid having to assign any properties to space he concentrates on the media that fill it. First he defines the place (*topos*) of a body: it is the shape of the boundary of the medium next to it (*Physics*, 211b). As the body moves the medium rearranges itself and the place changes. Occasionally Aristotle mentions space (*chora*); it is the totality of places that bodies can occupy. But the definition of place already contains a difficulty. Suppose a ship is anchored in a river (Duhem 1913, vol. 1, p. 199). The medium rearranges itself as it flows past the ship but the ship does not move. Is its place changing? Aristotle's answer is not quite clear, but it seems that since motion and rest are states absolutely defined, we must change the definition. "The place of a thing is the innermost *motionless* boundary of what contains it," i.e., in this case, the banks of the river (*Physics*, 212a). But now of course the boundary no longer fits the thing like a glove, and the definition no longer fits our notion of place.

As time went on writers pointed out more weaknesses in Aristotle's definition of place. What, for example, is the earth's place? It is at the center of a number of spheres that might be taken as boundaries except that none of them is at rest. And as John Philoponus asked, what does it then mean to say that the sphere of stars rotates if it is not contained by any boundary? Philoponus's resolution of his paradox presents us with a definition that will later be useful for scientific purposes:

> Space is not the limiting surface of the surrounding body . . . it is a certain interval, measurable in three dimensions, incorporeal in its very nature and different from the bodies it contains, for its nature is incorporeal. It is pure dimensionality, and indeed as far as matter is concerned, space and the void are the same. (Duhem 1913, vol. 1, p. 317)

I have already mentioned Philoponus's idea that a flying arrow is kept moving by the impetus from the bowstring and not by the action of the surrounding air, and I have shown that it contributed to the development of ideas which led in a continuous line to Newton's dynamics. His idea of space was just as essential for the same goal but the path was less direct, since as long as the universe was regarded as a vast mechanism of circles and spheres one tended to talk about them and there was little occasion to discuss cosmic space abstractly. A clockmaker has no need to think about the properties of the space in which he builds his clock, since it is the structure of the mechanism that determines the spatial relations and

[1] For help with Aristotle, see Duhem (1913, vol. 1, ch. 4), Jammer (1954, ch. 1), Bochner (1966, ch. 4).

the actions of its parts. The decisive moment came when Kepler threw away the machine and set the planets moving in ellipses specified not with respect to any structure of spheres but with respect to the background space exactly as Philoponus describes it. And the reason the planets follow these ellipses was not that they are parts of a contraption but because of forces originating in the sun and described by mathematical laws. Never mind that he got the details wrong. Kepler arrived at a point from which he had an entirely new view of the physical universe, of the role of mathematics in describing it, and of what constitutes an explanation of its phenomena. He did this entirely alone, and having done it remains almost alone at the summit of human genius.

In what arena do the performances of Euclidean geometry take place? During most of history the answer would have been, look around you. It is the space in which you live. During the nineteenth century a number of mathematicians, notably Karl Friedrich Gauss, began to wonder how one could know whether that answer is true or false. Euclidean geometry is generated out of points and lines. A point is an idealization of a familiar notion: this spot here. A line is a similar idealization: if I sight from point A to point B, the line of sight is the straight line between them. I assume that a beam of light defines a straight line. But suppose (as we now know to be the case) that the path of a beam of light is not straight in the Euclidean sense, i.e., that if one were to replace Euclid's mental triangles with large triangles defined by three beams of light, then the more accurately one measured, the more surely one would find that the various theorems on triangles do not hold exactly. This being true, what standard of straightness other than light should be used to relate Euclid's lines to real lines in space? We now know there is no way at all to do it; that is what it means to say that space is curved, or noneuclidean. More will be said about this in Chapter 14; it is mentioned here only to stress that the moment one tries to use the theorems of geometry to make statements about the world one has to introduce new hypotheses of a physical nature. They may very well be false, and some probably are. Einstein remarked that "as far as the propositions of mathematics refer to reality they are not certain, and as far as they are certain they do not refer to reality" (1954, p. 233).

These considerations about the relation of geometry to the world are of relatively recent date, but it should be obvious to anybody who thinks about it that, taken by themselves, geometrical theorems just sit there isolated from sensory experience, and that if they are to be used in physics and astronomy something must be assumed about what they have to do with the behavior of matter and with our sense impressions.

If we assume that a beam of light is straight, and for most purposes this assumption is good enough, we are assuming something about the way light travels through space, an assumption that therefore belongs to the science of physics. Kepler makes another one, that a body subject to a force F moves with a velocity that is proportional to F. To put it in the language that physicists use, Kepler assumes that the body knows when it is moving and when it is not, and that if the force is removed it knows it should stop. This is the same kind of knowledge that tells a beam of light how to travel in a straight line. The idea expressed in these metaphorical terms is of crucial importance for understanding physics. If concepts of space are to be used in a nontrivial manner the assumptions of geometry—Euclidean or not—do not fully characterize space; there must be further assumptions of a dynamical nature, involving time and perhaps involving forces and moving bodies, before we have said what role it will play in our thinking, or, more succinctly, what it is. No matter that Kepler's law of motion was wrong. When Newton proposed one that worked, the principle was the same: force does not produce motion; it changes motion, and a body has to know how it is moving with respect to space so that if forces are taken away it can continue to move in the same direction at the same speed. The *Principia* begins with definitions that tacitly assume a space loaded with dynamics, and in the Scholium that follows them Newton makes it explicit:

> Absolute space, in its own nature, without relation to anything external, remains always similar and immovable. Relative space is some movable dimension or measure of absolute spaces; which our senses determine by its position to bodies; and which is commonly taken for immovable space; such as the dimension of a subterraneous, an aerial, or celestial space, determined by its position in respect of the earth. . . . Place is a part of space which a body takes up, and is according to the space, either absolute or relative. . . . Absolute motion is the translation of a body from one absolute place to another. (1934, p. 6)

The discussion is very careful and continues for some pages, clarified by a description of an experiment that Newton says he has tried. Hang a bucket at the end of a long rope. Turn it around till the rope is strongly twisted, then fill it with water and start it turning so as to let the rope unwind. At first the bucket turns but not the water, and the surface of the water remains flat; but after a few moments the motion of the bucket communicates itself to the water and the water itself begins to turn. As it does so the water starts to recede from the center of the bucket and rise against the sides, so that the surface is no longer flat. This shows (to use physicists' language again) that the water in the bucket knows when it is rotating, and that it does not learn this from the bucket itself, for there is a moment at which the bucket turns but the water does not. It learns it because it is aware of its motion with respect to absolute space; there is,

Newton concludes, such a thing as absolute rotation that is in no way reducible to the spatial relation of the particles of water to nearby objects.

All this was enough to say about space, but Newton's religious enthusiasm carried him further. Biblical passages such as "in Him we live and move and have our being" led Newton into previously untrodden theological territory as he sought to establish the relation of absolute space to God. For example,

> does it not appear from Phaenomena that there is a Being incorporeal, living, intelligent, omnipresent, who in infinite Space, as it were in his Sensory, sees the things themselves intimately, and thoroughly perceives them, and comprehends them wholly by their immediate presence to himself? (*Opticks*, Query 28)

Galileo has already mentioned the *sensorium*. Since it was imagined as the locus of sensations in the brain, Newton is saying that it is *as if* the entire world were inside the brain of God. Leibniz did not let this pass. In the first letter of what became the Leibniz–Clarke correspondence he writes, "Sir Isaac Newton says, that space is an organ, which God makes use of to perceive things by. But if God stands in need of any organ to perceive things by, it will follow, that they do not depend altogether upon him, nor were produced by him" (Alexander 1956, p. 11).

It is useless to follow the further course of this argument through the correspondence, since it is used only to score debaters' points, but real science is involved in Leibniz's attack on Newton's absolute space. First, he objects to making space a substance (in the Aristotelian sense), as Newton does by endowing it with properties beyond those required for geometry. Leibniz prefers the Aristotelian definition as the totality of the places that solid bodies can occupy. The only spatial relations about which meaningful statements can be made are those between solid bodies, and geometry is the totality of such statements. If one introduces absolute space, questions can be asked such as, "Why did God not put the material universe one foot to the right of where he did put it?" (this opens an argument on the principle of sufficient reason, which we need not follow). But if space is what Leibniz says it is, the question is without meaning and cannot be asked. In his fifth letter he makes his point by an analogy: geometry is a mental ordering of spatial relations analogous to a genealogical tree that represents family relationships by a spatial ordering, but nobody would say that it is more than a convenient representation.

Leibniz's argument is attractive, for it is economical of hypotheses and it suffices for the purposes of geometry, and yet it is wrong, for the passage in the *Opticks* he is attacking concerns dynamics, and Leibniz's space is not a possible home for dynamics. Newton could quantitatively explain an immense range of dynamical processes while Leibniz could not explain anything at all and could only object. If we overlook his theological speculations Newton stood on solid ground. His results followed from simple and unambiguous hypotheses suggested by experiment, and yet there is still something unexplained. Suppose I sit on a carrousel outdoors at night, making some observation with a pendulum that enables me to know whether the carrousel is rotating or not. Whenever the pendulum indicates rotation, if I look up I will see the stars appear to turn overhead, but when it does not, they do not turn. There are two unrelated ways to detect rotation: I can look either at the pendulum or at the stars. They agree exactly and yet according to Newtonian definitions they have nothing to do with each other. Newton's absolute space appears to be anchored to the stars, or else the stars to it, but there is nothing in the hypotheses to suggest why either should be so. In Chapter 14 we shall see how Einstein tried to close the gap.

Time and the Perception of Time

It is customary to begin any discussion of this topic with Saint Augustine's question, "What is time? If nobody asks me I know, but if someone asks me, clearly I don't know" (*Confessions*, XI.14). Well, the same could be said of the word "green," and so the situation is not extraordinary. The difficulty arises from trying to do what should seldom be done: select a word and then find *the* definition of it. Language works the other way. As Karl Popper says, definitions should be read from right to left. Identify a concept or cluster of apparently related concepts that seems to have a kind of unity and connection—it seems to function similarly in your thought each time it is encountered—and then find a word for it. So for one person time is the number read from a clock, while for Thoreau it is the stream he goes a-fishing in. Father Time with his hourglass and sickle stands outside the door to remind us that each life must end. For Aristotle, as I have said, time is the number of movement with respect to before and after. None of these identifications is either right or wrong, but they can't all be used at once.

Take for example Aristotle's careful words about time quoted at the beginning of the last chapter. Things move, i.e., they change. If we want to say that a change occurs quickly or slowly we must have some quantitative notion in mind, and for this we need a "number of movement." This is the kind of thing a clock shows and in our science, by common consent, it is *the* thing a clock shows. We need this number for the scientific study of nature as well as for the functioning of complex societies, and by common consent we call it time. On the other hand, as everybody knows, time passes. How quickly does it pass? Evidently, with each second that passes time gets one second later. Therefore it passes at a rate of one second per second. Having said this we have said nothing. How much is a dollar worth? Exactly one dollar. To answer the question properly we must recognize that questions of value cannot be answered in monetary terms. We have to talk about what a dollar will enable us to do. To define a second, we must talk about what happens in a second. *If* time is a number then it doesn't move; as Aristotle remarked it is not described as fast or slow, and if it moves, or passes, it is not a number with respect to before and after. For the time that moves we ought to use a special word like Bergson's *durée*; we don't and confusion results, but it is a creative confusion for it forces us to examine our ways of thinking. Let us try to develop an idea of time that is adequate for scientific purposes and then see whether it is adequate to describe our experience of life.

Scientific time is a number on a scale, and just as the meaning of numbers on the scale of distance is (before we become too sophisticated) defined by the properties of a tape or a yardstick, so the scale of time is defined by a clock. Here we take advantage of a remarkable fact that perhaps no one would have anticipated: one kind of clock is enough. Suppose that as is customary today we define our time standard by oscillations of a certain kind of caesium atom. Now using this clock let us time some other periodic motions: the swing of a pendulum, the earth's rotation, the movements of the planets. Though we had no a priori reason to expect it, there is agreement between these various markers of time, exact except for tiny discrepancies of which most can be understood as the result of changes in the environment, and none of which shakes our conclusion that, to an accuracy that is very great and perhaps perfect, a single time scale suffices for all of nature. We may say then, if we like, that time is what clocks measure.

The logic of time is very simple although a little subtle, based as it is on a combination of subjective experience and laboratory measurement. It is an example of what I call universe-assisted logic: although there is much about the universe that we do not know, the universe is such that it allows us to make certain useful definitions and draw certain conclusions.

Isaac Newton was the first to understand the universality of time. Just as with space, he contrasts two kinds of time:

> Absolute, true, and mathematical time, of itself, and from its own nature, flows equably without relation to anything external, and by another name is called duration: relative, apparent, and common time, is some sensible and external (whether accurate or unequable) measure of duration by the means of motion, which is commonly used instead of true time; such as an hour, a day, a month, a year. (1934, p. 6)

This definition of flowing time would be a strange beginning for a mathematical treatise that was about to introduce dynamical laws expressed as equations involving a letter t denoting Aristotle's number, which does not flow, but the *Principia* is not such a treatise, and no time variable or letter t ever appears. The only numbers in the *Principia* relate to distances; everything about time is implicit. I think Newton is saying that we are conscious of the succession of our own sensations and thoughts, and also of motions and events in the world around us. Some of the motions and events are analyzed by the science of the *Principia*. When this is done it appears that the time they measure is the same as that which we experience in our inner lives: we perceive two successive hours by the clock as being of roughly equal length. Thus the clock moves "equably," and so therefore does the stream of time that in Newton's metaphor carries it along.

Newton's physics assumes that there is a single absolute time that our various ways of perceiving and measuring reveal as common time. Today we regard Newton's physics as at best an approximation to the truth. The modern dynamical theory of elementary particles and fields follows very different principles, but Newton's concept of a single true, mathematical time still works as well as it ever did—better, in fact, since the measurements confirming it are so much more accurate. There remains a nagging doubt: is the scale really the same at huge astronomical distances as it is in the solar system, and can one, without stumbling into logical or scientific difficulties, say it is the same now as it was in the remote past? These questions are important to cosmologists and we shall have to face them in Chapter 18. For the moment, suffice it to say that the situation appears maximally simple: everywhere, always, we can use one scale of time and for that matter, one scale of distance as well. Owing to certain relativistic effects we have to be a little careful how we use them, but this does not alter the basic facts.

So what has modern physical science added to Aristotle's definition of time? Essentially nothing. Aristotle defined it as a number that enters into the description of motion. Newton supplanted Aristotle's fallacious dynamical theory with one that is for most purposes very nearly correct. He cleared up the definition a little without changing its sense. Modern physics revises the equations of motion and finds in them the most useful definition of time. As an authoritative text on relativity theory remarks, "Time is defined so that motion looks simple!" (Misner et al. 1973, p. 23). That is pure Aristotle. The special theory of relativity (see Chapter 14) brings a surprise: the time intervals thus defined are observer-dependent. One can no longer say meaningfully that it is the same time everywhere. But what relativity takes with one hand it gives back with the other, for in the general theory our expanding universe, considered as a vast machine, does indeed define a time variable that is the same everywhere—only this time is not the time measured by every observer. We know the universe only sketchily and measure it roughly and this conclusion may not be exact. Universal cosmic time is something that must emerge from our knowledge as a result and not enter it via a definition.

If time is what a clock measures, what is a clock? It is a device whose law of motion is known. Since laws of motion involve time it would appear that an elementary error of reasoning has been made, and this would be true if the logic of the definition were not aided by the universe. Both definitions contain and assume the truth of information about the world. It doesn't make sense to say that new knowledge can prove a definition to be false, only that it can make it useless.

How useful is it to say that time is what a clock measures? No matter what we happen to think time is, most of us use clocks. For purposes of physics the definition is adequate. Chemists believe that their science rests on physical principles, and so physicists' time suffices for them. Biologists believe that their science is a science of matter, but they believe this with various nuances and caveats, and for them the situation is not quite so simple. Psychologists, novelists, historians, and economists are aware that they deal ultimately with the behavior of living tissue and inanimate matter, but they are not required to follow the physicists' definition. Of course real human behavior is regulated by the alternation of day and night, of summer and winter, but music and poetry and fiction are not so constrained. And for one area of experience, whether scientifically studied or merely lived, the physicists' definition is clearly not adequate. It is not appropriate to any discussion of human consciousness, since it does not contain the idea of now.

The simplest examples show the situation most clearly. Consider the mathematical description of an object that moves at constant velocity v during a time t. The distance it has traveled is given by

$$d = vt.$$

When we interpret this formula we may think of time as continually advancing and distance as continually increasing, but the formula says no such thing, for t is any value of time and d is the corresponding distance. We might just as well consider t as continually decreasing, or assign values to it at random. There is nothing in this or any other formula of physics that expresses our sense of orderly progression. According to Cartesian understanding our body is a machine; our eyes are machines; they watch an object as it moves. The various machines function according to "true, mathematical time" and keep step with each other. But consciousness, it appears, cannot be modeled with a machine, for it picks out one moment, calls it the present, and attends only to it. One can say easily enough that any particular value of *t* can be taken as now and that would not be wrong, but it does not correspond to experience. If we attend only to what is happening around us and let ourselves live, our attention concentrates itself on one moment of time. Now is when we think what we think and do what we do. Now seems by every test to get later and later, even though, as we have seen, it is impossible to say clearly what we mean by that, measuring time in terms of time. Augustine, taking the present as real, persuades himself in a brilliant argument that past and future are mental constructions and that therefore intervals of time are measured only in the mind (*Confessions*, XI). But if we turn our attention away from present sensations and think about what we have been doing or intend to do, we easily trace events from future toward past, or consider them as moments in no particular sequence, and they seem quite real as we think about them. When we do that we have constructed physicists' time in our minds.

Rudolf Carnap recalls a conversation on this subject with Einstein:

> Once Einstein said that the problem of the Now worried him seriously. He explained that the experience of the Now means something for man, something essentially different from the past and the future, but that this important difference does not and cannot occur within physics. That this experience cannot be grasped within physics seemed to him a matter of painful but inevitable resignation. I remarked that all that occurs objectively can be described in science. . . . But Einstein thought that these scientific descriptions cannot possibly satisfy our human needs; that there is something essential about the Now which is just outside the realm of science. (Schilpp 1963, p. 37)

One of the largest problems in science is to say what the brain is doing when it makes thoughts. What I have said about the limitation of the physical point of view suggests that it is impossible to explain the operation of consciousness by any arguments based on the material structure of the brain without introducing at least one concept, now, and possibly others, that perhaps are not in disagreement with physics but merely foreign to it.

The development of Newtonian physics marked an immense advance in thinking about the natural world, but this was achieved largely by letting go of some of the old concepts that had long filled men's minds and concentrating on new questions it was possible to answer. The old concepts remained. Form was one of the most important but it has not been mentioned for some time. Though it can only sometimes be captured in the net of mathematics, it persists as an important element of scientific thought. We had better see what happened to it.

Form in the Natural World

The theory that there are forms or ideas that have a kind of existence independent of any particular mind, or even any mind at all, has (for some) a certain attractiveness, but it is very hard to incorporate it into scientific thinking. As we have seen, Plato was obliged to slide over the question of how Ideas come to be imprinted on the material world, and while Aristotle's formal cause is easy enough to explain in illustrative situations like the making of a bowl, it turns out to be very hard to show specific examples of this kind of cause operating in nature (though remember Gilbert's formal explanation of magnetism). The Neoplatonists, and finally Augustine, transformed the Platonic doctrine by situating the forms within the divine intelligence so that it was no longer necessary to ask how they operate in the world, though this simple and natural Christian interpretation weakened the hope that mankind also can possess these same ideas as a direct source of inspiration in truth, beauty, and goodness.

In the Apocryphal book of Wisdom, Solomon says to God, "Thou hast ordered all things in measure, and number, and weight" (11:20) and Augustine echoes: "Recognize God at once as the author of everything in which you see measure, number, and order. If you take these entirely away, nothing whatever will be left" (*On Free Will*, II.20). This amounts to a command: look at the world so as to see measure, number, order, and, he adds later, good in everything. There are beauties and symmetries in the body of an animal, in the form of a flower or a seashell, in the colors of a butterfly's wing that are plain enough to see, but as one looks out on the chaotic landscape of inanimate nature it is sometimes not obvious where the order lies. There is one small exception: the natural crystals of certain minerals. Neither Plato nor Aristotle mentions them at all, but as time went on they assumed importance as indicating formal principles at work during the Creation and, if one believed that minerals are always being generated in the earth's womb, ever since.[2]

In 1611 Kepler wrote a little book translated as *The Six-Cornered*

---

[2] There is an excellent book (Emerton 1984), which shows how the study of crystals helped scientists to reinterpret earlier philosophical concepts of form into modes of thought that have run through science since the late Middle Ages.

Snowflake (1937, vol. 4)[3] as a New Year's present for a patron, in which he lets his mind race over some notions of physical form inspired by the way snowflakes always crystallize in six-cornered shapes even though no two are alike.

From the beginning, Kepler hopes and expects to explain the shape of a snowflake as a manifestation of the same divinely ordered numerical principles that he later thought he had discovered in planetary motions. He reasons that the cause of the shape cannot be material, since the vapor out of which snowflakes condense has no shape. Is it then efficient, implying action from outside? Or is it formal, implying a principle that works within? Searching for analogies elsewhere in nature, Kepler mentions the hexagonal structure of honeycombs and the rhomboidal packing of seeds in a pomegranate, both of which show a tendency toward geometrical form and in neither of which is there any evidence of action from outside. He is left with the formal cause, a hypothesis reinforced when he looks at the forms of fruits and flowers and finds along with the numbers 4 and 6 the number 5, again and again—the numbers 4 and 6 can be explained in terms of space-saving efficiency, but 5 cannot. After a detailed analysis Kepler decides that it is the same with the six-pointed snowflake.

> So having examined every notion that occurred to me I conclude that the cause of the snowflake's six-cornered shape is the same as the cause of the ordered shapes and fixed numbers we find in plants. And since in these things nothing occurs except with the highest reason—which is not, of course, the reason that we reach with our discourse but rather that which has existed from the beginning in the design of the Creator, . . . I do not think that even in snowflakes these ordered figures occur by chance. There is a formative faculty that resides in the body of the earth and is conveyed by vapor just as spirit conveys the human soul. (Kepler 1966, p. 32)

Water vapor emerges from the earth carrying the formal principle "in the innermost recess of its being," and where other forms appear in plants and rock crystals they must similarly be transmitted by the materials of which they are made. Had Kepler believed in atoms, would he have guessed that in some subtle way at least a few of these forms are explained in terms of geometrical properties at the atomic level? I think he would not have been satisfied with such an explanation until he realized that atomic theory has some right to be called a form.

Only the eye of faith could see form in the tangled disorder of the Ptolemaic apparatus, but as it was revised by Copernicus and finally replaced

[3] There is a bilingual edition (1966), but I must regretfully warn readers against the translation.

by Kepler's ellipses traced in free space, new aspects of formal symmetry and simplicity became visible in nature and new instances in which measure, number, and order could be perceived by anyone who took the trouble to look.

I am using the word "form" to denote the kind of pattern that educated senses perceive in nature or elicit from it, something complete in itself and if not simple, at least orderly. It may be a pleasing pattern, like Kepler's ellipse, or the stacking of an unimaginable number of atoms, in an arrangement as regular as that of bricks in a wall, to make a crystal of quartz. And just as crystals and planetary ellipses manifest geometrical forms, so the harmonies of the diatonic scale, based on Pythagorean fractions, manifest auditory forms.

Perhaps forms are the units of understanding. We easily grasp the form of Kepler's ellipse (though to do so we must take on faith the result of years of patient observing and computation), but when we have done so we have barely begun to understand planetary phenomena, just as when we have identified the harmonic structures in a Mozart sonata our understanding of the sonata itself has barely begun. We still need to know why the planetary orbit is an ellipse. When we learn about the ellipses we understand that Kepler did more than find a more accurate way to predict planetary omens. He penetrated further than others into the mathematical nature of things and arrived where he could begin to see planetary motions in their true simplicity. Ellipses can easily be drawn with a pencil and two pins and a piece of string, but the planet does not construct its ellipse that way: there must be another way, using the sun. And of course there is, as we have seen. The sun is one pin, the planet is the pencil, and the force of gravity is the string. Aristotle asked how, if Plato's forms were considered to have some kind of super-reality on an immaterial plane, they could cause anything to happen in the world of sense. If it is put that way the question has no answer, but I am speaking of patterns we perceive in nature, and nature does not stay still. We see patterns of repeated events and speak of cause and effect. There are patterns in time as well as in space, and they account for happenings even if they cannot be said to cause them. They are forms of causation, and they are the laws of nature, mathematically expressed.

For Aristotle, as we have seen, nature is understood directly as its forms enter the mind. For us, nature is intelligible and form exists, but selecting, defining, and perhaps creating forms is a creative act. What kind of reality do they have? Are they out there, in nature, whatever that may be? Are they in our minds? That was what Kepler believed. In his last book, the *Harmonices mundi*, he writes that,

As the perceptible things which appear in the outside world make us remember what we knew before, so do sensory experiences, when consciously realized, call forth intellectual notions that were present inwardly; so that that which was formerly hidden in the soul, as under the veil of potentiality, now shines therein in actuality (1937, vol. 6, p. 226, trans. from Jung and Pauli 1955, p. 163).

Later, reflecting no doubt on his own inconceivable struggles to understand planetary orbits in an entirely new way, he argues more specifically that if there were not a preestablished harmony between archetypal ideas (his term) present potentially in the mind and the actual forms and harmonies of the planetary system it would have been impossible, from the available data, to arrive where he did. He quotes his favorite ancient author, Proclus, who had expressed the same idea, "The soul . . . was never a writing-tablet bare of inscriptions; it is a tablet that has always been inscribed and is always writing itself and being written on by Nous."[4] Then Proclus explains why this is so. In the *Timaeus* (34ff) Plato has already told us that the world soul was planned and constructed according to mathematical principles, and so these are already fixed in the individual soul that partakes of it.

In our own time Wolfgang Pauli (1900–58), one of the most brilliant creators of twentieth-century physics, has written a modern interpretation of the experience we have when we succeed in fitting idea to experience:

I would like, leaning on Plato's philosophy, to propose that we interpret the process of understanding nature as well as the happiness one experiences in understanding, that is, of becoming conscious of new knowledge, as a coming into coincidence of preexisting inner images in the human psyche with external objects and their relationships. (1961, p. 91)

Archetypes again. They seem to be the patterns in which we think. How do they enter our minds? It seems unlikely that we are born with them, but there are plenty of opportunities for us to form them, or absorb them from our culture, as we grow. I have already mentioned that Einstein also, examining his own thinking, concludes that form comes before concept: "It seems that the human mind has first to construct forms independently before we can find them in things" (1954, p. 266).

The foregoing remarks are murky and conjectural but there is a brighter side to them, for the formation of the conceptual structures (whatever they may be) that are used when we think is a subject suitable for experimental investigation. I have not been able to discover that much

---

[4] Kepler (1937, vol. 6, p. 221), Proclus (1970, p. 14). *Nous* here stands for the rational principle that pervades the universe.

is yet known about it, but I think it is clear that one should go to the laboratory, and not to the study, to decide what the word "idea" ought best to mean.

Second Pastoral

Same hill, same tree. Phyllis in the shade, reading. Corydon walks slowly up and joins her.

COR. It's disappointing, isn't it?

PHYL. Why? I thought things were starting to go well. Newton and all—

COR. Yes, but what do you learn from it? Those people with their marble faces and marble minds carving formulas out of marble that tell me more about their own egos than about what the world is. I know the formulas but I still don't understand anything. What were they trying to do?

PHYL. Draw diagrams.

COR. Plato?

PHYL. Especially Plato, but all of them.

COR. Where did they draw them?

PHYL. In the sky mostly; sometimes on paper.

COR. What's Plato's diagram?

PHYL. There is a bed up in the sky. Of course it's only the idea of a bed but it's all right to draw it because it's just a diagram; then there are ordinary beds, all different kinds, on the ground and you draw lines from the bed in the sky to the ones on the ground.

COR. How is the sky bed different from the ground ones?

PHYL. You'd have to make it much larger or much smaller.

COR. Smaller.

PHYL. All right.

COR. How about Galileo?

PHYL. He has mathematical diagrams. The language of the book of nature is written in squares and triangles and things. No numbers, apparently.

COR. Yes there are numbers.

PHYL. How do you mean?

COR. Well, look at Galileo's relation that tells how something falls, distance proportional to t^2. That's numbers; he even says so. Descartes shows that if you graph it it's a parabola. (*As he speaks he draws a parabola in the air.*) There's Galileo's diagram. The formula is the same

as a diagram. There's a geometry of falling apples. You have to introduce time into geometry.

PHYL.   Geometry has nothing to do with time.

COR.   This parabola has nothing to do with time.

PHYL.   You just said it is about distance and time. You said t^2.

COR.   But the relation it shows has nothing to do with time. Things fall
just the same now as they did when Thales fell into the well.

PHYL.   You're talking about forms. Are your forms more like a Chinese
vase or a juggling act? Please select one.

COR.   Ninety percent of the forms in nature look like juggling acts, but
the reason you say "form" is that you think you know the principle,
and the principle, after you have scraped off everything that doesn't
belong to it, looks like a vase.

PHYL.   Beauty and all? (*He doesn't answer.*) So there isn't any difference
between a formula and a form?

COR.   How about listening to the word. Formula equals little form.

PHYL.   I need to know the difference between a formula and a law.

COR.   Galileo found that if air resistance isn't important everything falls
with the same constant acceleration as everything else. If g is the acceleration we write $d = \frac{1}{2} gt^2$, though Galileo didn't.

PHYL.   Why not?

COR.   You could talk about a number but when you wrote it down it
was a length.

PHYL.   Why couldn't they put it on a graph then? Then it would have
been a length.

COR.   I don't know. There are some medieval graphs that do that. I don't
know why Galileo didn't. (*There is a pause*)

PHYL.   I asked you what a law is. Is what Galileo said a law?

COR.   He says "everything."

PHYL.   "Everything" makes it a law?

COR.   It means it wants to be a law. I think a law in science is like a
classic in books. A book isn't a classic by itself. It isn't a classic at all
until one day some prominent people decide it's a classic. It depends on
historical context.

PHYL.   You couldn't be more exact?

COR.   No. Galileo's law is just a special case of Newton's law. You write
the acceleration as d^2x/dt^2——

PHYL.   And a thing's weight is proportional to its mass so you can write
it as mg where g is some number, and then the mass times the acceleration is mg and m cancels so that everything accelerates with g and you
integrate the equation and get t^2. I just saved myself half an hour of
boring explanations.

COR.   Feel free to be as insulting as you like. You'll be more comfortable
that way and it makes you stick out from the crowd.

PHYL. I know. Everybody says so. So $d^2x/dt^2 = g$ is a form? It lives in the heaven above heaven? It doesn't seem to have much of the Beautiful or the Good about it.

COR. I don't know about the Good. That seems to have got lost somewhere, but it may have a little beauty. It's shaped like a parabola.

PHYL. If you throw a ball it moves in a parabola.

COR. Gravity's rainbow.

PHYL. And Newton's equations for planetary motion in that appendix are shaped like an ellipse. You know, I don't agree with that book. It says that forms exist in the mind and psychologists are going to tell us how. Parabolas don't exist in the mind, they just exist.

COR. Where?

PHYL. Everywhere in the universe. Ask any of the little green men out there about physics or geometry and pretty soon they will draw you a nice curve. Of course the one they draw won't be very beautiful, but everybody will understand that it represents some perfect shape.

COR. You have to have a little green man before you have a nice curve?

PHYL. You have to have them before there is a drawing of a nice curve. But the curve is there anyhow. Otherwise they wouldn't all try to draw the same thing. The curve is a form—I see now. The book asks Einstein where he gets the forms behind the forms he makes, and he says he makes them.

COR. That doesn't say anything. What does he make them out of?

PHYL. Things he sees. Like crystals and insects and mathematics. You can't do mathematics without finding circles and parabolas. I don't know how little green men draw them or if they draw them, but they're waiting to be found. Then when you find these forms again, in some new situation, you feel happy.

COR. What's a theory?

PHYL. It's a law along with all the explanation that tells what it means and why it's true. Maybe more than one law.

COR. It's not quite like that. A theory has a shape.

PHYL. Beginning, middle, and end?

COR. Assumptions, arguments, laws.

PHYL. If it has a shape, is it a form?

COR. Yes, but some other kind. It doesn't say in the book.

PHYL. I think he's got the two kinds mixed up.

COR. Be good enough to define form.

PHYL. Maybe we can do that later. Definitions should be read from right to left . . .

COR. I can see what all this has to do with explanation. I keep writing formulas and drawing diagrams until the person has had enough and gives up and says "Yes, I understand"; then I've won and we stop. But what has it got to do with understanding?

PHYL. I thought about that yesterday morning while I was waiting for a lamb to be born. I think that seeing a form *is* understanding. That's what Plato and Kepler and all those people said.

COR. Come on, you don't take Plato and Kepler seriously?

PHYL. I take them very seriously. I just don't take them literally.

COR. You can't mean that as soon as you see a form you think you understand. Suppose someone hands you a piece of quartz crystal and you see the form.

PHYL. Maybe I don't understand very much until they show me the theory, but I understand more than if I had never seen the crystal at all. I understand there is something to understand.

COR. And if Galileo shows you that distance is proportional to time squared?

PHYL. It's the same thing.

COR. You make it sound much simpler than it is. What do you do when you understand something?

PHYL. Me?

COR. Well then scientists, what do they do? Contemplate the beauty of it all? The book doesn't say.

PHYL. I suppose they go on being scientists and think up some more things to do. That reminds me, I cut out something for you the other day. (*She finds it.*) He says, "But they understood the cardinal principle of all science—that the profession, as an art, dedicates itself above all to fruitful doing, not clever thinking: to claims that can be tested by actual research, not to exciting thoughts that inspire no activity."[5] I suppose theories generate experiments more often than experiments generate theories.

COR. The book doesn't talk about that.

PHYL. It's not what it's about. (*There is a pause.*)

COR. I've been thinking about those diagrams. There's more to them than their shapes. What color are they?

PHYL. You mean what Pauli said about happiness.

COR. Also Saint Thomas. That's part of it.

PHYL. It's funny that none of the others mentioned happiness.

COR. Does form include color?

PHYL. Let's go swimming.

COR. Yes, but it's important.

PHYL. I know it's important but I don't know what to say. (*They go.*)

[5] Stephen Jay Gould, *New York Review*, February 27, 1986.

CHAPTER 13

A World of Bronze and Marble

> An immense impulse was now given to science, and it seemed
> as if the genius of mankind, long pent up, had at length rushed
> eagerly upon Nature, and commenced, with one accord, the
> great work of turning up her hitherto unbroken soil, and
> exposing the treasures so long concealed.
> —J.F.W. HERSCHEL (1840)

THIS chapter covers the period of almost two centuries after Newton's death, roughly 1700–1900, when there occurred a confluence of two streams of intellectual energy that had always flowed in separate channels: that of philosophy and that of practical invention. During the Middle Ages the practical men had never been much influenced by what intellectuals thought and wrote. They belonged to a different social class from the intellectuals, and though some were literate in their own languages they usually did not read Latin. When scientists like Stevin, Galileo, Descartes, Newton, and Boyle started to publish books in the popular languages before they appeared in Latin it showed that science was becoming more open to public understanding and less attached to scholarly tradition. We have come to the time when science began to be considered not as philosophy but as its offshoot, and was cultivated outside university walls.[1] When the Royal Society of London was founded under

---

[1] I will use "science" as we do today, but it was only at the end of the nineteenth century

243

Charles II in 1660, it consisted of nonacademic gentlemen and emphasized its distance from scholarly tradition by adopting as its motto *Nullius in verba*, loosely translated as "Don't take anybody's word for anything." It was an age of investigation that called itself an age of reason, and of course it was, but in an entirely different way from the earlier age of reason that was the Middle Ages. All reasoning is based on authority: that of the Middle Ages on God's authority, that of the age of Enlightenment on nature's.

During these two centuries the scientific principles that had slowly matured over two millennia steadily produced results. There were so many discoveries of fundamental importance that it is impossible to mention them all. We will see, though, that the vast increase in the scope of physical theory was not matched by comparable changes in modes of explanation. At the end of the period, as at the beginning, the aim was to explain everything in terms of Newton's laws. It will turn out that this worked very well for selected problems at the atomic level, thus greatly extending the scope of Newton's theory, but badly for phenomena involving electric and magnetic fields. In the following chapters we shall see that there is more to fundamental theory than Newton's laws.

In 1692 when Richard Bentley preached his sermon "A confutation of Atheism from the origin and frame of the world," in the preparation of which, as we have seen, Newton helped, I doubt that atheism posed a very serious threat to the established order. What Bentley could not very well say was that the old reasons for believing in God were beginning to give way. New ones must be found and they must be found in nature, for nature, interpreted in the light of common sense, spoke the simple truth. That there is a God few people doubted, but what he is and who he is and what are his true and essential attributes had become such complicated and divisive questions during the long centuries of Christendom that it was now necessary to refer to something called Nature, usually personified as a woman, representing a source of truth and rightness acting according to unspoken but rational principles, subordinate to her Creator but accessible to human understanding, like Wisdom in Solomon's book. It was not just for euphony that Jefferson, writing the Declaration, referred to "the laws of Nature and of Nature's God"; at least at that moment, for his purpose, he placed first the most certain and immediate source of political principles.

---

that this usage became general. Earlier the word lingered on from medieval philosophy to denote truth derived from first principles: one spoke of moral science as opposed to natural philosophy, and John Locke in his *Essay on Human Understanding* doubts that natural philosophy can ever become a science. When natural philosophy turned into natural science, moral science abandoned the field and became moral philosophy.

Newton Continued

On picking up the *Principia* we are struck by the number of figures it contains. They are not there just to make reading easier; the book's reasoning is geometrical and the figures usually show the essentials of a proof. But a figure that is useful for one geometrical proof is not very useful for another; every theorem demands a fresh figure, a fresh exercise of ingenuity. I have suggested that for Newton, as for the ancients, the idea of a symbol representing pure numerical quantity did not exist. It is hard to find evidence to the contrary. In the *Principia* the lines on a diagram represent only what they actually show, distances and directions, and nothing else. Even the principle of the parallelogram of forces, illustrated by a neat parallelogram, is stated: "A body, acted upon by two forces simultaneously, will describe the diagonal of a parallelogram in the same time as it would describe the sides by those forces separately." The lines do not represent forces at all; they show where the body would go in a given time. Newton's proof has nothing at all to say about the addition of static forces. In addition to the principles of natural philosophy that he explained, there were many more that he did not. The work must be carried forward. One reason why reading the *Principia* is a daunting experience is that we ask ourselves at every point, "Would I be able to solve for myself even the simplest problem that Newton has not solved here?" Newton translated the book of nature into a language a few could read but only he could speak. It was not obvious how, once Newton disappeared, the scientists remaining behind him would be able to do more than tidy up the details.

At the same time as this great work was slowly convincing its opponents and becoming lodged in the world's consciousness, mathematicians in England and Europe were developing the techniques of calculus that Newton and Leibniz had independently invented. The last edition of the *Principia* was published in 1726, the year before Newton died. Ten years later the Swiss Leonhard Euler (1707–83), perhaps the most talented and versatile mathematician who ever lived, the Mozart of mathematics, published his *Mechanica* (Euler 1911, Ser. ii, vols. 1 & 2). The Empress Catherine I of Russia, wife of Peter the Great, had established an Academy of Sciences at St. Petersburg as one of her first acts and had brought there a group of talented men from Western Europe to add intellectual light to her empire. The day Euler, aged twenty, arrived on Russian soil in response to her invitation was the day Catherine died, but the Academy survived stormy transitions. Except for a long sojourn in the court of

Frederick the Great at Potsdam, Euler spent the rest of his life in St. Petersburg and wrote most of his eight hundred publications there, half of them after he had become blind. Unlike Newton, Euler freely used mathematical symbols to denote any physical quantities he chose, and he expressed mathematical relations much as one would today. Instead of Newton's verbally expressed law of motion he writes simply $dp = fdt$ (dp is an increment in momentum, f is the applied force, dt is a short time interval [*tempusculum*] during which the force is applied; Euler uses other letters).

Following some of the arguments in the Mechanica (such as the proof that a planet attracted toward the sun by Newtonian gravity follows an elliptical path), one finds nothing that looks familiar, and the strangeness of the mathematical form arises from the unfamiliar things that mathematics is being asked to do. The momentum p is a property that inheres in the moving planet, directed along the tangent line at any point of the orbit. The variables are the lengths of various lines drawn on the diagram, changing as the planet moves—as if Disney had animated Euclid.

When Euler needs a continuously changing angle he represents it in terms of the changing lengths of lines. Even the magnitude of the gravitational force is represented by the length of a line. Two things are missing. The first is the familiar convention that represents the planet's position by its coordinates with respect to a set of axes fixed in space. John Bernoulli had used them to solve the same problem twenty-six years earlier (see Note F), but Euler, perhaps hoping to avoid Bernoulli's analytical acrobatics, introduces geometrical acrobatics of his own. The second missing element is any mention, verbal or mathematical, of time, and this is odd when one remembers that he had used it to express the law of motion and that the goal of the calculation is to find how the planet's motion proceeds as time goes on.

The idea of representing time explicitly in an actual calculation was so radical that after the *Mechanica* Euler needed another fourteen years to arrive at it. The paper he finally wrote is called "Discovery of a new principle of mechanics." It is mainly devoted to the very difficult question of the rotations of a rigid body, but he works out the theory by representing the body as an assemblage of point masses linked together, each one subject to Newton's law of motion. This would have been impossibly difficult with the old geometry, which had trouble enough following a single point, and so Euler started again. In a section titled "Explanation of the general and fundamental principle of all mechanics" he very carefully explains how to construct a rectangular coordinate system in space and shows how to use it. Although more technical than most of the passages quoted here, this one is of such importance for the history of scientific thought that I reproduce the first paragraph.

Assume an object that is infinitely small, the entire mass being concentrated at a single point and equal to M; assume also that this object has received an arbitrary motion and that it is being acted upon by arbitrary forces. To determine the motion of this object one has only to consider its distance from a motionless plane; let this distance at the present moment be x. Now decompose all the forces acting on the body according as their directions are parallel to the plane or perpendicular to it, and let P be the force resulting from this decomposition that is perpendicular to the plane and will therefore tend to carry the body further from the plane or closer to it. After an elementary interval of time dt, let $x + dx$ be its distance from the plane, and taking the element dt as fixed, let

$$2 M \, ddx = \pm P \, dt^2,$$

according as the force tends to bring the body further or closer to the plane.[2] And this single principle contains all the principles of mechanics. . . . Since this formula determines the body's motion towards or away from an arbitrary fixed plane, to find the true position of the body at any instant one needs only to locate it with respect to three planes at the same time, fixed and mutually perpendicular. (1911, Ser. II, vol. 5, p. 89)

Euler has found out how to represent directed quantities in terms of three orthogonal components and to use coordinates as an arbitrarily defined system of place-markers in the background space. Sixty-three years and an eon of mathematics after the *Principia*, we now have the first careful statement of what children are taught as "Newton's second law, $f = ma$." Let anyone who considers it a trivial transcription of Newton's statement or Euler's $dp = fdt$, note that Euler had first to disengage the notion of velocity from that of momentum and then to introduce a notation that gives to every possible point P not only a name but an address.

Returning to the *Mechanica*, there is a curious paragraph there that demands a few words. It strikes the modern reader as very odd that Euler introduces his statement of the law of motion not as an "axiom or law of motion," as Newton did, but as a theorem, and the plausibility argument supporting it ends with an unjustified Q.E.D. Corollary 2 of this "theorem" reads as follows:

It is therefore clear that this theorem is not only true but is necessarily true, since setting dc [the increment of velocity] equal to $f^2 \, dt$ or $f^3 \, dt$ leads to a contradiction. But since the celebrated Daniel Bernoulli regards these as

[2] The factor of 2 in the equation will look strange to people who know calculus, but here $d \, dx$ is $d^2x/2$. The 2 is the one that occurs in the second term of Taylor's series.

equally probable [he means mathematically plausible³], I would very much
like to see him prove it rigorously.

The implication is that Euler possesses a rigorous proof that any law
other than Newton's is inconsistent mathematically. This is surely not so,
but it is a tribute to the simplicity and naturalness of Newton's postulates
that many others fell into the same error. Jean LeRond d'Alembert, for
example, wrote a few years later that "The laws of equilibrium and
motion are necessary truths" (1758, p. 29). Perhaps all Euler meant is
that other assumptions lead to contradiction with *experiment*—for
example, that mechanical energy might not be conserved—but if so that
was a strange way to say it.

Joseph Louis Lagrange's *Mécanique analytique* appeared in 1788, and
after the books of Newton and Euler this one looks very tidy; there are
many pages of historical and explanatory writing, some neat and compact mathematics, and that is all. In the preface are the following words:

> There are no diagrams in this book. The methods I explain need no construc-
> tions, no geometrical or mechanical arguments, but only the processes of
> algebra [we would say calculus], performed in regular sequence. Those who
> love Analysis will see with pleasure that mechanics has now become part of
> it and will thank me for extending its domain.

The segment of the science of mechanics that deals with particles and
rigid bodies is then developed step by step, in general terms and using
procedures applicable to all calculations alike. The methods are just
under two hundred years old but they remain at the center of the stage,
for at present the most fruitful line of approach to the quantum theory is
through these methods.

Even though Lagrange's methods are those of calculus, diagrams might
have helped the reader, but there are none. To see why this is so,
remember Newton's demonstration (it is not a proof) of Kepler's law of
areas that is outlined in Chapter 11. The *Principia* is a gallery of ingenious
mathematical proofs; weak spots are rare. Newton states a proposition,
draws an appropriate diagram, perhaps fudges something, and an-
nounces the result. His dynamical theory contains truth but not a method.
Lagrange's *Mécanique* is a method. First, one looks to see whether the
system under analysis has a symmetry of some kind. For the system
composed of the sun and one planet this is clearly the case, for
the sun is (nearly) a perfect sphere and the attractive force it exerts

---

³ The reference is to a very sophisticated paper in which Daniel Bernoulli (1728), recognizing Newton's law of motion as a postulate and his proof of the law of the parallelogram
of forces as incomplete, asks what other postulates independent of Newton's would lead to
the same law.

is the same in every direction. It follows from Lagrange's equations that corresponding to any symmetry there is some dynamical quantity that remains constant. In this case it is angular momentum, the $mr^2\dot\theta$ in Note E, and as shown there, the law of areas follows as soon as we know it is constant. Thus one has the result without solving any sort of equation, for it follows at once from the existence of the symmetry.

Lagrange's proof has three obvious advantages over Newton's. One need not be a genius to invent it, for the connection between symmetry and constant ("conserved") quantities crops up all the time, and after a little exposure to the method one feels one understands *why* they are conserved. Lagrange's argument, in its general modern form, is at the core of modern physical theory, but in confronting the mysterious phenomena of matter at the most fundamental level physicists tend to turn the argument around: behind the flux of appearances certain quantities such as the masses, electric charges, and other identifying characteristics of elementary particles are seen to remain constant. The builder of theories makes a great effort to find the symmetries of nature that correspond to these constancies, and to incorporate them into the mathematics. There will be more about this in Chapter 17.[4]

Lagrange's methods work for a wide variety of calculations and the practitioner does not ordinarily need any great ingenuity in performing "the processes of algebra in regular sequence"—these processes often lead to an equation that is not easy to solve, but for that there are general methods too, and if they fail it is not the end of the matter for one can often approximate. Thus Lagrange's "there are no diagrams in this book" should be read as an affirmative statement: 101 years after the *Principia* first appeared, a calculation could at last stand on its own feet.

Final Causes?

To explain Snell's law of the refraction of light at the interface between two media the Newtonians and the Cartesians had a simple theory involving the assumption that light travels faster in dense media than in air (since it is wrong I leave it as an exercise). We have seen in Chapter 8 how using the contrary assumption Fermat generalized Hero's principle of least time to give a correct derivation of the law. The hypothesis of least time has a certain attractiveness—we compare the personified Nature with a skilled artisan who makes no unnecessary moves, but it

---

[4] Clio reminds me that though the connection between symmetry and conservation is clearly implicit in Lagrange, it was not until 1869 that Jacobi called attention to it as a general principle (Jacobi 1884).

cannot have been stated right, for the Cartesians at once pointed out the exception shown in Figure 13.1: light passing from P to Q obviously reflects from the most distant part of the curved mirror and follows the path that takes the longest time. There was another objection: physics, especially optics, was still generally regarded as a subject to be brought under the rules of geometry, and geometry has nothing to do with time. Leibniz, for example, who agreed with the Cartesians, expressed the optical law in the form of a minimum-principle with the hypothesis that light follows the "path of least resistance," not least time, concocting a definition of resistance that led to the desired result.

As set down in Euler's *Mechanica*, the general procedure for solving problems in dynamics can be illustrated with a simple example, the trajectory of a ball. If I throw a ball I decide where I will stand, how hard and in what direction I will throw it, and when I will throw it. No theory controls my choices, but once the ball leaves my hand law takes over; the ball goes where it must, regardless of anything I may do thereafter. This being the case, there does not seem to be any place for a minimum-principle in mechanics. Light travels at a fixed speed in any medium, but if I want the ball to go more quickly from P to Q I have only to throw it harder.

In 1747 Pierre-Louis de Maupertuis (1698–1759), a French polymath who became president of Frederick the Great's Academy of Sciences in Berlin, managed to show that in the impact between two perfectly elastic or perfectly inelastic objects the motion can be analyzed with what he called the principle of least action: the motion after impact minimizes a certain mathematical quantity involving the masses and the squares of the velocities.

Friends of this kind of argument joined the critics in pointing out the extremely radical nature of any minimum principle: it seems to bring back the Aristotelian notion of final cause. The formulation of such a principle goes like this: "If a particle travels from P to Q in a given time, the minimum principle will tell you what path it took." This is not quite the way one thinks in dynamics, for when I throw a ball its point and time of arrival are not given in advance. In the proposed formulation it is as if they *were* given in advance, as if the minimum principle, in determining the ball's path from its initial and final points, illustrated in a tiny example how God's plan for the world actually works.

In his *Essaie de Cosmologie* (1752) Maupertuis contrasts his position with that of rationalists who want to eliminate God from natural philosophy. Maupertuis himself is not satisfied with the various metaphysical proofs of God's existence; here at last is the opportunity for a mathematical proof.

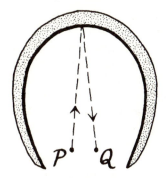

FIGURE 13.1. Sometimes when light is reflected it follows the longest path.

Until now, almost the only aim of mathematics has been to serve the physical demands of the body or the useless speculations of the mind: one hardly thought of using it to demonstrate or to discover truths outside the domain of numbers. . . . Let us see if we can use this science in a happier way. It will prove the existence of God with the transparency that characterizes mathematical truth. . . . I could have started from the laws of nature, written in mathematical form and confirmed by experiment, seeking to learn of the wisdom and power of the Supreme Being, but since the discoverers of these laws based them on hypotheses which were not purely mathematical and hence are not susceptible of rigorous proof, I thought it surer and more useful to derive those laws from the attributes of an all-wise and all-powerful Being. If the laws I find in this way are those obeyed by the Universe, is not this the strongest proof that this Being exists and is the author of those laws? (1758, vol. 1, pp. 21–23)

Richard Bentley would doubtless have considered this argument naive, and it may even strike some modern readers that way, since both lines of reasoning equally rest their proofs of God on the uncertainties of experiment. Voltaire, anonymously reviewing a collection of Maupertuis's works, offered a different objection:

A mathematical theorem is a necessary truth. The three angles of a triangle equal two right angles because things cannot be otherwise. But the necessity of things stands precisely in opposition to a God who is infinitely powerful and infinitely free. Necessity excludes choice, and it is in God's choice of means that the great mathematician Newton found a most striking verification that there is a God who created and who governs. (Voltaire 1879, p. 539)

We have encountered Newton's argument and Leibniz's attack on it in Chapter 11. Voltaire's objection to Maupertuis is the one with which Pope Urban crushed Galileo.

The ensuing philosophical debate got lost in a struggle over priorities in which Voltaire became involved. As a friend of one of Maupertuis's opponents, and being largely ignorant of the finer scientific points, he turned his terrible invective and ridicule on Maupertuis in a display of wit, the *History of Doctor Akakia and the Native of Saint-Malo* (1879, p. 559) that nearly destroyed the reputation of a gifted and far-sighted scientist. While this was going on, however, Euler and Lagrange developed a more general and useful principle of least action and showed that it provides the simplest mathematical foundation for Lagrange's dynamical method. One assumes a mathematical expression for the action characteristic of the system under discussion. It depends on the beginning and end points, the time taken, and the path followed. The principle states that of all paths that might imaginably be followed, the one actually followed is the one that minimizes the action. There is something very attractive in a formulation of natural law in the form of a principle of economy from which mathematical results follow deductively. This is not the same as showing that such a law is logically or mathematically necessary, but it gives insight as to why it rather than some other law is the case. Many people thought then, as they think now, that quite apart from issues of theology, this is the right approach to a fundamental understanding of nature.

There remains the question of the final cause: is it really an assumption of the theory? Nowadays we would say no, that the proper formulation is: "A ball is thrown from P. *If* it arrives at Q after a given interval of time, I can tell you what path it followed. Since this is so for any possible Q and any reasonable time, the principle describes all trajectories from P without favoring any particular one." There remains, however, a nagging question. If P, Q, and the time are chosen, how does the ball know in advance which path to follow in order to minimize the total action calculated along its path? I will give you the answer now, though it will not emerge into daylight until Chapter 16. The ball doesn't know in advance; it tries them all.

The Unexpected Dividend

Other than mathematics, Newton's main interest in the *Principia* was celestial mechanics, and he devoted more space to it than to any other part of dynamics. Euler and others studied new applications over a great range of problems in elasticity, the motion of fluids, and mechanisms of all kinds. Between 1798 and 1827 Pierre Simon de Laplace (1749–1827) published his *Treatise on Celestial Mechanics*, an encyclopedia of calculations relating to the six known planets and their satellites, to the shapes of rotating planets, and to tides in the earth's oceans. But all this time the atoms that compose our world went about their daily business without anybody wondering how they do it. Until the early nineteenth century nobody except Daniel Bernoulli, to be mentioned in a moment, asked whether the laws so carefully worked out for large objects might also apply to things so small. When the work finally began, its course was more clouded by errors and misunderstandings than any other part of physics ever was—but they are not our business here, and I shall ignore them as much as possible as I explain things.[5]

In 1738 Daniel Bernoulli (1700–82) published a treatise *Hydrodynamica* on the motion of fluids, on the forces they exert, and on the design of water pipes and fountains. In the middle of it he invented the molecular theory of gases. Let us assume that a gas is not a homogeneous substance but rather that it consists of particles "practically infinite in number," moving rapidly in otherwise empty space (Figure 13.2), and since gases are extremely compressible we assume that almost all the space is empty. The pressure exerted by such a gas is not a uniform force but the averaged result of many tiny impacts per second as the particles collide at random with the containing walls of the cylinder. Nothing in our experience of air encourages any such assumptions—atomists like Newton and Gassendi and their predecessors back to Democritus all assumed that air consists of particles in empty space, but nobody thought the particles are far apart or moving fast.

Bernoulli asks us to suppose that the weight P is pushed down the cylinder so as to halve the volume without changing the average speed of the particles. There will now be twice as many particles per unit volume in the cylinder and hence twice as many impacts per second on the surfaces. The pressure will thus be twice as great, and in general the product of pressure and volume will remain constant. The fact had already been discovered experimentally by Robert Boyle and was by that time well known. In Boyle's law the temperature of the gas is assumed constant. In Bernoulli's argument the particles' average speed is assumed not to change. Temperature must therefore depend on the average speed.

Suppose now that the average speed of the particles is doubled without increasing their density. The pressure will increase for two reasons: each impact will be twice as strong as before and there will be twice as many of them: the new pressure will be four times what it was. Thus pressure is proportional to the square of the particles' average speed or, more simply, to their average kinetic energy. Bernoulli relates this quantity to temperature; much later it was shown that with reasonable precision and a proper choice of zero on the temperature scale the two are proportional.

[5] For the facts see Truesdell (1968), Brush (1976).

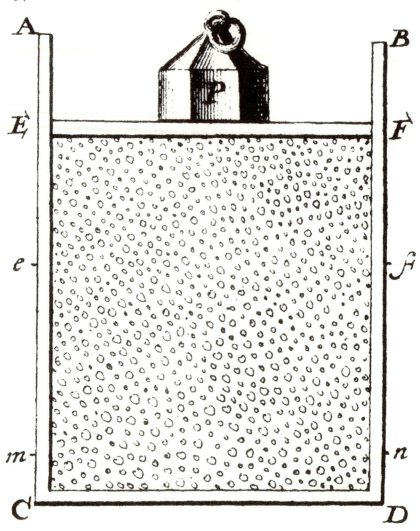

FIGURE 13.2. To illustrate the kinetic theory of gases. From Daniel Bernoulli's *Hydrodynamica* (1738).

Real gases, if they are made cold enough, condense into a liquid or a solid, and the pressure drops abruptly (Figure 13.3). But a magic gas that does not condense would cool to the point marked o at which all motion would cease, and could not be cooled any further. Long after Bernoulli, experiments showed that for different kinds of real gases, this extrapolated zero is at very nearly the same point, about 273 degrees below zero Celsius; it is absolute zero. We need a name for theories based on Ber-

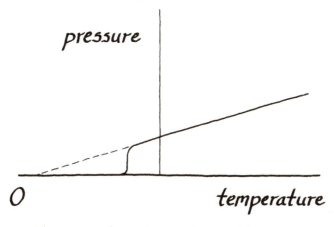

FIGURE 13.3. The pressure of a gas decreases linearly with temperature until condensation occurs. The extrapolated point is absolute zero.

noulli's model: they are called kinetic theories, *kinesis* being Greek for motion.

Bernoulli's hypothesis concerning gases was only one of several put forward at the time and it neither deserved nor received much attention, for it depended on exactly the thing Newton had warned of, the making of fancy and implausible hypotheses to explain a very small amount of data. Before the proportion of fact could be increased there had to be quantitative studies of heat and temperature in the different states of matter. The atomic theory had to be clearly stated and atoms had to be distinguished from molecules. It had to be established that heat is not, as many people thought, a fluid substance but just a measure of molecular motion, that it makes sense to say this even of the atoms and molecules in a solid— they vibrate without going anywhere—and finally, that heat and mechanical motion are fundamentally the same. We have seen that Galileo had an idea of this kind in *The Assayer* in 1623, and it is plain in Daniel Bernoulli's hypothetical model. If heat is introduced the particles move faster and their kinetic energy is increased, but a qualitative hunch was not enough: all this had to be properly stated and proved.

That took more than a century and many people were involved. It turned out that the fundamental quantity is energy, a quantity unknown to Newton. Leibniz had noticed some simple examples of mechanical energy, and the word had been introduced, with something like its present meaning, in Thomas Young's *Lectures on Natural Philosophy and the Mechanical Arts* (1807). During the 1840s James Prescott Joule, the heir to a Manchester brewery who had time, money, skill, and inexhaustible patience, made a series of measurements tending to show that there is an exact equivalence between mechanical energy and heat (Joule 1884), and in 1850 appeared a long memoir, "On the mechanical equivalent of heat," summarizing his work. The idea is that if a certain mount of mechanical energy, as might be measured by the descent of a heavy weight through a certain distance, is dissipated in frictional heat so that some object, usually a vessel of water, warms up, the amount of heat developed always bears the same ratio to the energy dissipated. If one believes that all matter is made of moving atoms this is easy to believe: the ordered mechanical energy of the descending weight is transformed into disordered motion of the atoms. Few believed that however, and for others the result was hard to accept.

Knowing the equivalence, Joule could calculate the average kinetic energy of a given mass of molecules and hence their speed. For hydrogen gas at room temperature he found it is about 2000 meters per second, and this explains a striking feature of the mechanical equivalence of heat: that for a vigorous expenditure of mechanical energy one gets not much heat. Rub your hands together hard for a while; they get warmer but do not begin to smoke. It takes quite a large windmill to light and heat a house. The point is that the molecular speeds are so large: to warm a sample of hydrogen by only 10 degrees requires as much energy as throwing the whole sample at a speed of almost 50 meters per second.

For scientists who found the kinetic model of a gas hard to take seriously it did not help to be told that the molecules move several times faster than a rifle bullet. Now we are used to being served popular science garnished with numbers that are very large or very small. Scientists of the mid-nineteenth century were not used to it. It smelled of nonsense and most of them wanted no part of it. Still, the evidence accumulated.

Julius Robert Mayer (1814–78) graduated in medicine from the University of Tübingen in 1838, with a mediocre record as a student. Failing to obtain a hospital appointment, he signed on as ship's doctor in a sailing vessel carrying bricks to Java. While the ship lay at Surabaya one of the sailors contracted a fever and Dr. Mayer, following the usual practice, bled him. When the vein was opened the doctor was astonished to see that the blood looked red like arterial blood rather than purple as it should have. As he reflected on this observation it occurred to him that in the hot climate of Java the sailor used less oxygen in his blood to maintain his body temperature, and thinking further about the consumption of oxygen in muscular tissue when work is being done he was led to the idea that blood carries something that is the equivalent of both heat and work; later he called it energy and seems to have been the first to use the word in such a general sense—heat and mechanical energy are only two of many forms that energy can take. His writings (see Mayer 1862, Lindsay 1975) contain a quantity of metaphysical material and weighty conclusions balanced on small evidence and they desperately need a rational terminology, but in their main lines they have been vindicated. They were not convincing at a time when even Joule's beautiful experiments attracted little attention. Mayer was ignored and even ridiculed, but toward the end of his life the tide swung the other way. He was awarded the Poncelet Prize in Paris and the Royal Society's highest honor, the Copley medal, and the king of Württemburg decreed that he might write *von* before his name. A vagrant medical fact rounds out this brief account: the color of veinous blood is the same in Java as it is in Germany (Ober 1968). What is different is the quality of sunlight.

My point in telling this story while leaving so many others untold is to illustrate an essential fact. Since Galileo, physics had made great progress by concentrating its attention on things that could be seen and felt and weighed. To introduce into the discourse of physics an entity such as energy that lacks these qualities and then say it is not only real but constant in value through many transformations went squarely against the principles of the best science, and we should not be surprised that it was many years before the idea gained currency. The mid-nineteenth century was a time of tremendous scientific advances—perhaps more fundamental scientific discoveries were made then in medicine, physiology, biology, chemistry, and physics than at any time since, but for that very reason it was a time of confusion, and people held tenaciously to old ideas even as they explored new ones.

Central to the acceptance of energy as a physical quantity was the best established example, the equivalence of heat and mechanical energy. As I have mentioned, this can be understood if one thinks of heat as mechanical energy of invisible molecules, but lacking good evidence, that hypothesis was as fanciful as the rest. What was needed was evidence.

James Clerk Maxwell (1831–79) came from a Scottish family and was educated in Edinburgh until he went on to Cambridge. When he was fourteen his first little scientific paper came out, on the construction of a class of oval figures (using two pins and a piece of string) and on their properties. In 1860, while teaching in Aberdeen, Maxwell wrote a theoretical paper "Illustrations of the dynamical [i.e., kinetic] theory of gases" (1890, vol. 1, p. 377), in which he showed how to calculate some properties of gases that had not been discussed before: their viscosity and heat conductivity and the rate at which one gas diffuses through another. He was particularly intrigued by the result that the viscosity of a gas is independent of its density. Everyone expected that viscosity would increase with density, but Maxwell showed that for a gas of freely moving particles to be viscous at all the molecules must transport momentum from one part of it to another, and that when the density of molecules is increased the mean free path (the average distance a molecule travels between collisions with other molecules) is reduced; the effect cancels the effect of more molecules and the viscosity remains the same. The experimental data relating to this prediction were not very good, but a few years later Maxwell and his wife Katherine made careful measurements that verified it. There were at this time almost no firm numbers relating to the atomic scale of things, but from the measured viscosity it was possible to derive a figure for the mean free path: $1/447,000$ inch ($= 5.7 \times 10^{-4}$ mm) for air at room temperature. Maxwell's conclusions concerning the heat conductivity and diffusion of gases could be more easily verified, and they were. From that time, people who could follow the arguments began to take the kinetic theory more seriously.

This story of the kinetic theory of gases is appended to an account of Newtonian mechanics because Maxwell's calculations, like Bernoulli's, assume that molecules, however small and inaccessible to observation they may be, obey the laws of Newtonian dynamics; if they did not, the answers would have been different and wrong. I have written in Chapter 11 that the mass of Saturn is about 10^{27} times greater than that of one of the pendulums with which Newton experimented. The mass of an air molecule is about 10^{25} times less. Thus the range of masses for which by 1860 Newton's laws were reasonably well verified covered 52 orders of magnitude. Today Newtonian theory is used without comment in analyzing the motions of galaxies in a cluster of galaxies. A galaxy has a mass about 10^{14} times greater than Saturn's and the 52 orders of magnitude become 66. Can the range of validity of Newton's generalization be extended downward from the air molecule? We shall see in Chapter 15 that it cannot.

Thermodynamics

The name thermodynamics was invented in about 1850 for the science that tells how to derive mechanical power from heat, as in a steam engine. For many years after the time of James Watt (1736–1819), people tried to increase the efficiency of steam engines—that is, to increase the amount of mechanical energy they put out per ton of coal consumed—but the efforts seemed to bump against an upper limit, and engineers were anxious to find a way past it. There is no way past. In an engine, very hot steam issues from a boiler, drives the engine, and is then expelled into the surrounding air at a lower temperature. Since that lower temperature is not absolute zero (in fact, it is often quite hot), the steam carries heat energy with it that is paid for but not used; hence the limited efficiency of the engine. The flow of energy, in the engine or anywhere else, is always from a warm place to a cold place and never, unless you buy a refrigerator and plug it in and pay the bill, from a cool place to a warm place; and thus energy that takes part in any process that can change its temperature becomes steadily less available for useful work. All the heat energy from the kitchen stove stays in the kitchen for a while, but unless you put the pot where the flame is you can't cook with it. Rudolf Clausius (1822–88) found a measure for the increased unavailability of thermal energy and called it entropy; his magisterial paper of 1865 closes with a couplet that summarizes what he has learned:

Die Energie der Welt ist constant.
Die Entropie strebt einem Maximum zu.

The energy of the universe is constant; the entropy streams, or strives, towards a maximum. These two principles, imagined to envelop the entire universe, are known as the first and second laws of thermodynamics.

If you think about it, the second law is not very surprising. Temperature measures the speed of molecules. It is easy to imagine how a fast molecule can hit a slow molecule and speed it up, but how often will a slow molecule speed up a fast one so as to make it go still faster? Try it with billiard balls or marbles; it just isn't very easy. As Newton wrote, "Motion is much more apt to be lost than got, and is always upon the decay" (*Opticks*, Query 31).

Death

Heat passes from hot places to cold places and entropy streams toward a maximum. The world's energy remains constant but becomes steadily less available for doing anything: heating something or lighting it or making it move. The laws of thermodynamics push us into a corner. What's gone is gone. The Victorians woke up to the realization that for humanity, no matter what else happens, some day will really be the last. In 1862 William Thomson (1824–1907), later Lord Kelvin, wrote,

Within a finite period of time past, the earth must have been, and within a finite period of time to come the earth must again be, unfit for the habitation of man as at present constituted, unless operations have been, or are to be performed, which are impossible under the laws to which the known operations going on at present in the material world are subject. (W. Thomson 1882, vol. 1, p. 511)

Thomson is properly careful with his provisos, but we can see that to him they offered small possibility of escape, and by now it is an axiom of biology that no living system escapes the second law. Later, in a popular lecture, he gave some numerical estimates for the life of the sun. He works out the numbers in front of his audience and concludes that "it would, I think, be exceedingly rash to assume as probable anything more than twenty million years of the sun's light in the past history of the earth, or to reckon on more than five or six million years of sunlight for time to come" (1889, p. 369).

These are very disconcerting numbers: the past too short for geologists and too long for fundamentalists; the future much shorter than the past. The same science that had progressed so rapidly and seemed to offer a future safe from ignorance and danger and want now promised that relatively soon all would end in ice. Charles Darwin spoke for his generation when he wrote to a friend that

> even personal annihilation sinks in my mind into insignificance compared with the idea or rather I assume certainty of the sun some day cooling and we all freezing. To think of the progress of millions of years, with every continent swarming with good and enlightened men, all ending in this, and with probably no fresh start until this our planetary system has been again converted into red-hot gas. *Sic transit gloria mundi*, with a vengeance. (Darwin and Seward 1903, vol. 1, p. 260)

A hypothesis was wrong. Kelvin did not know about nuclear reactions in the sun and got the wrong numbers. It now seems certain that our future will be about a thousand times longer than Kelvin thought and that we shall end in fire rather than ice, but perhaps we can comfort ourselves with the thought that earth's inhabitants at that late date will probably include no species that we would recognize as human.

Light and Color

Thus far the story of the great rush forward has focused on mechanics and what we now realize to be its offshoot, the science of heat. It is time to see what happened to the ancient science of light, the first corporeal form. In doing so we shall also touch on the sciences of electricity and magnetism, which were in their infancy when we last encountered them, and see how they work into the story.

After Galileo had pointed his telescope at Jupiter's moons the astronomers who studied them expected to find them revolving around Jupiter periodically, like our own moon. They were surprised to see that their periods seemed to fluctuate, so that they crossed the planet's disc sometimes earlier, sometimes later than expected. In 1676 the phenomenon was explained by Ole Roemer (1644–1710), a Danish astronomer working at the Paris Observatory, as an example of what we would now call the Doppler effect. The orbits are indeed periodic, but we observe them from a planet that is moving alternately toward and away from Jupiter, and so the finite speed of light makes the orbital frequencies seem to vary. Roemer deduced that light takes about 22 minutes to cross the earth's orbit (a better figure is 17 minutes), but his paper (1676; see Magie 1935) does not mention a speed. I think I know why. Roemer had opponents. The idea that light is a condition instantly established in a transparent medium had lasted better than most of the rest of Aristotle's physics. The speed Roemer would have written down still seems to most people ridiculously large. From astronomical data then available, some of which Roemer had helped to measure, a number of the order of 130,000 miles per second would have followed (Boyer 1941). Since he was arguing against the Aristotelians he would hardly have handed them a number like that to beat him with. Robert Hooke calculated that if light moves this way it crosses the earth's diameter in less than a second and remarked that, "if so, why it may not be as well instantaneous I know no reason" ("Lectures on light," in Hooke 1705, 1; see also Cohen 1940).

But others were convinced by Roemer's argument. In the *Principia*, a few years later, Newton corrects Roemer's 22 minutes to 15; there and in the *Opticks* he discusses the propagation of light in terms of the motion of a stream of particles after rejecting the possibility that it might be explained as a wave in the ether. The argument (*Opticks*, Query 28) is simple. We know the properties of waves from our experience of sound and of waves in water. Waves travel around corners. If my friend speaks to me from the next room I can hear her but not see her. There are, however, some difficulties with Newton's explanation, seen clearly in many of the experiments described in the *Opticks*. For example, when light hits a glass surface some of it is transmitted and some reflected. One would expect that a stream of particles would all follow the same path. If they are alike, and moving alike, why should some be reflected and others go on? To explain what happens Newton endows them with "fits of difficult and easy reflection," and in Book II, Proposition 12, he suggests how this might occur: the rays of light may excite vibrations in the ether similar to the vibrations of sound in air, which would sometimes aid and sometimes oppose the passage of the light. "But whether this hypothesis be true or false I do not here consider."

Further, a sharp edge does not cast a shadow that has a sharp separation between light and dark; carefully studied, the edge is seen to consist of a few parallel alternations of light and dark in which color can be seen. Figure 13.4 shows the shadow cast by a needle. Newton tries to explain this effect in Query 3: "Are not the Rays of Light in passing by the edges and sides of Bodies, bent several times backwards and forwards, with a motion like that of an eel?" Clearly the author is at a loss, and throughout

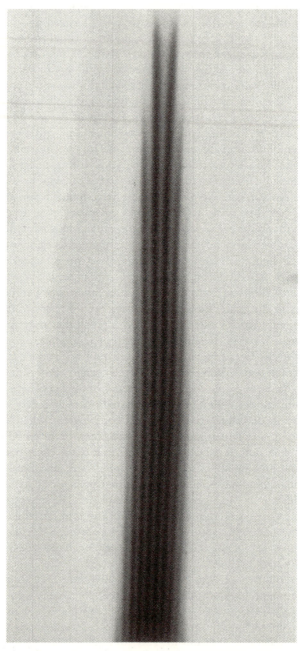

FIGURE 13.4. The shadow of a needle in monochromatic light from a point source.

the *Opticks* it often seems that in the seventeen years since the *Principia* he has forgotten his strictures against hypotheses. In the *Principia*, however, and on his side of the Leibniz–Clarke correspondence, he tried to keep hypotheses out of a deductive theory. Here he had no theory to propose, and so he allowed himself to explain the observed facts with arguments that are often far-fetched and even contradictory but which might perhaps, in other hands, become convincing.

The reason why Newton insisted on particles may, I think, be seen in the results of his experiments on color. Writing in 1671/72 to Henry Oldenburg, secretary of the Royal Society, he recalled that

> in the beginning of the year 1666 (at which time I applyed my self to the grinding of Optick glasses of other figures than *Spherical*), I procured me a Triangular glass-Prisme, to try therewith the celebrated *Phenomena* of *Colours*. And in order thereto having darkened my chamber, and made a small hole in my window-shuts, to let in a convenient quantity of the Suns light, I placed my Prisme at its entrance, that it might be thereby refracted to the opposite wall. It was at first a very pleasing divertisement, to view the vivid and intense colours produced thereby; but after a while applying my self to consider them more circumspectly, I became surprised to see them in an *oblong* form; which according to the received laws of refraction, I expected should have been *circular*. (Newton 1959, vol. 1, p. 92)

Newton, in that same year that he discovered so much else, was on the way to discovering that a prism does not produce its colors by coloring light previously white, but rather that white light contains all colors, and the prism, by refracting different colors through different angles, spreads them out so that they can be seen separately. He proved this by using a second prism to recombine the colors into white, and by a series of brilliant experiments using colored paints and powders. What could be more natural, then, than to suppose that a beam of white light is a mixture of particles corresponding to the different colors, and that some of them are deflected more by the prism than others?

In writing of planetary motions Newton has nothing to say about Kepler's Pythagorean harmonics, but they turn up in the *Opticks* when he seeks to define the difference between one color and another. As anybody who has looked carefully at the continuous spectrum of white light knows, it is not easy to decide just where one color leaves off and the next one begins, and if we did not know the sequence "violet–blue–green–yellow–orange–red," it might even be hard to say how many different colors there are. For Newton there were seven colors in the spectrum, not six; he inserts "indico" between violet and blue. Various authors have suggested that Newton's marvelous eyes actually perceived indigo as a distinct color, but in Book I, Proposition 3, he tells us the real reason: he wants the colors to correspond to the seven notes of the musical scale. Newton projects a spectrum onto a sheet of paper and draws eight lines on it defining the seven colors. He then compares the separations of the lines to the intervals in an octave. When the *mi* is flatted to make a minor third the correspondence is good, and the indices of refraction corresponding to the different colors fit very well into the Pythagorean ratios. Nature is indeed conformable to itself.

In 1690 Christiaan Huygens (1629–95) published his thoughts about light in a small *Treatise on Light*, in which he maintains that light is a disturbance that moves through ether as sound moves through air. He speaks of waves, but they seem to be of no particular wavelengths, and it would be more accurate to say that they are like a sequence of little ripples. Euler, in 1768 and probably before, seems to have been the first person to propose that just as the wavelength of a sound wave determines the pitch we hear, so the wavelength of light determines the color we see (Euler 1911, Ser. III, vol. 11, p. 46; vol. 12, p. 4; trans. 1842, vol. 1, p. 112; vol. 2, p. 68). This is quite correct, but the assertion was not supported by experiments, and so great was the authority of Newton that a century went by before the wave theory became firmly established.

Thomas Young (1773–1829) was one of the brightest of all British intellects. As a boy he learned the modern and classical languages including Hebrew, Persian, and Arabic. He studied medicine and set up practice in London but was never very successful; apparently he insisted on telling his patients that medical judgments are only probabilities. Having inherited some money he became for a few years a professor at the Royal Institution, where he gave demonstration lectures and made experiments in physics and physiology.

In 1801 Young delivered a lecture before the Royal Society entitled "On the theory of light and colours." He starts with a tribute to Newton (whose *Opticks* was then ninety-seven years old) and, after this brief bow, says that he will argue for the wave theory. According to this hypothesis light striking the retinal substance at the back of the eye excites vibrations in it and

the frequency of the vibrations must be dependent on the constitution of this substance. Now as it is almost impossible to conceive each sensitive part of the retina to contain an infinite number of particles, each capable of vibrating in perfect unison with every possible undulation, it becomes necessary to suppose the number limited, for instance, to the three principal colours, red, yellow, and blue, of which the undulations are related in magnitude nearly as the numbers 8, 7, and 6; and that each of the particles is capable of being put in motion more or less forcibly, by undulations differing less or more from a perfect unison; for instance, the undulations of green light being nearly in the ratio of $6\frac{1}{2}$, will affect equally the particles in unison with yellow and blue, and produce the same effect as a light composed of those two species; and each sensitive filament of the nerve may consist of three portions, one for each principal colour. (1802, p. 20)

With minor modernizations of the physiology, this absolutely original proposal becomes a sketch of the current theory of color vision. The cone cells of the retina are indeed of three kinds distinguished by dyes of three different colors, and our sensation of color is determined by the extent to which the three are excited. Why did Young choose the number three? He doesn't say. It turns out to be the minimum possible number to do the job and so, not surprisingly, it is the one chosen by nature, which affects not the pomp of superfluous causes. The lecture continues with a sober and careful account of the wave theory of light and the various phenomena it explains.

The aftermath is less edifying. Henry Brougham (1778–1868) was a clever young Scotsman seeking to rise in the world. By the time he was twenty he had communicated to the Royal Society some unimportant ideas on light and color entirely in the Newtonian tradition. When in 1802 the *Edinburgh Review* was founded he became one of its most frequent contributors, certainly the most vitriolic, and he took the trouble to pour ridicule, anonymously, on Dr. Young: "As this paper contains nothing which deserves the name, either of experiment or of discovery, and as it is in fact destitute of every species of merit, we should have allowed it to pass. . . . But we have of late observed in the physical world a most unaccountable predilection for vague hypothesis daily gaining ground. . . . We wish to raise our feeble voice" (Anon. 1803). Brougham's heaviest fire falls on Young's hypotheses. He does not deign to mention the theory of color vision but takes fresh offense at every mention of waves in the ether. A reader of Brougham would never guess that Young's exposition of the wave theory was based on careful experiments.

What was the reason behind this savagery, and what had gone wrong with British science? The dead hand of Newton had imposed two doctrines on his successors: that the theory of light must finally be a theory of particles and that as I have quoted earlier from the *Principia*, hypotheses have no place in experimental philosophy. In this philosophy "particular propositions are inferred from the phenomena, and afterwards rendered general by induction," and I have suggested that Newton, when he could not do this, allowed himself to hypothesize freely in the *Opticks*. His successors saw the distinction less clearly, tended to follow Newton's precepts rather than his example. If it didn't matter what light is, if quantitative laws were the only goal and the mind didn't need to be aided with models, the rule they followed would be a useful guide to research. But a wave and a beam of particles are very different things, and propositions and questions referring to one are not relevant to the other. At the stage at which physics then stood, a hypothesis had to be adopted, and so just as the Copernican model was tacitly built into theory of planetary motion, Newton's hypothesis of particles got built into British ideas about light. In Chapter 15 we shall see that wave and particle models, though very useful, should not be taken so literally as to exclude each other, but such a degree of tolerance did not exist anywhere in 1800.[6]

Young's experiment, which exhibits the wave nature of light, is sketched in the printed version of his lectures where he discusses phenomena of color (1807, vol. 1, p. 464). Figure 13.5 is his illustration of the idea: waves in step with each other (with light, this is most easily arranged by deriving them from a single source) are formed at points A and B. As they move toward the screen on the right there are lines (seen as dark if one looks edgewise at the page) along which the crests of the waves from one source coincide with the troughs from the other. The experiment can be done with water waves or on a smaller scale with light, and it is found that along these lines there is no disturbance of the medium—crests and troughs cancel. Thus in the optical experiment there are alternate light and dark regions on the screen, the points C, D, E, and F remaining dark. This fact is hard to explain if light is imagined as a rain of little particles. We shall need Young's formula for calculating wavelengths in Chapter 16, and it is derived in Note G. Thus Young, who may or may not have known of Euler's hypothesis made a generation earlier, was able to do experiments that verified it.

After Young resigned his professorship he turned to his many other interests. He deciphered the inscription in demotic Egyptian script on the Rosetta Stone, held various public offices, and wrote several books on medicine. Henry Brougham went on to become Lord Chancellor. His personal style, however, continued unchanged until he died in his ninetieth year.

Some time after the events described, in a letter to a European savant, Young proposed an explanation for the phenomena of polarized light. He found that he could save the phenomena by assuming that the vibrations of the ether occur sidewise like the waves in a rope that is shaken from side to side, rather than along the line in which the wave travels like the vibrations of sound (Young 1855, vol. 1, p. 380). This discovery raised a question much larger than the one it settled. A fluid cannot transmit transverse vibrations. There has to be some kind of rigidity. Jelly will do it but not a fluid. Apparently Young was filling the universe with jelly. But why are we not aware of it as we ride on the earth at such a great speed? And how can the planets make their way through such a substance year

---

[6] For more on this controversy, see Cantor (1975).

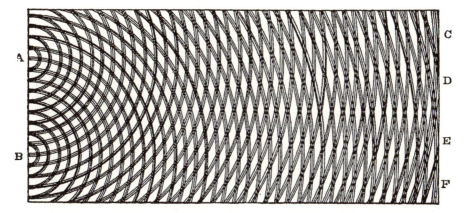

FIGURE 13.5. Interference pattern produced by light from two coherent sources. From Thomas Young's *A Course of Lectures on Natural Philosophy* (1807), Vol. 1, plate 20.

after year without finally spiralling downward into the sun? Thus was inaugurated a program of theoretical study that occupied brilliant minds in Britain, France, and Germany for a century: to explain how the ether can be substantial and insubstantial at the same time, how it can be an elastic material and yet offer no resistance to objects passing through it. Ignoring the mechanical effects on moving bodies a satisfactory theory of light was possible. Ignoring the phenomena of light it was possible to make plausible explanations of why we move through the ether without noticing it. But to harmonize the two seemed just as difficult at the end of the nineteenth century as it had at the beginning. A new idea was needed.

Electromagnetics

At the time of Newton nobody knew any more about magnetism than Gilbert had known a century earlier, and what was known of electricity amounted to little more than phenomena produced with a hairbrush on a cold morning. The great advances in knowledge occurred during roughly the century following 1750, and at the end of this time one knew that electricity is some sort of substance that like energy is not seen, weighed, created, or destroyed. There are two kinds of electricity, arbitrarily called positive and negative. The electric substance can be made to flow through metal wires, and when it does it gives rise to the same magnetic forces that one observes near a lodestone. If a piece of lodestone or other magnet is moved close to a loop of wire it causes electricity to flow in the wire. (The quantity of electric substance is called the electric charge, and when it flows, the amount transferred past a point in one second is called the current.) Generally speaking, action at a distance was then under the Newtonian interdict as it remains today: nobody in his right mind thinks that two objects can influence each other across a distance if there is nothing whatever in between. The job of transmitting the forces of electricity and magnetism—since they are so closely related one speaks of electromagnetism—was given to the ether, though whether it is the same ether as is involved in the transmission of gravity and light was not clear.

This was not all just talk. By 1860, if one stuck to the verifiable facts and did not worry about the ether, there was a fairly complete body of quantitative, verifiable knowledge concerning electricity and magnetism and their interconnections. There were mathematical formulas that correctly accounted for the observations.[7] They described them but did not, in the general view, explain them. How does it all actually work?

As the experimental picture of electromagnetism became clearer it was necessary to have a word to denote the state of things in the apparently empty space surrounding a magnet or a charged body. The word "field" was chosen. Nothing is assumed about the physical nature of the field when one speaks of the forces it exerts or the other things it does, but people generally thought that the field is a manifestation of some state of tension or motion in the ether that would not be there if the agent producing it were taken away. At length, in December 1861, James Clerk Maxwell wrote to William Thomson (later Lord Kelvin) that he had a mechanical explanation of it all.

> I suppose that the "magnetic medium" is divided into small portions or cells, the divisions or cell walls being composed of a single stratum of spherical particles, these particles being "electricity." The substance of the cells I suppose to be highly elastic, both with respect to compression and distortion and I suppose the connection between the cells and the particles in the cell walls to be such that there is perfect rolling without slipping between them and that they act on each other tangentially. (Larmor 1937, p. 34)

He describes it in excruciating detail, a working model of the ethereal machine that saves all the phenomena. The model characterized the field by partial differential equations; the technicality is necessary because equations of that kind refer to all parts of space at a given moment. They

---

[7] The theories were complicated and the equations sometimes fearsome. We now know that this was because most of the formulas were only approximate and different parts of the theory were based on different approximations, so that their connections were not apparent. It is much simpler when it you know the correct laws and can do it all exactly.

usually have an enormous variety of solutions, and sometimes there are solutions that tell you more than you knew when you set them up. Maxwell's equations turned out to possess a class of solutions in which the electric and magnetic fields detach themselves from the charges and currents that produce them and travel off into space in the form of electromagnetic waves, as far as you please. Figure 13.6 shows the electric and magnetic fields in a simple wave.

A strange thing about Maxwell's theory is that although it was based on a detailed model of the ether almost no numbers relating to the model ended up in the equations: nothing about the sizes of the cells or the masses of the particles forming the walls—nothing but two experimentally measured numbers, k and μ, which tell how strong an electric field is produced by a given electric charge and how strong a magnetic field is produced by a given current. The speed of the electromagnetic waves predicted by Maxwell's equations turned out to be a universal number, the same for all waves, equal to $\sqrt{(k/4\pi\mu)}$. The quantities were fairly well known at the time and Maxwell deduced that the speed is

$$c_{electromagnetic} = 310,740 \text{ kilometers per second (km/sec)}.$$

A recent French measurement of the speed of light had given

$$c_{light} = 315,070 \text{ km/sec}.$$

"I made out the equations in the country before I had any suspicion of the nearness between the two values of the velocity of propagation of magnetic effects and that of light, so that I think I have reason to believe that the magnetic and luminiferous media are identical" (Larmor 1937, p. 35).

By 1865 Maxwell had written up his results in a paper that occupies seventy-two pages of the *Scientific Papers* (1890). It sets the stage with remarks about the ether; then the equations begin and ether is left behind. Carefully he shows how the equations explain the known electric, magnetic, and electromagnetic phenomena; finally come the new electromagnetic waves. He compares their velocity with that of light:

> The agreement of the results seems to shew that light and magnetism are affections of the same substance, and that light is an electromagnetic disturbance propagated through the field in accordance with electromagnetic laws. (1890, vol. 1, p. 580)

Notice the change in language. In 1861 light waves traveled in the ethereal medium; now there is only vague talk of a substance, and light is a

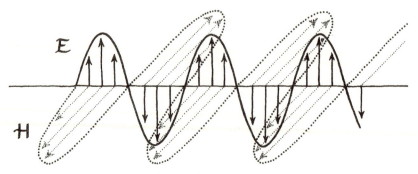

FIGURE 13.6. Electric and magnetic fields in plane-polarized light.

disturbance in the field. Maxwell has become aware that because his equations explain everything without mentioning any dynamical properties of the ether, it is difficult or impossible to find out anything about the ether from studying electricity and magnetism. All we can ever know are the fields; the ether is inaccessible but luckily it is also superfluous. Having built his house Maxwell had taken down the scaffolding, but of course most physicists felt that he had pulled out the foundation.

According to Maxwell's theory every electromagnetic wave originates in the motion of electric charges. Light issues from matter; therefore a radiating atom contains moving electric charges. One ought also to be able to generate electromagnetic waves by creating alternating currents in the laboratory, but the quantity of radiation depends on the frequency of oscillation, and the currents must oscillate quickly. Developing the experimental techniques took a while but in 1887, in Karlsruhe, Heinrich Hertz (1857–94) was able to produce Maxwellian waves and show that they move at the speed of light. In Germany this was hailed as the greatest experimental discovery of the century but the British physicists were less impressed, since they had already been largely convinced by Maxwell's theory. Hertz's waves were the first radio waves. The development of that technology needs no comment here.

Reading about Maxwell's theory in a modern text one gets the impression that there must have been handshakes, champagne, and a brass band. Nothing could be less true. To most physicists of the time the great task of physics was to unify the different laws of physics into a single theory. A good many people thought that not only was ether *a* fundamental substance; it might be *the* fundamental substance. Ether was also the medium whose motions gave rise to electromagnetic phenomena, and therefore the theory of such motions, based, one supposed, on Newtonian mechanics, appeared as the Holy Grail of physics. The flaw in Maxwell's theory was that though it undoubtedly squared with the facts it contributed little to this great project, and most writers of books felt that their main task was to complete Maxwell's ideas by making up mechanical models to explain what the equations left unsaid. Up until about 1910 most physicists spoke of ether as naturally as they took their morning coffee, even though by that time attempts at a detailed theory were collapsing in fatigue and defeat.[8]

It is interesting to note how the response to Maxwell's theory followed the pattern of the debate between Leibniz and Clarke. The British and German ether theorists, like the Leibnizians, considered that Maxwell had found the basic mathematical relations between the field variables but that he had explained nothing, and they redoubled their efforts to explain electromagnetic effects mechanically. Only there was this difference: in the eighteenth century the British were content with a mathematical theory that saved the phenomena while the Europeans sought deeper explanations; now most of the diehard ether theorists were British.

In 1889 the physicist Oliver Lodge (1851–1940) published a popular book, *Modern Views of Electricity*, in which he explained the ether theory to the public. Figure 13.7 shows Maxwell's rotating cells on each side of a current-carrying wire. The saw represents the current. Move it and the positive charges in matter spin one way, the negative charges the other. We learn in school that when a current changes a magnetic field is set up that opposes the change. That is the effect of the inertia of the wheels. When one changes the speed of the saw it requires some effort, and a little time, before they respond. In 1905, as we shall see, Einstein put to bed the idea of the ether as a mechanical system. Most of those who had given their lives to it were old and discouraged but they swam on. In 1905, in a reprint of his *Elementary Lessons in Electricity and Magnetism*, Sylvanus P. Thompson told his readers that "It must never be forgotten that the **electric force acts across space in consequence of the transmission of stresses and strains in the medium with which space is filled**" (Thompson 1905, p. 76). Ether was a logical necessity: if there is a wave, something must be waving. If radiant energy moves from the sun to the earth, ether must convey it. Action through empty space was still unthinkable and few people saw a field as an alternative to emptiness. Note that the arguments against emptiness were different from Aristotle's, but they were just as strong.

Among the most eminent of the French *savants* was the physicist and historian Pierre Duhem, whose *Aim and Structure of Physical Theory* first appeared in 1914. Here he extols the "straitness of mind which makes the Frenchman eager for clarity and method, and it is this love of clarity,

---

[8] The intellectual currents of this period are traced with many references in Jungnickel and McCormmach (1986, vol. 2).

FIGURE 13.7. Maxwell's mechanical explanation of the production of a magnetic field by an electric current. From Oliver Lodge, *Modern Views on Electricity* (1889). The serrated rod represents electric charges. When a current flows the rod moves, setting in motion the wheels marked +, which produce the effects of a magnetic field. The wheels marked − (which turn in the opposite direction) serve only to set the more distant wheels in motion.

order, and method which leads him, in every domain, to throw out or raze to the ground everything bequeathed to him by the past, in order to construct the present on a perfectly coordinated plane." He turns to the British:

> In the treatises on physics published in England, there is always one element which greatly astonishes the French student: that element, which nearly invariably accompanies the exposition of a theory, is the model. . . . Here is a book [referring to the French translation of Lodge] intended to expound the modern theories of electricity and to expound a new theory. In it there

are nothing but strings which move around pulleys, which roll around drums, which go through pearl beads, which carry weights; and tubes which pump water while others swell and contract; toothed wheels which are geared to one another and engage hooks. We thought we were entering the tranquil and neatly ordered abode of reason, but we find ourselves in a factory. (Duhem 1962, p. 70)

In pages that will surprise and delight the modern reader, Duhem (following Pascal) distinguishes between two kinds of mind, those that are strong and narrow and those that are broad and weak. Napoleon's mind, alas, was of the second sort and it is "endemic" in Britain. The haughty rejection of the aid offered by models and by experiments performed only in the mind may help to explain why French physicists contributed so few fundamental physical insights in the century after the death of Ampère in 1836.

As the years went on, physicists learned to think in terms of the electromagnetic field and to consider that it has an independent physical existence just as much as a planet or a stone. It exerts forces and transports energy (the sun's heat, for example). It can sometimes be felt—put your hand into the sunlight, yet you feel nothing near a magnet unless it is tremendously strong. The field can be seen if it happens to oscillate at a frequency that affects our eyes; at lower frequencies it carries the radio traffic. As we shall see, Einstein showed that it even has weight. Being spread out everywhere in space it is not like a material object. It satisfies Maxwell's laws of motion and not Newton's.

By 1900 physics was penetrating the unseen world. Energy and entropy had been defined and studied, the kinetic theory of matter brought atoms under the domain of Newtonian theory, the electromagnetic field was understood and at least the younger workers felt comfortable with it. One elementary particle, the electron, had been discovered and some of its simpler properties were known. Now we are almost another century further along. How do those ideas stand today? It is scarcely an exaggeration to say that though the territory of physics has broadened, what lies in the old domain remains as it was. The laws of dynamics have been extended to cover phenomena unthought of a century ago, but within the old domain Newton's laws are an accurate approximation. The kinetic theory has been extended, but the theories that agreed with experiment a century ago agree quite well today. Ideas of energy and entropy are somewhat wider now but the couplet that Clausius wrote in 1865 is still the law. Maxwell's electromagnetic equations, published in the same year, continue unaltered, though the quantum theory has changed our idea of the meaning of a field. Even the ether survives, though in disembodied form and under other names.

Those theories were built to last, and they have lasted. What has not lasted is the sense of security they engendered in their founders. In 1900 Lord Kelvin could tell an audience at the Royal Institution "The beauty and clearness of the dynamical theory, which asserts heat and light to be modes of motion, is at present obscured by two clouds" (W. Thomson 1882, vol. 4, p. 531; 1904, Appendix B). One of these was that he saw no way for the earth to move freely through the elastic jelly of the ether, and the other was a thermal anomaly that was soon to be blown away by quantum theory. A sunny sky with two clouds. Now there are many. The scope of investigation has broadened, and every answer provokes new questions. As Karl Popper wrote,

The more we learn about the world, the deeper our learning, the more conscious, specific, and articulate will be our knowledge of what we do not know, our knowledge of our ignorance. For this, indeed, is the main source of our ignorance—the fact that our knowledge can only be finite, while our ignorance must necessarily be infinite. (Popper 1962, p. 28)

Two Theories of Relativity

The supreme task of the physicist is to arrive at those universal elementary laws from which the cosmos can be built up by pure deduction. There is no logical path to these laws; only intuition, resting on sympathetic understanding of experience, can reach them.

—ALBERT EINSTEIN

UNTIL recently it was axiomatic that the universe sets absolute standards of rest and motion. For most of history the standard of rest was the earth. After Copernicus it was the sun, and as people began to realize that there are other suns not necessarily at rest with respect to our own, the standard of rest became the ether. But nature seems to conspire to keep us from measuring or even detecting the effects of motion with respect to the absolute standard. If it had been easy there would have been no doubt as to whether the earth, the sun, or both were in motion, and if the ether is the bearer of electric and magnetic interactions one might have expected that they would be affected by our motion through the ether.

Consider, for example, what happens if we move a loop of wire in the neighborhood of a magnet: a current begins to flow in the loop. This is easy to explain. The wire contains electrons that are free to move along it. When we move the wire we move the electrons, and an electron moving in a magnetic field experiences a force. This causes the electron to flow in the wire; the current can be calculated; theory and experiment agree.

Now hold the loop still and move the magnet. A current flows. Again it is easy to explain: at a given fixed point the field of a moving magnet varies in time. By Maxwell's theory, a changing magnetic field gives rise to an electric field. This field acts on the electrons to propel them around the loop. Again, perfect agreement. But wait: when we say we hold something still are we really holding anything still? We live on a grain of cosmic dust forever whirling through space. By what right do we speak of stillness? Ptolemy is dead. And of course it doesn't matter to the experiment. Whether we hold the magnet and move the wire or hold the wire and move the magnet makes no difference to the observed effect. The only difference is in the theory, for completely different arguments are used to explain the same effect in the two cases. It is with thoughts about this ridiculous situation that Einstein began his first paper on relativity.

That paper (1905c) took Einstein only a few weeks to write. It solved the puzzle of the two explanations of the same effect and had a profound effect on subsequent physical theories, but in his mind it raised more questions than it answered, for it still did not say how one should speak about absolute motion. Generalizing the theory so as to partially take care of that took another ten years of very hard work. Only now, seventy years later, are we beginning to understand how further generalizations can be made that will accomplish still more of Einstein's program. A little will be said about that in Chapter 17.

What Are the Properties of Ether?

As we have seen in the last chapter, ether theories got involved with complicated and purely imaginary hypotheses precisely because there were no observations that measured any physical property of ether and so might have helped decide between one theory and another. The only results were negative. Ether offers no perceptible resistance to the passage of the planets, and so if it is a fluid it is very thin. It does not scatter light, and so it is not made of atoms. Even on a very clear day the sky is blue. This is because air molecules scatter the sunlight into our eyes, making it appear to come from the sky rather than directly from the sun, and the intensity of sunlight is correspondingly reduced. A simple calculation showed that over the vastness of space scattering by an atomic ether would extinguish the light from the stars, but there they are.

If the ether stands still in space, with the sun and other stars standing still in it or nearly still, while the earth travels around the sun at some 30 kilometers per second, we should be able to detect an ether wind. Not that it could exert any force on material objects, of course—that would long ago have sent us spiralling toward the sun. Apparently ether slips between atoms like wind between bare tree trunks. But if it does not interact with matter it obviously interacts with light, for it conveys light, as water waves are conveyed by the sea. The ether wind must be detectable in an optical experiment.

In the decades following 1880 several very clever and subtle optical experiments sought to surprise the ether into revealing its presence. The most notable was that of Albert Michelson and Edward Morley, carried out in Cleveland (1887); over the years it was repeated with more and more refinements (Swenson 1972) even when nothing at all was detected. The older physicists, nourished on ether since childhood, were bewildered and dismayed and they put forth a string of hypotheses to save the phenomena. The younger generation accepted the situation more easily. Optical experiments only confirmed the message of our senses: there isn't any wind. A letter written by Einstein in 1901 (Pais 1982, p. 132) mentions an idea for an improved way to measure motion through the ether, but in the next year or so he seems to have become convinced it would not be worthwhile. Later he could not recall that he had even heard of the Michelson-Morley experiment when he wrote his first paper on relativity theory. Experiments continued into the 1920s but by that time, except in a few backward-looking minds, the question had been swept away in the tide of events.

The Special Theory of Relativity

Albert Einstein (1879–1955), son of an unsuccessful electrical contractor, was born in Ulm and received his early education in Munich. In 1896, after failing once, he was admitted to the Swiss Federal Institute of Technology. There he lived on the fringes of the Institute as some university students do today, attending class selectively, studying material of his own choice in his own way and making friends with other outsiders. In 1902, not finding any better job, he began work as a technical expert, third class, in the Federal Patent Office in Bern, which he said afterwards was ideal for someone like himself—it filled his day with an occupation that was moderately interesting but not demanding enough to produce fatigue or to preclude a little thinking on the job, and it left evenings and weekends for hard work. It was during his third year there that he wrote the three papers that quickly established his reputation. The next chapter will discuss the papers on molecular motions and the nature of light; here I will explain some of the thoughts and results in the paper called "On the electrodynamics of moving bodies" (1905c). The most detailed account of the development and reception of the special theory of relativity is given by Miller (1981).

Einstein begins by calling attention to the farcical situation described

at the beginning of this chapter, in which two entirely different explanations are given for the same phenomenon depending on obviously foolish
distinctions between rest and motion. He points out the real conclusion
to be drawn: "the phenomena of electrodynamics as well as of mechanics
possess no properties corresponding to the idea of absolute rest."
Mechanical experiments are less precise than optical and electrodynamic
experiments, but the lack of a mechanical effect is striking if you think
about it. Throw a ball a few feet into the air; the earth's motion carries
you and it dozens of miles sidewise before you catch it again, but nobody
notices the fact. The conclusion is that already expressed by Virgil in the
line about the ship (Chapter 7) and Oresme with his flying man (Chapter
6), and it is at the center of Copernicus's argument: you can say that one
thing is moving with respect to another but you can say no more, for the
universe provides no standard of absolute uniform motion or rest.

Since the proposition is important, let me say it in other ways. Suppose
we are shut up in a large, windowless box and set moving *uniformly*, i.e.,
in a straight line at constant speed through the universe. Then the proposition states that no observation or experiment could tell us anything
about the motion of the box. Experiments test the laws of nature, and so
another way to say it is that inside the box the laws of nature always hold
in exactly the same form, provided the motion does not change. The
restriction to uniform motion is necessary. If the box turns a sharp corner
or comes to a sudden stop, of course we will know it. Please note that this
last sentence is part of an argument. In writing it I do not guarantee that
it is true. For purposes of the theory being described, the special theory of
relativity, it is assumed to be true. Later in this chapter we shall see that
in the general theory it is assumed not to be true. The proposition I have
stated in several ways was named by Henri Poincaré the principle of relativity (see, for example, Rogers 1905, p. 607), and it is the first hypothesis of Einstein's theory.

Einstein's second hypothesis is that, as all experiments indicated and
still indicate, Maxwell's theory of the electromagnetic field is really correct, and if the two hypotheses are combined we are forced to some
strange conclusions.

In the last chapter we saw that Maxwell's theory gives an unambiguous
value for the speed of light. Let us suppose there is a flash of light moving
somewhere in the world and you and I, each in our own box, calculate its
speed. Since by hypothesis we both use the same equations we will both
arrive at the same result. But that is odd, because if I were driving a car
and you were sitting by the road, for example, and we both looked at the
same bicyclist, her speed with respect to you would certainly not be the
same as her speed with respect to me. Why is this not equally true of a
flash of light? Well, it just isn't; later in this chapter I will mention one of
many experiments that support the conclusion that the speed of light in a

vacuum is the same for all observers. This conclusion enables us to understand some strange consequences of the theory, for much of our signalling is done by light, and we reach new conclusions when we take this property into account.

Figure 14.1(a) shows a moving flatcar with a light mounted exactly in the middle. I am on the flatcar, you are beside the track. When the car is about opposite you I flash my light. The question is, does light from the flash reach the two ends of my car at the same moment? From my point of view, no question could be simpler to answer. Since the light is mounted in the middle, and since light travels toward the right at the same rate as toward the left, then of course it reaches the two ends simultaneously. But to you, Figure 14.1(b), the situation is different. You assign light the same speed that I do, but while the rear end of the car is drawing closer to the point from which the light was emitted, the front end is moving further away. Therefore the light reaches the rear of the car before it reaches the front. We can argue about this all day but the conclusion follows from the premises: two events that are simultaneous for me are not simultaneous for you.

This fact may be surprising but it has equally surprising consequences. Suppose, for example, that I am carrying a pole just as long as the flatcar. If I set it parallel to the track and invite you to measure its length, I must warn you that you may use any means you like but you must be sure to locate the two ends of the pole simultaneously. If you measure the position of one end at one moment and the other at some other moment you will get an answer, to be sure, but it will not be the length of the pole. To this you agree, but if we decide to make an optical measurement using two flashes of light we will not agree on simultaneity. If you follow my directions for making the measurement you will get an answer that you believe is wrong, since in your view the flashes will not be simultaneous and you will be measuring the positions of the two ends at different times. By instructing me how to flash the lights you can design a procedure that will deliver two flashes that are simultaneous for you but not for me. It appears that we cannot agree on the length of the pole.

Suppose, however, that I hold the pole upright. Now we can easily agree on its length, since the height of the pole does not change and we do not care when top and bottom are measured. Motion affects the results of measuring both times and poles, and we must separate the two influences before we can understand what is happening.

Relativity of Time and Distance

The structure of our language, and the things we hear and read, suggest to us that time is a sort of moving fact spread out all over the universe, in

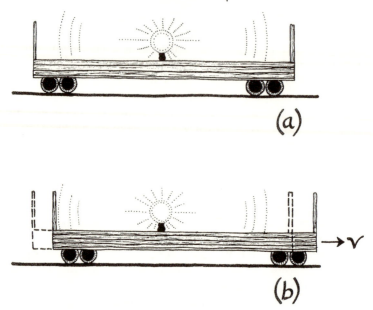

FIGURE 14.1. The relativity of simultaneity. A light flashes at the center of a moving flatcar. For an observer on the car the flashes arrive simultaneously. For an observer on the ground the light reaches the back of the car before it reaches the front.

the sense that if it is three o'clock here it is also automatically three o'clock everywhere else, account being taken of the time zones that cover the earth. It follows from the postulates of relativity that things are not quite so simple and we must look a little more closely, helped by simple-minded conceptual experiments like the ones just described. The arguments to follow do not come from Einstein's paper but they are in accord with it.

Figure 14.2 shows an arrangement suitable for the comparison of clocks. You mount your box on a moving flatcar and project a flash of light to the ceiling and back, measuring the elapsed time with your watch. Evidently it is

$$t' = 2L/c,$$

where c represents the speed of light and the prime on the t means that it is as measured on your clock. From my point of view, things look as shown in the second drawing. The distance L is the same, as we have just seen, but plainly the path is longer and so the time is greater. How much

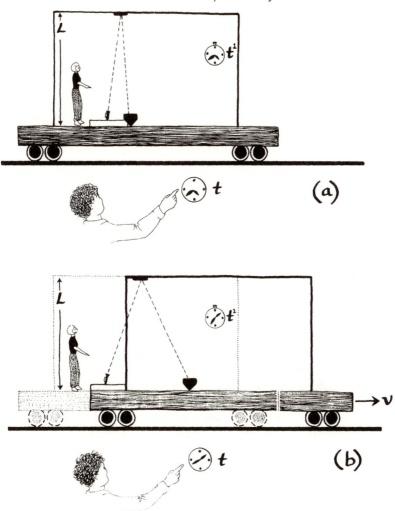

FIGURE 14.2. Relativity of time. Another conceptual experiment on a moving flatcar. For the observer on the ground the light path is longer and therefore the flash takes longer to arrive. Since the two observers assign different time intervals between the same two events, their watches do not agree.

greater is told by a simple formula derived in Note H which we can write neatly as

$$t_{\text{across and back}} = \gamma t', \qquad \gamma = \frac{1}{\sqrt{(1 - v^2/c^2)}}.$$

It is more important to see the idea than the formula. You and I are thinking about the same two events, the emission and reception of a flash of light and we are using our watches, in imagination, to measure the interval between them. The interval is determined by how far the flash travels at its fixed speed. Obviously, for me it travels farther than it does for you. Therefore for me, measuring by my watch, the interval between the same two events, emission and reception of the flash, is longer. This is what the formula says. Standing on the solid ground and observing your moving watch, I see that its reading is less than the reading of my watch. This means that the moving watch is slow. Of course, as I have already pointed out, I only think I am standing still; from another perspective I am whirling through space. It doesn't matter. All that matters is that your watch moves with respect to mine. But this remark opens a paradox: if I find that your watch is slower than mine, then if you perform the same measurements you will find my watch slower than yours. How can my watch be both slower and faster than yours? Not possible, if one thinks of the measurements as delivering unambiguous news about the relative rates. Quite possible, if a measurement simply delivers its own result. A contradiction would result if the same measurement, repeated, yielded the opposing results, but these are measurements of two different objects, your watch and my watch, made by different observers, you and me, in different states of motion. There is no logical reason why they have to give the same result, and they don't.

Now, in order to compare lengths, I change the experiment a little, as in Figure 14.3. As before, $t' = 2L'/c$, but note the prime on the L to indicate that I do not assume we agree on the length of the light path. The time I measure is given by another formula from Note H,

$$t_{\text{down and back}} = \gamma^2 2L/c.$$

We already know the difference in rate between the two clocks: $t = \gamma t' = \gamma 2L'/c$, so this is

$$\gamma^2 2L/c = \gamma 2L'/c \qquad \text{or} \qquad L = L'/\gamma.$$

Since γ is always greater than 1 this means that the length a "stationary" observer assigns to a moving object, when the length is measured parallel to the motion, is less than the length assigned by the observer who travels with it. As mentioned earlier, the length measured perpendicular to the motion is the same for both observers.

The postulates have yielded three results that ought to be tested: the universality of c, the slowing of moving watches, and the contraction of moving rods. To test the first, it is relatively easy to set a source of light

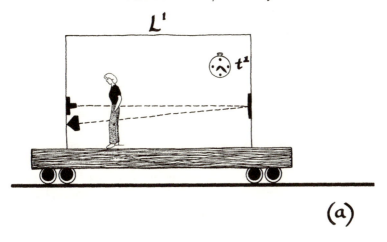

(a)

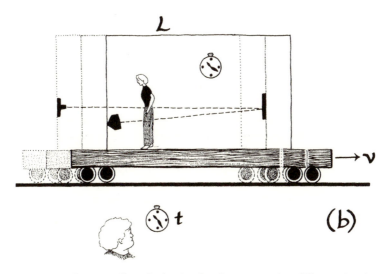

(b)

FIGURE 14.3. Relativity of length. Again, the observers assign different time intervals. They now find that they disagree also as to the length of the flatcar.

into motion. This was done a number of years ago with some particles called neutral pions that break up into two photons, which for our purposes can be considered as two flashes of light. The pions were created in one of the high-energy laboratories. They were moving very fast, with speed 0.99975c (as we shall see in a moment, no material particle can be accelerated to the speed c), and the speeds of the flashes of light directed forward and backward with respect to the pions' motion were measured.

They did not differ from the usual c by as much as one part in 10,000 (Alväger et al. 1964; see also Brecher 1977). Setting the observer into motion at comparable speeds is more difficult, but the principle of relativity tells us that only the relative motion matters. The Michelson–Morley experiment was designed to detect absolute motion, and the negative results of it and its much more precise successors (see, for example, Brillet and Hall 1979) show that the principle of relativity is fairly securely established.

To test the slowing of watches it happens that nature gives us some watches that move very fast. Most of the particles that are created when an energetic particle hits a piece of matter are short-lived. The impacts occur naturally, when atomic nuclei at the top of the atmosphere are struck by cosmic ray particles from outer space, or they are produced at particle accelerators. Particles created in this way live for varying times before they disintegrate, but each species has a well-defined average lifetime and some of these have been measured when the particles are moving quickly as well as when they are at rest. The lengthening of the lifetime always agrees with Einstein's formula (see, for example, Bailey et al. 1977, Kaivola et al. 1985).

Nature is less generous with flying yardsticks than with flying watches, and even the slowing of a particle's decay is a rather special effect. At this point one might wonder whether relativity deserves its reputation if the effects are visible only in such unimportant ways. In fact, relativity theory is used all the time in physics, but not those parts of it that relate to clocks and yardsticks. Most important are the dynamical parts. The considerations relating to space and time are certainly correct, or very nearly so, and they have made us realize that intuitively simple notions of space and time contain subtleties we would not have expected, but they are rarely encountered in practice. Now we must see what happens to the laws of motion.

Dynamics in Spacetime

Relativistic dynamics starts with the equivalence of mass and energy. This emerges almost automatically from a mathematical treatment, but its meaning, and the hypotheses on which it is based, are perhaps most clearly seen in one of Einstein's simple arguments. Figure 14.4 shows a conceptual experiment of the same style as the ones we have been studying (Einstein 1950, p. 116). A block of material whose mass is m floats in interstellar space and we float near it. Someone directs two flashes of light at it, one from each side. Each flash carries an energy we will call $E/2$, which it deposits in the block, leaving it with a little more

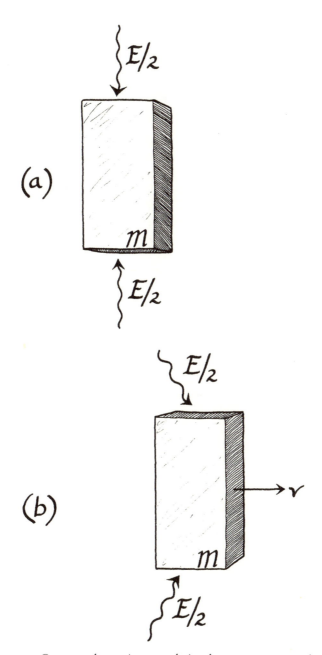

FIGURE 14.4. Conceptual experiment to derive the mass–energy equivalence.

energy than it had before. Perhaps the block gets warmer, perhaps the energy helps a little plant inside it to grow. Anyhow, we assume that the energy is conserved, that it enters the block and stays there. The flashes also carry momentum, and in Maxwell's theory the momentum of a flash of light is calculated by dividing its energy by c. Since the two flashes are equal and oppositely directed their total momentum is zero, and the block remains at rest.

Now we repeat the imaginary experiment with one change having nothing to do with the block or the flashes: we give ourselves a small velocity v toward the left. As we see it, the block now moves toward the right with unchanged velocity v during the entire experiment. Also, as the two flashes approach the block at speed c, their velocity has a rightward component v, which is v/c times c. Since momentum is parallel to velocity each particle's momentum $E/2c$ will also have a rightward component v/c times as great, $Ev/2c^2$, which it delivers to the block.

Let us assume that momentum, like energy, is conserved. Before the flashes arrive the total momentum of block and flashes is

$$p_{\text{before}} = mv + 2(Ev/2c^2) = mv + Ev/c^2.$$

What is the momentum after the flashes arrive? Though we might be tempted to say that it is again mv, that would be inconsistent with our assumption that momentum is conserved. It must be greater than before, and, since the velocity is unchanged, the mass must be greater. Let us call the new value M, so that the new momentum is Mv. Equating them gives

$$mv + Ev/c^2 = Mv \qquad \text{or} \qquad M = m + E/c^2.$$

Thus, under the assumptions made, any increase in the energy of the block increases its mass. What I specially like about this illustration of Einstein's is that the conservation laws of energy and momentum are so simply and clearly used. The relation between mass and energy is a verifiable fact only within the ordinary conceptual framework of dynamics.

I have called the argument an illustration because referring to such a special situation it can hardly be called a proof. But, like the conclusions regarding clocks and rods drawn from the imaginary experiments considered earlier, this one turns out to be generally true. One is at first inclined to say that the added energy turns into mass, but that is an unnecessarily complicated hypothesis. Rather, the relation should be interpreted as saying that energy and mass are different manifestations of the same thing—that traditionally, not understanding this, we came to measure energy in one kind of unit and mass in another, just as some people might measure a length in inches and others in centimeters. To translate from inches to centimeters we multiply by 2.54, but we would not say that at a certain moment inches turn into centimeters. Similarly, to pass from mass units to energy units we multiply by c^2. These cautionary remarks are here because there is a flood of printed confusion about the simple content of Einstein's relation $E = mc^2$.

If I heat my egg or wind my watch they get a little heavier, but because c^2 is such a large number the increase is immeasurably small. There are cases, though, in which it is detectable. An alpha particle is a nucleus composed of two neutrons and two protons, but its mass is not equal to the sum of the four masses; it is about 5×10^{-26} grams lighter, and the difference is not hard to measure. To interpret this mass defect, as it is called, we say that to transform an alpha particle into two neutrons and two protons we must take it apart by doing work against the forces that hold it together. The mass defect, multiplied by c^2, gives the amount of this work. Considerations like this belong to the everyday routine of nuclear physics.

If I roll a ball across the floor I exert a force and move my arm; I do work on the ball, and the energy that it acquires in this way is called kinetic energy. Because I have given it some energy I conclude that the moving ball is more massive than it was before I touched it. But not everyone agrees that the ball is now more massive than it was. Suppose my friend walks along the floor at the same speed as the ball. For her the ball is not in motion and its mass for her is the same as it was for me before I started it rolling. Thus the mass of the ball, like its diameter, is a relational quantity. In this respect it joins quantities like momentum and kinetic energy which are obviously relational. Perhaps you wonder whether *all* quantities in physics are relational. No. The amount of electric charge on a charged object, for example, has the same value no matter who measures it, and the same is true of the speed of light.

By exactly how much does the mass of a thing increase if we set it in motion? The calculation is not difficult but since it involves a little calculus it goes into Note I. The result, though, is simple: using the γ defined earlier,

$$m = \gamma m_o,$$

where m_o represents the mass the object has when it sits on a scale, not moving; this is called the rest mass. Clearly the mass m increases without limit as the velocity tends toward c, and since, as we have seen, energy is proportional to mass, this means that to accelerate any material particle to speed c we would have to supply it with infinite energy. Therefore c is an upper limit for speed, to be reached only by particles that are truly immaterial, i.e., have m_o exactly zero. The photons of light may very well be immaterial (Williams and Park 1971), and perhaps also the particles called neutrinos. (The possibility that some particles might be *created* with speeds larger than c has been considered, but experiments have failed to find any.) Speeds close to c have been produced in the laboratory. If we were to stage a race from here to the moon between an electron from the Stanford Linear Accelerator and a flash of light, the light would win, but by only a few centimeters. The gap can be further narrowed but it cannot be closed.

The relation between mass and rest mass leads to another useful result. Square both sides of the last relation and write the result as

$$m_o{}^2 = m^2 - (mv)^2/c^2,$$

or, writing the momentum mv as p,

$$m^2 = m_o{}^2 + p^2/c^2.$$

This shows how to calculate the extra mass (or energy) a thing gains from being set in motion with momentum p.

The last formula has a special case that will be needed in the next chapter. Multiply both sides by c^4 and write mc^2 as E:

$$E^2 = m_o{}^2c^4 + c^2p^2.$$

A massless particle is one whose rest mass m_o is zero or effectively zero. For such a particle the last formula gives the simple relation between energy and momentum,

$$E = cp.$$

The particle of light, called a photon, is massless and so this formula gives the momentum of a beam of light in terms of its energy, but it cannot be used to prove that photons exist because as I mentioned in proving the mass–energy relation, Maxwell's theory of electromagnetic waves yields exactly the same result.

The relations just written down illustrate an important principle, but before stating it I will show an example of it. Again, let us think about an imaginary experiment. You are equipped with a light source that shines in every direction. I come rapidly past you on my train and just as I reach the spot where you are you flash your light. As you see it, the light from the flash will spread out on the surface of a sphere whose radius, after a time t, will be equal to ct. Represented in terms of coordinates x, y, z, the distance of any point on the sphere from the center is given by the three-dimensional form of Pythagoras's theorem,

$$x^2 + y^2 + z^2 = c^2t^2,$$

or, in a form that will be easier to talk about,

$$c^2t^2 - x^2 - y^2 - z^2 = 0.$$

Now, how will I represent the same spherical surface? I cannot anticipate that my x', y', z', or t' will be the same as yours, but since both of us see the same sphere expanding in exactly the same way I define my sphere by exactly the same formula,

$$c^2t'^2 - x'^2 - y'^2 - z'^2 = 0.$$

Though the values of the letters differ, this particular quadratic expression has the same value, zero, for both observers.

Now let us look again at the fundamental relation that unites energy, rest mass, and momentum, writing it this time as

$$E^2 - c^2(p_x^2 + p_y^2 + p_z^2) = (m_oc^2)^2.$$

If you and I both look at some moving object and I am in motion we assign different values to E and the components of momentum, but m_o is the object's rest mass—it tells how massive the object thinks it is, and it has a fixed value. Comparing the last two expressions we see that they are similar in form: a certain quadratic expression in four variables, carrying one positive sign and three negative ones, has the same numerical value when calculated by two observers in motion with respect to each other. There is an analogy to this in ordinary geometry. Consider a point on a circle of radius r (Figure 14.5). The coordinates of the point satisfy $x^2 + y^2 = r^2$. Now locate the same point using coordinates tilted with respect to the first. The coordinates will be different but they will still satisfy the same relation since the point is still at the same distance from the center. The operative concept here is invariance: when the coordinates are rotated the expression $x^2 + y^2$ remains invariant.

The results we have just found can be expressed in the same way: under the change of temporal and spatial coordinates that occurs when one coordinate system moves and turns with respect to another the quadratic expressions written above remain invariant. In the case of simple geometry the invariance expresses a sort of homogeneity among the dimensions of space: space is space; coordinate axes are only a conventional

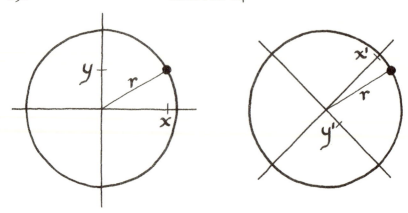

FIGURE 14.5. Since the circle is invariant under rotation, it is described the same in the two coordinate systems.

way of labeling its points and it does not matter how they are chosen, for the expressions representing geometrical relations within the space are unchanged by a reorientation of the coordinates. The theorems of geometry are no different in Australia from what they are in Iceland, even though Australian and Icelandic coordinates are oriented differently. Now we have the same situation in four dimensions, where the new dimension is time, or, more exactly, *ct*. Once more, invariance expresses homogeneity, indifference as to choice of coordinates, in—we need a fresh word. The word is spacetime.

You have been taken through this last example in order to make a single point: the idea of spacetime is not a trivial one. In the eighteenth century it was a commonplace that whereas the arena of geometry is a space of three dimensions, that of mechanics, where motion is added, requires the additional coordinate of time and the science of mechanics was said to be geometry in four dimensions. This amounts to pasting a time coordinate onto a space of three dimensions, but relativistic spacetime is a different idea, for here the time dimension occurs on the same footing as the other three and its separate identity is lost when a change of coordinates is made. *Except*, it is marked by a difference in sign. Thus it is always clear which coordinate represents time. Look for the positive sign. And this corresponds to an unmistakable fact of experience: we never have the slightest difficulty in distinguishing time from space. Our sensory and concept-forming systems are sensitive to the difference in sign, and even though Maxwell's equations and the relativistically revised formulas of dynamics are invariant under rotations of coordinates in spacetime, there is never any doubt as to how to pick out the direction of time.

Spacetime was the invention of two mathematicians, Henri Poincaré (1905) and Hermann Minkowski (1908). They saw that the mathematical symmetry of Maxwell's and Einstein's equations went beyond any symmetry of our perceptions, but probably it is just for that reason that Einstein only slowly became interested in spacetime, under the pressure of his work on the general theory, for initially he distrusted abstract mathematical arguments and used them sparingly. Minkowski, on the other hand, saw new mathematics waiting to be born, and his address in 1908 to a German mathematicians' meeting at Cologne opened thus:

> The views of space and time which I wish to lay before you have sprung from the soil of experimental physics, and therein lies their strength. They are radical. Henceforth space by itself, and time by itself, are doomed to fade away into mere shadows, and only a kind of union of the two will preserve an independent reality. (Lorentz et al. 1923, ch. 5)

After these familiar but stirring words Minkowski showed his audience how to draw four-dimensional representations of physical processes and how to represent graphically the transformation from the coordinates appropriate to one observer to those appropriate to another. His words serve to emphasize the strange dichotomy between the mathematics of relativity and the physical perceptions on which they rest, the first situated in spacetime and the second in space *and* time. Spacetime seems to be a mathematical entity useful for certain calculations but irrelevant to the description of human perceptions. If so, its position in the theory is equivocal. What is it, anyhow?

Is It Real?

With all the talk about measurements that come out differently for different observers, one may well ask whether anything about the phenomenon itself is really different or whether the effect is merely one of perspective. If I look at a boat end-on and then from the side the appearance changes but nothing has changed about the boat. On the other hand, a very fast particle decays noticeably more slowly than the same particle when brought to rest. That appears to be an objective difference, but is it too only an effect of perspective?

The answer is different according as we decide whether we are going to situate the process in space and observe it with a clock in hand or to situate it in spacetime. In the first case the difference is clearly objective. In the second, the whole process is represented *sub specie aeternitatis*; nothing changes because the temporal course of the process is part of the representation. (Writing of Parmenides in Chapter 2, I mentioned that Parmenides' world of discourse can be regarded as comprising the history of the world as well as its spatial extent—as representing the world in spacetime. That is the representation I am talking about here. Later, after Chapter 12, I heard Phyllis arguing the timeless nature of these representations of time.) Observers in different positions and states of motion will view from different directions this reality which exists in spacetime independently of any choice of coordinates. But in order to represent the event as it appears to them they must choose coordinates that distinguish space from time in a way that answers to their own perceptions. This determines what perspective they adopt and how the event appears to them. What in spacetime is an effect of perspective becomes an objective difference when viewed in the way that is forced on us by our ways of representing experience.

Finally, we might wish to return to the verbal battle that pitted Leibniz against the Newtonians. For Newton, space and time are prior to the created universe; we have seen him daring to situate them in the sensorium of God. For Leibniz they arise out of our experience with things and events in the world and are completely determined thereby. I pointed out that Leibniz's space is a good one for geometry but a bad one for defining distances and motions. The question that devastates it is "How do you distinguish a good yardstick or a good clock from a bad one?" Evidently the introduction of spacetime leaves the situation unchanged. A Leibnizian version of spacetime furnishes an economical mode of description but it is sterile if one wants to formulate dynamical laws in it, while Newton's version is still rather like a thing, composed of absolutes with prescribed properties that appear necessary for the formulation of dynamical laws but still cannot be touched by the question "How do we know they exist?" I mention this old question only because from here on spacetime will become more and more Newtonian, like a thing.

The ideas described up to this point spring from Einstein's paper, "On the electrodynamics of moving bodies" (1905c). In 1916 he published an extension of the theory (described below) that explains gravitation in terms of curved spacetime. He called it the general theory of relativity, and the one we have been discussing accordingly became known as the special theory.

The Principle of Equivalence

In the years following the publication of his paper of 1905 Einstein remained bothered by a paradox he saw in it. The principle of relativity on which it is based says that if we travel in uniform motion through space in a closed box, no experiment performed in the box will give us any information about the motion. But let the box accelerate and we know at once. Thus it seems that the metaphysical apparatus of Newton's absolute space has not been banished along with the ether, and that it is just as necessary in the new theory as it was in the old one.

In a slashing attack on Newtonian absolutes the Austrian physicist Ernst Mach (1838–1916) denies that an argument of this kind carries any necessity (1883, ch. 2). He points out that the universe is full of stars and that it is useless to speak of moving boxes as if they were alone in space. It is an experimental fact that if the box rotates or accelerates its motion with respect to the stars, travelers inside would know it. Anything we may say about motion with respect to absolute space, of which we know nothing, we may also say about motion with respect to the stars, which we know well; therefore absolute space is a superfluous hypothesis.

It was in 1907 that Einstein, pondering this and related matters while still working in the Patent Office, had what he later called the happiest thought of his life. It will be easier to talk about this thought if we once more put an observer into an imaginary box that floats in space. In this box nothing has any weight, and if the box moves uniformly with respect to the stars its inhabitant would not know it. Suppose that while its inhabitant was playing with some of her floating possessions, the box is given a gentle acceleration in some direction which we may call upward. As the box begins to move with respect to the observer and her playthings she notices that the floor is getting closer, and after a moment there is contact. Not knowing that the box has been set in motion she deduces that a force has suddenly attracted everything in the box toward the floor, and once settled on the floor she feels the force as we all have felt the extra weight when an elevator rises. She knows nothing of the origin of the force but notices that it has a striking property: it gave everything in the box exactly the same downward acceleration. Then she remembers having read that there is such a thing as gravity, even if she has never experienced it, and that, as John Philoponus and Simon Stevin and Galileo had in turn discovered, under its action all objects acquire the same downward acceleration. She reasons that a gravitational field has somehow been created inside the box.

If one experiments only with material objects the effects of a uniform gravitational field are absolutely indistinguishable from those observed in a uniformly accelerating box. But what about experiments with light? The observer looks up Maxwell's paper; it says nothing at all about how gravitational force affects an electromagnetic field, and she can only guess. The simplest guess is this: the effects of gravity on light, as on matter, are indistinguishable from those observed in an accelerating box. If this is so, then just as the principle of relativity states that it is absolutely

impossible to distinguish rest from uniform motion, one can formulate a
principle that says it is absolutely impossible to distinguish between the
local effects of gravity and those observed in an accelerating system.[1]
After some preliminary explorations of this idea (1907, 1911, trans. in
Lorentz et al. 1923), Einstein formally named it the principle of equivalence (1912) and suggested that it has testable consequences.

Figure 14.6 shows a consequence that illustrates the principle. It is
designed to show the effect of gravity in bending a beam of light. Let a
flash of light be projected across the room, hitting the wall at A. Then let
the room be given an upward acceleration equal to the ordinary acceleration of gravity, g, and a second flash be projected. It takes a time $t = l/c$
to reach the wall, during which the acceleration of the room carries it
upward by a distance

$$s = \frac{1}{2} g t^2 = \frac{g l^2}{2 c^2}.$$

Put in any reasonable numbers and the result comes out much smaller
than the diameter of an atomic nucleus, impossible to measure, but it
provides an answer in principle to the question of how a gravitational
field deflects a beam of light. For the effect to be detectable the gravitational field would have to be much stronger and the room much larger
than the ones we know. In 1911 Einstein showed how the effect might be
observed using the immense size and gravitational field of the sun. Wait
until there is a solar eclipse and then observe the apparent shift in position
of a star whose ray into the telescope grazes the sun's surface (Figure
14.7). The calculated angular displacement, α, is very small, only 0.83
seconds of arc, but it might be measurable. Luckily, no one tried; the
calculation was wrong. The true effect is twice as large, but this did not
appear until four years later when the general theory was finished. It has
now been verified by observation.

Another imaginary experiment leads to the conclusion that gravitational fields affect the motion of a clock. Imagine a room of height h, in
which a source of electromagnetic waves has been installed at the ceiling
and a detector on the floor. Let us compare the frequency of a flash of
light received at the floor with the frequency given it by the source. If the
effect of gravity is the same as that of accelerating the whole room
upward, the floor will be rising at an increasing rate while the light is in

[1] The word "local" is necessary because if one were to measure ordinary gravitational
forces in a room very accurately one would find that because the floor of a room is closer to
the center of the earth than the ceiling is, the forces are a very little stronger at the floor.
This effect cannot be mimicked by accelerating the room.

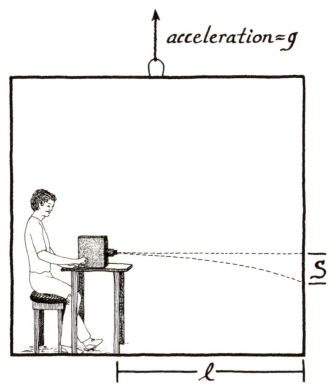

FIGURE 14.6. The principle of equivalence. Light projected toward the wall of a room that is accelerating upward *or* in a gravitational field should bend toward the floor.

transit. The transit time is (nearly) h/c. If g is the acceleration, the floor will acquire an upward speed gh/c during this time. Whenever the detector of a light wave moves toward the source with velocity v, the frequency detected increases by the fraction v/c; this is an example of the Doppler effect. In this case the fraction is gh/c^2.

Thus the number of vibrations received at the floor each second is consistently greater than that emitted at the ceiling. The effect is very small but it poses a puzzle: suppose that the room is a tower 20 meters high and that yellow light is used. During the course of an hour 34 more vibrations will be received at the floor than were emitted at the ceiling. Where did they come from? Since the waves travel continuously from ceiling to floor the question has no answer. We must have said something wrong. Assuming the principle of equivalence and the elementary physical ideas involved we are left facing the assumption that the clock that measures the frequency at the floor runs at the same rate as the one that measures

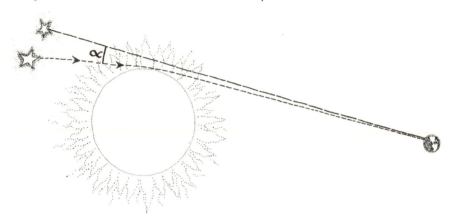

FIGURE 14.7. An effect of a gravitational field. Einstein predicted that light from a star passing close to the sun should be bent so that the star appears displaced in the sky. If the observation is optical it can be made only during an eclipse of the sun, but radio waves from quasars can be observed at any time.

the frequency at the ceiling. If that is not the case, if the clock at the floor runs more slowly by a fraction gh/c^2, then the number of vibrations counted during a given clock interval will be the same at both places.

As early as 1907 Einstein had arrived at this conclusion and predicted that (reversing the light's direction of motion) light produced by a given atomic mechanism in the sun will reach us at a frequency lower than that which would be produced by the same mechanism on earth. The effect is very hard to measure in the sun because at its tumultuous surface the currents of gas move so fast that the Doppler effect masks the effect sought. There are, however, small stars called white dwarfs at which the gravitational force is much stronger and the gases much calmer, and in some of these the shift toward lower frequencies is clearly seen.

In 1972, in one of the sweetest experiments of our era, J. C. Haefele and R. E. Keating, of the Caterpillar Tractor Company in Peoria, Illinois, flew twice around the world in commercial airliners, once eastward and once westward, accompanied by some very accurate caesium maser clocks. Before and after each trip they compared the readings on the traveling clocks with the accurate time signal from the Naval Observatory. There were two reasons to expect that the readings would differ: the traveling clocks and the Observatory clock were in different states of motion and also their average altitudes differed. (Remember that the earth turns rapidly; this adds to the speed of an eastward flight, while with respect to the nonrotating universe the airplane in a westward flight actually flies backward.) Luckily for the experiment, the special-relativity effects of motion are of about the same size as the effects of height, and by comparing data for the two flights (1973) the experimenters could measure both effects at once. Theory, calculated from the airplane's log, is compared with observation in Table 14.1. The effects described by relativity theory are real effects. The clocks really are affected. A more precise version of this experiment (Vessot et al. 1980) verified the theoretical prediction with an accuracy of 70 parts in a million.

Observations confirming the principle of equivalence cast new light on the question raised at the beginning of this section: what does it mean that nature does not tell us whether we are moving uniformly through space, but tells us at once if we begin to accelerate? We see that the assumption underlying the question is false; nature does not tell us we are accelerating, for the same effect can be produced by a gravitational field. What is a gravitational field? Our imaginary experiment provides part of the answer: it is an influence produced in the neighborhood of a massive object that affects the rate of a clock.

Einstein seems to have arrived at this point and then turned his attention for several years to finding a rational explanation for some of the quantum phenomena to be discussed in the next chapter. Only at the end of 1910 did he finally confess that he could not do it (Pais 1982, ch. 10) and return once more to problems of gravity.

The Twin Paradox

As soon as the special theory was published a number of people drew attention to a disturbing consequence that became known as the twin paradox. Suppose there are twins Alice (A) and Bertram (B), and that A goes on a long space voyage at high speed, say a tenth of the velocity of light, while B stays home. According to the results we have derived above, A's clock is slower than B's, but A's clock is not necessarily a mechanical clock. Her body is also a clock, and when it arrives home it will be younger by every test than B's. This is odd but not paradoxical. The paradox is in the objection that if the effects of absolute motion are unobservable and only relative motion can be detected, one might just as well say that the earth with B on it went away from the spaceship and came back, so that A would be the younger. Thus the argument seems to require A on her return to be both older and younger than B.

The key that unlocks the twin paradox is the fact that A is obliged to move nonuniformly during at least part of her trip, while B does not accelerate at all. A detailed analysis taking this acceleration into account (Darwin 1957) shows that A does indeed return younger, just as predicted by a naive application of special relativity. Now recall that the mechanical

TABLE 14.1
Time Gains and Losses in Flights around the World

| | Time gains (units of 10^{-9} seconds) | |
	Eastward	Westward
Predicted		
gravitational effect	144 ± 14	179 ± 18
effect of motion	-184 ± 18	96 ± 10
total	-40 ± 23	275 ± 21
Observed	-59 ± 10	273 ± 7

effects of accelerated motion can be mimicked by the effects of a gravitational field and that a gravitational field affects a clock. Can the difference in the twins' ages be explained as an effect of the field? Figure 14.8(a) shows the situation as viewed by earthlings: A travels a distance L at speed v, turns around, and comes back. By B's reckoning the time required for the journey is

$$t_B = 2L/v.$$

The formula for the slowing of moving clocks states that the time as measured by A is

$$t_A = \gamma t_B = \frac{1}{\sqrt{(1 - v^2/c^2)}} t_B \approx (1 + v^2/2c^2)t_B,$$

where the last expression is an approximation that is very close when v is much less than c. The difference between the times is thus

$$t_B - t_A = vL/c^2$$

(for any forseeable space trip, it seems reasonable to expect that this is a very small number).

Now let us look at the same process from the standpoint of A, who watches B disappear into the distance, reverse, and come back [Figure 14.8(b)]. There is this difference between the experiences of A and B: A seems to experience a force halfway through B's journey, while B, naturally, does not. Assuming that A moved while B stood still this was easy

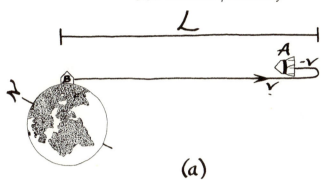

(a)

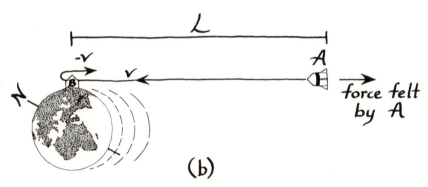

(b)

FIGURE 14.8. Resolution of the twin paradox. When A returns home from a space voyage she will have aged less than B. If B considers himself at rest he explains the fact by the kinematic effect. If A considers herself at rest the explanation involves the gravitational effect as well, but the result is the same.

to understand as a consequence of the reversal of her motion, but if A believes she is not in motion she does not interpret this feeling as having been produced that way. Instead, she concludes it must be the result of a mysterious gravitational field that acts for a short time t_a and then ceases. When A calculates the reading on B's clock the effect of this field must be taken into account. Let a be the acceleration whose effect A feels. At this moment B is at a distance L and higher in the imagined gravitational field than A; therefore the rate of B's clock, by the formula derived above, is increased by an amount aL/c^2, and the clock's gain in the time t_a during which the field exists is

$$t_{\text{gain}} = (aL/c^2)t_a = 2vL/c^2,$$

the last equality reflecting the fact that the acceleration a acting during a time t_a serves to reverse B's motion from v to $-v$. The rest of the calculation goes through as before. According to A, B's clock will have lost vL/c^2 as a result of its motion, but will have gained twice as much while the "gravitational field" was present. The result is a net gain of vL/c^2, so that B returns older than A, in full agreement with the earlier calculation.

This argument was intended to show how Einstein's hypothesis does away with Newton's absolute space. The observed effects and the laws responsible for them take the same form in both coordinate systems, but what is explained as an effect of acceleration in one of them is explained as an effect of gravity in the other. At the same time, the argument shows something else: that gravity is not just another force along with electricity and magnetism; it seems to play some special part in the description of nature.

The General Theory

Thus far we have been able to draw conclusions from the principle of equivalence by inventing little experiments and arguing from them as if they could be carried out, but that is not the same as constructing a theory. In 1911 when Einstein returned to the problem of absolute space he had two beliefs to guide him: there were experiments that seemed to show that the formulas of his earlier theory of relativity were correct, and he had confidence in the principle of equivalence according to which the gravitational field surrounding a massive body must manifest itself (among other ways) by slowing a clock. These were the only new physical notions. The rest of the theory had to be constructed from conventional notions, from hypotheses, and from mathematics.

What emerged after four years of very hard work was a theory in which the presence of gravitating matter distorts measurements of space as well as time—one says that it produces a curvature of spacetime. The idea of gravitational force is not part of the theory. Its magnitude can be calculated if one likes, but the curved paths of planets and fly balls are not attributed to any force. Rather, they flow from a minimum principle closely analogous to that of Hero and Fermat: an object traveling from point P to point Q in spacetime (meaning that, as before, the times as well as the places of departure and arrival are specified) follows the path *through spacetime*, not through space, that using an appropriate definition of length is either longer or shorter than any other. This is, in Aristotelian terms, its natural motion. There is no talk of forces unless one seeks to impede the motion. If I hold an apple in my hand I am blocking its natural motion and that requires a force, but if I drop the apple it experiences no force until it hits the ground.

To describe the curvature, Einstein had available to him a body of mathematics relating to curved spaces laid down by Karl Friedrich Gauss and Bernhard Riemann in the nineteenth century and skillfully developed by several Italian mathematicians. Without this preparation Einstein, for whom pure mathematics never came easily, would hardly have been able to produce what he proudly named the general theory.

At the core of the general theory are equations that tell how gravitating matter produces curvature in spacetime. Let me give a very rough analogy. Consider a sheet of rubber, stretched tight and fastened around its edges to a frame. Now push the sheet at some point. The surface will become curved, in a way determined by the magnitude and direction of the push, by the size and shape of the frame, and by the particular properties of rubber. Experts in the theory of elasticity know equations from which, when these things are specified, they can work out the shape that the stressed sheet will assume. The calculation of curvature in spacetime contains the same ingredients except that here one is dealing with curvature in a manifold of four dimensions rather than two, so that visual imagination is of little help. One must specify the nature and location of the gravitating object that produces the curvature, make suitable assumptions as to the conditions around the edges of the region, and assume that spacetime has some properties of elasticity analogous to those of rubber that enable it to regain flatness, if it was initially flat, far away from the point at which it is disturbed.

I write this analogy precisely because it raises unpleasant questions. How can spacetime be like rubber? How can it be like any kind of a thing, when it is composed of the dimensions in which we experience the world? Well, it is more than that. It is the exerter of weight, the bender of planetary orbits; in fact it is a field, and fields have properties. One property of spacetime is curvature. The spacetime we experience is curved enough so that it makes big stones hard to lift but not enough so that rays of light are appreciably bent, and we most easily think of it as flat and explain weight by introducing the hypothesis of gravity. Would it not be easier to say that gravitation is a field that inhabits spacetime? No; that is the old idea. In the general theory the field *is* spacetime.

A rubber sheet has dynamic as well as static behavior. If you move your finger at the point of contact waves will spread out, and in the same way the dynamics of curved spacetime allows wave motions that carry momentum and energy. Remember Maxwell's theory, which at first represented electric and magnetic fields as states of stress and motion in the ether and predicted electromagnetic waves. Einstein's spacetime also obeys equations of motion, and they predict that a rotating system like a double star will radiate gravitational waves. In recent years a double star has been found in which the energy loss can be observed, and it is as calculated by the theory (Taylor et al. 1979, Weisberg and Taylor 1984). Many attempts have been made to detect gravitational waves arriving at the earth from outer space, but none so far has succeeded (see Davies 1980).

The general theory yields planetary orbits that are almost but not quite the same as Newton's. The differences are mostly too small to detect, but one is important: a tendency for planetary ellipses to rotate slowly around the sun instead of standing still as Kepler and Newton thought they did. Actually, each orbit rotates quite a lot, dragged around by the combined influences of the other planets. When the extent of this dragging is calculated using Newtonian theory there is a small residual effect that is not accounted for, and is most conspicuous in the orbit of Mercury where it amounts to a swing of 43 seconds of arc per century. This is exactly the correction that relativity gives to the Newtonian theory.

There is also the effect already mentioned, in which gravity pulls on the beam of light from a distant star as it passes the sun (Figure 14.7). Einstein's first calculation, taking into account only the curvature of time, gave an angle of about 0.83 arc seconds. In the completed theory spatial curvature contributes an equal amount. The precession of Mercury's orbit had been known long before Einstein, so that it was possible to argue (and some did) that he had cobbled the theory together so as to account for it. But the predicted deflection of light was an effect that had never been measured, and when very delicate observations carried out by British astronomers in the eclipse of May 1919 verified the prediction, newspapers announced the end of the Newtonian era. Much against his will Einstein became a public figure, and in science, in his own word, something of a monument.

Nothing has happened in the last seventy years to force changes in Einstein's ideas, but we now see those ideas in a wider context. Apparently general relativity is an approximation to a more complete theory of spacetime that we do not yet possess. Maxwell's electrodynamics fits beautifully into the modern quantum theory of fields, and the resulting quantum electrodynamics is the most precise physical theory we have. General relativity does not fit. Consider what this implies for scientific progress: a theory that agrees perfectly with experiment may stand as a memorial to human genius, but it does not tell us what to do next. Quantum electrodynamics is almost complete and apparently almost perfect. General relativity is not. Efforts to find a theory of spacetime that agrees with Einstein's in a certain approximation and at the same time fulfills the requirements of quantum theory have led to a theory called supergravity.

It is far from complete but people who work with it are very optimistic. It just might work, and if it does it will unify at least several of the fundamental fields into a single theory. That is what Einstein wanted to do.

Finally, what about the ether? It is often said that Einstein put an end to it, but that is not what he thought. In 1920 he delivered a lecture, "Ether and relativity," at the University of Leyden. His closing words were:

> Recapitulating, we may say that according to the general theory of relativity space is endowed with physical qualities; in this sense, therefore, there exists an ether. According to the general theory of relativity space without ether is unthinkable; for in such a space there would not only be no propagation of light, but also no possibility of existence for standards of space and time (measuring-rods and clocks), nor therefore any space-time intervals in the physical sense. But this ether may not be thought of as endowed with the physical quality characteristic of ponderable media, as consisting of parts which may be tracked through time. The idea of motion may not be applied to it. (1922, p. 23)

The situation has changed little since Einstein said these words. Nobody speaks of ether any more, but to say that the apparently empty space between two objects has no properties at all would contradict everything we know. Empty space is now full of interacting fields; a little about what we think these fields are will be told in the next chapter.

One More Dimension

The general theory explains the action of gravity by curving the spacetime in which we live. There are other kinds of force. Have they too some explanation in terms of geometry? It would seem that we have used up the geometrical possibilities of spacetime in explaining only one kind of force. What more can be done? Einstein spent the last forty years of his life searching for ways in which the general theory could be extended so as to explain the forces of electricity and magnetism. His efforts were not very fruitful, but one, which he worked on for a while but finally abandoned, has recently turned into a major industry. This is the Kaluza–Klein theory.

In 1919 a German mathematician named Theodor Kaluza (1885–1954) wrote to Einstein about an idea for combining the theory of electromagnetism with that of gravity by representing electromagnetism as curvature in a fourth dimension of space, so that spacetime would have five dimensions (Kaluza 1922). He showed that Maxwell's equations can live beside Einstein's in such a spacetime, the two not really unified but at least sharing the same domain. At once a question leaps to the mind: if space has four dimensions why haven't we known it all along? In 1926 Oskar Klein (1894–1977) clarified Kaluza's work and introduced a new hypothesis: we do not see the extra dimension because it is rolled up into a little cylinder. How this can happen is illustrated with an analogy. A sheet of paper has two dimensions. Roll it up tightly into a narrow tube like a soda straw. Seen from a distance it resembles a line, a one-dimensional domain, but on closer inspection we see that it has thickness as well. Klein's hypothesis is that the thickness in the fourth spatial dimension is too small for us to perceive. Where is the extra dimension to be found? Cut the paper tube at any point in its length: the circular cross section is there and the extra dimension measures distances along its circumference; it is present at every point along the tube. I think only a trained mathematical imagination sees how this simple construction can be generalized to more dimensions, but writing down the formulas is quite straightforward. Perhaps the trouble with the Kaluza–Klein theory is that in it Einstein and Maxwell lie side by side without a murmur of protest. Nothing needs to be changed in either theory, and therefore no new physics results from the combination. Lacking a microscope that shows us the extra dimension, we have a theory whose truth or falsity cannot be tested by experiment. In Chapter 17 we shall again encounter the idea that the spacetime in which physics plays its games has more dimensions than four, the extra ones being everywhere present but rolled up, too small to detect.

Einstein moved on from Kaluza–Klein and proposed a sequence of theories designed to unify gravitation with electromagnetism, but today they are not considered to be among the promising lines of approach. One reason for Einstein's failure was that, as we shall see, he did not trust the quantum theory that was being developed at the same time and therefore did not avail himself of the insights it offered. Another was that he chose the wrong directions in which to generalize the mathematics of spacetime. Today the situation looks very different from the way it did in 1955 when Einstein died. The problem of generalization is not solved; there is no definitive theory, but geometry is being extended in new ways. It seems Einstein was right in linking physics to geometry but wrong, after 1915, in the paths he chose to follow. In Chapter 17 I will try to explain some of the new ideas.

CHAPTER 15

Very Small and Far Away

Kick at the rock, Sam Johnson, break your bones:
But cloudy, cloudy is the stuff of stones.
—RICHARD WILBUR

EINSTEIN'S work was a useful lesson for people who thought they understood at least the simpler aspects of physical reality—who thought they knew, for example, that the mass and size of a thing are fixed quantities, and that it makes sense to say it is the same time everywhere in the world. But even if relativity theory introduced some cautionary rules for the interpretation of measurement and some new philosophical notions, it dealt with the things of the world in conventional terms. The truths of special relativity were solidly grounded in experimental fact; Einstein needed only to get rid of certain universally held prejudices about space and time in order to construct his theory by careful argument from what was known. The general theory required much more than that; it took a great leap of imagination to see the fabric of spacetime as a dynamical system and not just the stage where events take place. But even before Einstein took that leap, the most obvious fact to be explained, gravity, was quantitatively understood.

The inner nature of atoms is a truth that lies in the abyss. We have no direct experience that reveals that truth, nor is there any straight road leading to it. Every experiment needs interpretation in the light of one theory or another, and at the beginning of this century the most direct approach was by the relatively crude technique of bombarding atoms with electrons and studying the light that was produced. It is as if one tried to learn the structure of a machine, say a typewriter, by listening to the noise it makes when it is thrown downstairs. But the difficulty was even greater than that, for if a typewriter is taken apart each of its pieces turns out to be of ordinary material, specially shaped. Of course everyone expected that this would be the case when atoms were finally understood: it would turn out that they are structures made of little bits of matter bound by certain forces, too small to see with any microscope, but still one could imagine making some magnified model of it to show what it is like. That is exactly what turned out to be a false hope.

For more than a decade after Ernest Rutherford proposed that an atom is like a little solar system, people tried to understand it by scaling it up in their imaginations as one understands the solar system by scaling it down. At length the founders of quantum mechanics realized that this was not to be—that just as ordinary notions of nearly flat spacetime do not tell us how to think about its curvature over cosmic distances, our experience with marbles and tennis balls does not prepare us to think about electrons. We have found that each realm of nature must be represented to our minds in terms appropriate to its scale of magnitude, and even commonplace ideas like place and motion must be redefined for each radically new scale. It is certain that the process has not finished. Particle theory forces us to reconsider even such elementary concepts as the dimensionality of space when the scale becomes small enough, so that the considerations of this chapter on atoms must be thought of as applying only to objects of intermediate size. Still, they required a remarkable amount of penetration and ingenuity, and the challenge they offer to our familiar notions of reality is very great. They suggest the extent, but not the nature, of the conceptual revolutions that will be required as we move further down the scale.

Discreteness in Nature

The events of this chapter begin late in the year 1900 when Max Planck (1858–1947), professor of physics at the University of Berlin, decided to try to understand the electromagnetic radiation that we can feel being given off by a hot object or, if the object is as hot as the sun or the filament of a light bulb, that we can both feel and see—it is called black-body radiation. By ingenious trial and error Planck put together a formula that correctly related the observed color and intensity of the radiation at different temperatures; then he thought very hard to see what had to be assumed if he was to derive it.

Concerning what Planck assumed (1900) there are a majority and a minority opinion. The majority opinion is found in most books that discuss the history of twentieth-century physics (for a concise statement, see M. J. Klein, in Weiner 1977). It has Planck assuming that when an atom emits radiative energy at frequency v, it emits it not in a smooth wave but in multiples of a definite quantity of energy, proportional to the frequency of the radiation. The proportionality is now expressed by the celebrated formula

$$E = hv,$$

where h, the constant of proportionality known as Planck's constant, is a very small number, for this all happens on a microscopic scale of magnitudes.[1] Planck later called this amount of energy a quantum, a word that in late Latin referred to a portion of something, and which was commonly used in Germany in that sense.

Many who read Planck's original papers carefully must be surprised (I know I was) that Planck never makes any such claim explicitly, though it is perhaps possible to read it into his words. Instead, he explains how he borrows an idea of Boltzmann's to define the internal states possible for an atom—though at that time nobody had more than a vague idea of what they are. The minority position (Kuhn 1978, 1984), to which I subscribe, is that this computational technique is Planck's discovery. He said as much in a letter written long afterward, in 1931: "This was a purely formal assumption and I really did not give it much thought except that, no matter what the cost, I must bring about a positive result" (Hermann 1971, p. 24). That is hardly the way one would recall the making of a precedent-shattering physical hypothesis. That hypothesis, I think, was made by Einstein.

Einstein's paper (1905a) was concerned with consequences that followed from Planck's formula. It concluded that the observed facts are most readily understood if the radiation is pictured as consisting of massless particles, later called photons, each carrying energy hv and a corresponding momentum. Einstein did not try to explain how his picture squared with the accepted picture of radiation as waves in the electromagnetic field, which had already been abundantly verified. To emphasize that he was not making a hypothesis that light really consists of particles, but only that certain facts are more readily understood if one thinks of it that way, Einstein titled his paper "On a heuristic point of view concerning the creation and transformation of light." With the unusual word "heuristic" he suggested only that the point of view might be useful in

[1] Numerically, $h = 6.626 \times 10^{-27}$ erg-sec. An erg is a very small amount of energy; one might think of the energy expended by a flea walking one millimeter up a vertical surface.

further work. At the end of the paper he pointed out that picturing radiation as a rain of photons makes it easy to understand the photoelectric
effect, which is the emission of electrons from a metal surface when light
strikes it. According to his picture an electron is emitted only if a photon
hits it. This has immediate consequences: there should be no time delay;
electrons should be emitted as soon as the light hits the surface. Doubling
the intensity of the light should double the number of electrons without
changing their average energy, whereas since the energy of a photon is
proportional to the frequency of the light, increasing the frequency
should increase the energies of the electrons that get knocked loose. This,
it turned out, is exactly what happens.

During the next decade delicate and exact experiments verified each of
these predictions, but nobody knew what to do with the paradoxical
notion that had generated them. The equally radical theory of relativity
had been adopted by most active physicists with surprising speed, but
these same people were generally agreed that the photon concept was a
mistake. As late as 1909, while lecturing at Columbia University, Planck
gave a lucid account of his ideas without mentioning light-quanta at all
(1915), and in 1913 the letter from Planck and other distinguished physicists that recommended Einstein for membership of the Prussian
Academy concluded with the words: "That he may sometimes have
missed the target in his speculations, as, for example, in his hypothesis of
light-quanta, cannot really be held too much against him, for it is not
possible to introduce really new ideas without sometimes taking a risk"
(Pais 1982, p. 382). Nevertheless, in 1922, this was the only work that
was cited specifically when Einstein was awarded the Nobel Prize. Apparently the members of the Royal Swedish Academy had not quite satisfied
themselves about relativity (Pais 1982, ch. 30).

As time went on the hypothesis of light-quanta—whether or not one
thought of them as little bullets—began to be taken seriously, and in a
few years their existence was again demonstrated in a particularly convincing way. Arthur Compton (1892–1962), in St. Louis, performed an
experiment of which the obvious explanation was that X-ray photons
bounced off electrons as one billiard ball bounces off another. Each particle was assumed to carry energy and momentum, and Compton's calculation assumed only that these quantities are conserved in the collision.
Both the deflected X-ray and the recoiling electron could be studied, and
every measurement was explained by the literal and simple-minded picture of a miniature billiard shot (Compton 1922). Nothing was lacking
to prove that light consists of particles. On the other hand, Young's
experiments showing that light is a wave had been refined and repeated
during more than a century, and the evidence was still perfectly convincing. The situation was perhaps unprecedented in science: simple

experiments, requiring no subtleties of interpretation, yielded clear but entirely contradictory conclusions. It was hard to imagine that either could be wrong or that both could be right. I shall tell presently how modern physics resolves the paradox, though not to the satisfaction of absolutely everyone.

Atoms Are Real

I have already suggested that for people with the proper amount of scientific skepticism in their characters it was not easy to believe the claim that matter is composed of fast-moving atomic particles. Studies of the thermal properties of gases by Maxwell and others had made the idea more plausible, but there was always the possibility that the same facts could be explained with less radical assumptions. At the beginning of the twentieth century physicists were experiencing a crisis of conscience. Some had decided that in attributing physical reality to the electromagnetic field—assuming that it exerts force and carries energy—as well as in giving plausible explanations of physical and chemical phenomena in terms of atoms that no one had ever seen, science was building its house on sand. A small number of eminent workers, including the chemist Wilhelm Ostwald and the physicist Ernst Mach, went so far as to repudiate the entire atomic doctrine, except possibly as a computational strategy, but for most people, viewing the great range of phenomena that the atomic theory explained, this was too much. The difficulty was that atoms are so small that even now they can only be photographed through very sophisticated microscopes in very special situations. At that time even the true believers assumed that atoms live in a country so far from our own that they are forever inaccessible to direct observation. Still, there might be some means by which observation could carry the observer at least a little way toward the distant land.

In 1905 Einstein wrote yet another paper (1905b) to add to the ones I have discussed. In it he studied some consequences that follow if one assumes that liquids and gases are composed of fast-moving atoms. In this paper and in his doctoral dissertation published soon afterward (1906) he suggests that even if atoms are too small to see, there might be objects at an intermediate level of size that are small enough to be influenced by the fluctuating movements of atoms but large enough to be observed. As an example, he shows that a particle suspended in a liquid and observed with a microscope would be seen to dance around under the influence of random fluctuations in pressure that are to be expected if the liquid consists of atoms in rapid motion. Though he did not know it when he worked out the theory, the phenomenon was already well known to microscope users as Brownian motion. The ceaseless trembling of grains of pollen suspended in water had sometimes been interpreted by biologists as a sign of life, but Einstein's work showed that it resulted from molecular fluctuations. During the next decade measurements were made that completely supported the theory's predictions, settling forever the question that Leucippus had raised: is the hypothesis that matter is made of atoms necessary for explaining the phenomena of nature?

Inside an Atom

In the 1880s, largely through the work of Joseph John Thomson at Cambridge, it became clear that atoms somehow contain within them negatively charged particles that are very much less massive than themselves. These soon became known as electrons. Since an atom as a whole (unless something has been done to it) is electrically neutral, this implies that an atom is composed of at least two constituents, and that what is not electrons bears a positive charge. Early in 1911, reasoning from some experiments done in his laboratory at Manchester, Ernest Rutherford (1871–1937) deduced that almost all of an atom's mass is concentrated in a very small, positively charged object at its center, which soon became known as the nucleus. Figure 15.1 shows the principle of one of the experiments. Alpha-particles emitted by a radioactive substance were directed toward a very thin gold foil (alpha-particles are small nuclei of helium, thousands of times more massive than electrons, and they move very fast). It was expected that they would probably act like bullets hitting a board, passing straight through with diminished speed, but in fact many were deflected from their original paths by a degree or so, which was hard to understand, while a few were deflected through very large angles. "It was almost as incredible," Rutherford wrote later, "as if you fired a fifteen-inch naval shell at a piece of tissue paper and the shell came right back and hit you." Only interaction with a small and much more massive body could have turned the alpha-particles around. Rutherford guessed that an atom consists of electrons orbiting a massive nucleus as planets orbit the sun. Assuming that the alpha-particles interact with the nucleus with a purely electrical force, Rutherford (1911) calculated the statistical distribution of deflections through various angles that was to be expected, and agreement with the measurements seemed to validate his model.

This was one kind of evidence that related to the structure of atoms. There was other evidence, but it was not so easy to understand. There was the chemical knowledge that had been accumulated over the centuries, plainly full of information if only one knew what it meant, and there were studies of the light given off by atoms of various gases and vapors,

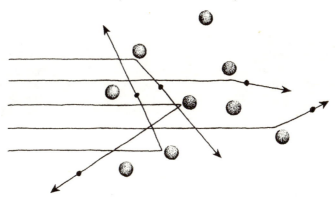

FIGURE 15.1. The discovery of the nucleus. Alpha particles bounce off nuclei in gold foil.

which showed, in the spectroscope, characteristic groups of wavelengths. These were hard to interpret because they showed very little order or regularity, either in wavelengths or intensities, though there was one exception, hydrogen, in which the wavelengths of the lines seemed to follow a regular pattern (Figure 15.2). This pattern seemed to indicate some mechanical regularity in the atom which produced it, but interpretation was far from obvious.

In 1911 Niels Bohr (1885–1962), a research student from Denmark, arrived in Manchester to work with Rutherford. There he learned of the new atomic model and began to think about it, probably more seriously than the proposer had, and in a few months he put together some ideas that explained the exact pattern of the hydrogen spectrum and began to clarify an enormous range of other phenomena.[2] Up until then the diversity of the physical and chemical properties of the elements had been simply part of the marvelous variety of nature. Regularities were known (summarized in Dmitri Mendeleev's arrangement of the elements in order of increasing mass), which showed that certain properties recurred periodically; now Bohr was able to explain these periodicities qualitatively, at about the level of a present-day high-school course in chemistry. For the first time it began to be clear that there is a system, easy to visualize even if it left some questions unanswered, that embraced all the known elements and predicted the occurrence and basic properties of several that were not yet known.

Bohr's theory was certainly not a fundamental theory in the sense that Newton's and Maxwell's theories are fundamental. Like the theory of photons it was heuristic, borrowing some bits of Newton, Maxwell, and

[2] For the history of these developments, see J. L. Heilbron, in Weiner (1977).

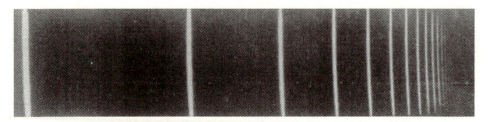

FIGURE 15.2. The spectrum of hydrogen, from red to near-ultraviolet.

Einstein and rejecting the rest according to rules justified only by the results. In Newtonian dynamics an electron circles around the nucleus in an orbit determined by the various fortuitous influences to which it has been exposed. According to Maxwell's theory it will not stay in any orbit for long, since accelerated electric charges emit radiation and so it will spiral into the nucleus, emitting more and more radiation at higher and higher frequencies. Bohr saw that if the Newtonian idea were taken seriously no atom would preserve its orbital motion for more than a moment, and that Maxwell's ideas about radiation, while fully validated on the scale of laboratory apparatus, cannot possibly apply to Rutherford's model of the atom. Instead, he put together a theory out of his own ideas and others that were in the air at the time. In this theory an atom can exist only in its ground state, the state in which it is ordinarily found, and some specific short-lived states of higher energy. It can be driven into an excited state by impacts of photons or other particles, and when it loses energy to return to a lower state it gives off a photon carrying exactly that amount of energy. Thus only certain photon energies are ever seen (identified by their corresponding frequencies or wavelengths), and this explains the appearance of the spectrum. How a transition actually takes place and how the photon is emitted are questions concerning which Bohr's theory has nothing to say. The reason why the spectrum of hydrogen is so simple is that hydrogen has only a single electron. The spectra of other elements are more complicated because the motions of the electrons are more complicated—they interact with each other as well as with the nucleus.

Bohr assumed that the motions of electrons in their orbits are governed by Newton's laws of motion, and their attraction to the nucleus and repulsion from each other are governed by the inverse-square law of electrostatic force. Each electron is limited to certain allowed orbits by a limitation on the possible values of its angular momentum. In the simple case of a circular orbit it amounts to

$$mvr = nh/2\pi, \qquad n = 1, 2, 3, \ldots,$$

where m is the electron's mass, v its orbital velocity, r is the radius of the orbit, h is Planck's constant, and the different possible values of n correspond to different possible states (Bohr 1913).

There was also a tacit assumption that for five long years got in the way of applying the theory to heavier atoms. The analogy of the structure of atoms with that of the solar system was so vivid and so rich in deep and extraneous philosophical implications that people thinking about atomic structure forgot that whereas planets attract each other by their gravitational force, electrons repel each other electrostatically. Thus, while the planets tend to get close to each other by arranging their orbits in the same plane, the electrons try to get as far as possible away from each other by spreading out in three dimensions. "The mere analogy of Bohr's orbits with those of the planets around the sun, repeated over and over in the literature, served as a psychological block, for a half dozen years, to a consideration of the possibility of three-dimensional systems" (A. Landé, in Yourgrau and van der Merwe 1971, p. xix).

It is quite easy to calculate the wavelengths of the spectra of hydrogen and singly-ionized helium (which also has a single electron) that follow from Bohr's assumptions, and the results are excellent, but the mathematics becomes far more difficult as soon as one turns to more complicated atoms, and in fact for a decade not even the normal helium atom, with two electrons, was within reach. In 1923, when it was finally mastered by Born and Heisenberg, the results disagreed grossly with experiment, but even before then clouds were gathering, for there were subtler disagreements, especially with the theory of the Zeeman effect, which is the splitting of spectral lines in an atom immersed in a magnetic field. The situation was not simplified by the ad hoc nature of the assumptions Bohr had needed to make to derive his results. Since they were not derived from anything, any of them could be wrong. Obviously the hydrogen atom was a special case for which Bohr's theory happened to work quantitatively; for the rest, it gave little more than a pictorial insight—but what an insight! Just as Newton had been able to extrapolate the results of measurements performed on his pendulums into a theory of planetary orbits on a scale 10^{12} times larger, so, at least in the hydrogen atom, the same Newtonian principles were now giving something like insight into the behavior of systems 10^{10} times smaller. Clearly nature is very consonant to itself—though not perfectly so.

We may note here the reemergence of integers into physics after a long time. As we now know, the integers $n = 1, 2, 3, \ldots$ occur in Bohr's angular momentum formula for exactly the same reason that they occur in the formulas governing Pythagorean harmony—they count the ways in which wavelike disturbances can be set up in a region of finite size. The early exponents of the theory were not slow to point out the resemblance.

"What we are nowadays hearing of the language of spectra is a true 'music of the spheres' within the atom—chords of integral relationships, an order and harmony that becomes ever more perfect in spite of the manifold variety" (Sommerfeld 1923, p. viii).

For calculating properties of atoms heavier than helium Bohr's modified version of Newtonian dynamics turned out to be essentially useless, and there was little in the experimental material to tell theoreticians which way to change their thinking. The group of young people that gathered around Bohr in Copenhagen must have sensed that the Newtonian theory had reached the limit of its development, for their response to the failures of Bohr's theory was not to patch it up with new hypotheses (and Newton's ghost smiled) but to doubt all of them—particles, orbits, the whole picture, for by what experimental magic could one ever hope to say whether some statement about a particle orbiting inside an atom was true or false? Physical thought should move further from pictorial models and closer to the observational material. But while these thoughts were being thought in Denmark and in Germany, an idea arose in another quarter.

Waves of Matter

The mist of centuries had begun to disperse. Mankind had become aware that there is order behind the diversity of elementary substances, to the point where one could even begin to predict the existence and properties of elements that had so far eluded discovery. Bohr was given the Nobel Prize in 1922, and people all over the world discussed rationally what goes on in the interior of the atom. An atom consisted of a nucleus surrounded by orbiting electrons, its diameter was measured in units of 10^{-8} cm, its mass in units of 10^{-24} g. No imagination, however skilled, can grasp numbers like these. It helps to tell little stories: if a glass of water were dipped from the ocean and the atoms somehow marked and scattered into all the oceans of the world, another glass, dipped anywhere at random, would most likely contain several of the marked atoms. As to the atom's constituents, the electron was very tiny, its mass almost two thousand times less than that of the hydrogen nucleus. The electron bore a negative electric charge and was, of course, usually visualized as a sphere. In 1925 it was discovered that every electron spins on its axis, as the planets spin going around the sun, but with the difference that all electrons spin at exactly the same rate. It was not long before new developments in atomic theory showed that this is a very naive way to look at the phenomenon of spin, but for thinking everyday thoughts the notion of a little spinning ball is still serviceable.

The nucleus—what was that? "Massy, hard, impenetrable spheres," Newton's words come to mind. Very small but finite in size—Rutherford's measurements had revealed sizes of the order of 10^{-12} cm, so that an atom's nucleus occupied only about 10^{-12} (a million-millionth) of its total volume. A few years later Rutherford accomplished the transmutation of elements that mankind had long dreamed of by splitting a nucleus of nitrogen into two smaller nuclei (accordingly in 1919 he was awarded the Nobel Prize for chemistry, a transmutation that surprised him very much). "Splitting the atom," "smashing the atom" were phrases in the Sunday newspapers. Mankind was being treated to an entirely new view of nature. And in the midst of the party sat the guest of honor, the quantum, veiled in mystery. No one knew what lay beneath the veil. They had asked it whether light was a wave and it had answered "Yes!" and only that. They had asked it whether light was a particle—again, "Yes!" leaving them more puzzled than before. A particle has energy; a wave has frequency of vibration. A formula connects them: $E = h\nu$. It is unimaginable that anything can be both a wave and a particle—a particle exists in one spot while a wave spreads over space. What was this strange number h which, in a formula no one claimed to understand, bridged the unbridgeable gap?

Prince Louis de Broglie (1892–1987; the title is Napoleonic and the name rhymes with *feuille*) got his training in France, where in those days physicists performed beautiful experiments and regarded speculative theory as an Anglo–German occupation. Isolated from the discussions going on in the main research centers, he thought about the gap. The idea that struck him was so simple that nobody had thought of it: instead of worrying about the pathological behavior of light, why not, nature being consonant to itself, assume that the behavior is not pathological at all? What if every particle is also a wave? The scope of the problem is enlarged and the problem certainly is not solved, but it is transformed. In de Broglie's mind (Price et al. 1973, p. 12) the quantum looked like a particle sitting in the middle of a wave, but just as Maxwell's electrodynamics finally bore no trace of the ether theory from which it sprang, nothing finally depended on de Broglie's mental picture.

The arguments that de Broglie presented to the world in two preliminary papers (1923) and his doctoral thesis of 1924 (published in 1925) are based on the theory of relativity and to some extent (it is not very clear) on the mental image he held. For our purposes his central contribution was a formula giving the relation between the momentum p of a particle and the wavelength λ of its associated wave, analogous to the earlier relation between energy and frequency,

$$\lambda = h/p = h/mv,$$

where for p is substituted the usual expression for the momentum of a moving object. De Broglie applied this formula to an electron moving around a nucleus. Assume a circular path, and assume that the wave makes a pattern that is closed on itself, as in Figure 15.3. The circumference of the circle is a whole number of wavelengths:

$$2\pi r = n\lambda, \quad \text{where } n = 1, 2, 3, \ldots.$$

By de Broglie's hypothesis, this is

$$2\pi r = nh/mv,$$

whence

$$mvr = nh/2\pi,$$

and this, as if by a miracle, is the formula that Bohr had had to guess in 1913 in order to derive his formula for the energy levels of hydrogen.

De Broglie's thesis goes on to explore the remarkable consequences of the wave–particle idea in the realm of mechanics. If the laws of classical dynamics as worked out by Newton, Lagrange, and the other founders of theoretical mechanics really describe the behavior of a wave, then there must be a correspondence between those laws and the laws of optics that assume a wave from the beginning. De Broglie finds the connection in a comparison of two principles I have already mentioned: Fermat's principle of least time, and Maupertuis's principle of least action. They both say the same thing. Since then, optics and the mechanics of a particle have been branches of a single mathematical theory (see, for example, Park 1979).

Not many people saw the thesis of an obscure French physicist, published in a journal that was not much read, but when Einstein received a copy he was impressed. He thanked the sender, remarking that de Broglie had lifted a corner of the veil, and sent copies to a number of his friends.

Already in his first paper de Broglie had suggested that the wavelength of an electron could be measured by techniques used in optics. This was done by several people during the next few years, yielding photographs that show a wavelike diffraction of electrons from crystals, almost indistinguishable from pictures made with X-rays. Later, in more difficult experiments, diffraction patterns were obtained using beams of hydrogen atoms and even entire hydrogen molecules. For heavier atoms and molecules direct measurements like these become impossible for rather subtle reasons, but as we shall see, the indirect evidence is overwhelming that

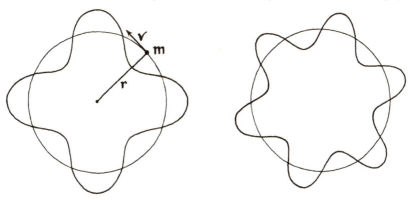

FIGURE 15.3. De Broglie's waves fitted onto circular orbits.

not only is the wave nature of matter a universal property, but the wavelength is exactly as given by de Broglie's simple formula.

All this clever reasoning does nothing whatever to resolve the fundamental puzzle, what is the connection between the wave and the particle? De Broglie had a mental picture that was shared by some others, but nothing that was known, to him or to anyone else, suggested a way of finding out what the connection really is.

Quantum Mechanics

I have already mentioned the reasons why in the mid-1920s the people around Niels Bohr became convinced that the quantum theory was coming apart. A new theory was needed, but not too new, since Bohr's theory gave right answers as well as wrong ones, and also it fitted into Newtonian theory in a beautiful way: if one allows oneself to consider a hydrogen atom in which the electron orbit has grown to the size of a bicycle wheel (n is very large), the classical theory of Newton and Maxwell will suffice to tell how it will radiate energy, but so will Bohr's theory. The two theories are utterly different in their description of what is going on, but Bohr's theory, with its apparatus of stationary states, single photons, and quantum transitions, gives results that, for such large systems, are the same as those of the classical theory that has electromagnetic radiation flowing from the neighborhood of an accelerating electric charge. This precious harmony between the new and the old must not be lost.

Very carefully, in the summer of 1925, Werner Heisenberg (1901–76) put his hands into the mathematical machinery of the existing quantum theory and began to make subtle changes so as to move it away from the pictorial image of orbiting particles in which it had started. He and his colleagues had reluctantly become convinced that those orbits lead nowhere and that the mental picture must be given up. He wrote to Wolfgang Pauli in 1925:

I am really convinced that interpreting [Bohr's theory] in terms of circular and elliptical orbits in *classical* geometry makes not the slightest physical sense, and my whole effort is to destroy without a trace the idea of orbits that cannot be observed and to replace them with something better. (Pauli 1979, vol. 1, p. 231)

The new theory (Heisenberg 1925; discussed in Park 1979), soon named quantum mechanics, with which Heisenberg replaced the Bohr theory seems at first sight to have nothing at all to do with it. There is no mention of particles or orbits; gone is anything that can be described as a process in space and time. Instead, the electron is an entity that produces certain effects, nothing more. It is more like a verb than a noun. It is hard to describe the new formalism without ascending to technicalities, but its most characteristic feature is that the numbers it uses to designate the physical states of an atom are whole numbers. Nothing varies continuously as the position of a moving particle does; everything jumps, but not in the space and time we know. And yet the new theory had the same harmonious relation as its predecessor with the principles of classical physics: whenever they can be properly applied to the same problem they give the same answer.

A number of the mathematical methods Heisenberg used, especially the replacement of continuous variables by discrete ones, had already been tried, but Heisenberg changed the relation between physical concept and mathematical symbol that up until then had been taken for granted in physical theory.

In order to say anything at all about the dynamics of a physical system one must introduce descriptors of some sort, and in a mathematical theory they are mathematical quantities represented by symbols written on a piece of paper. In the new theory, if one is talking about an oscillator one must say, somehow, that something is swinging back and forth. Thus one must talk of both its position, which changes, and its speed of motion, which also changes, but without introducing numbers for either, since numbers would define an orbit, and this was just what Heisenberg intended to leave out. He escaped from this dilemma by using mathematical symbols as if they were words, not numbers. The letter x stands for the idea of position but not, as heretofore, for the numerical value of a distance measured along an axis called x. In the same way, p stands for momentum. They occur in symbolic formulas representing the relations that exist between the corresponding numbers in classical dynamics, but the rules for calculating with them are not the rules for numbers. They can be added and multiplied, but it turned out that the product of two quantities often depends on the order in which they are written down; for example,

$$xp - px = ih/2\pi,$$

where i is $\sqrt{-1}$. This feature of the theory, understood a little better with the passage of time, persists today. In 1925 it looked very strange, not least because for the first time in history an imaginary quantity was being introduced into the basic mathematical structure of a physical theory. These mathematical quantities that are not numbers have a mathematical representation in terms of arrays of numbers called matrices, which are rather cumbersome to calculate with, but it was in this form that Heisenberg first encountered them, and the theory soon became known as matrix mechanics.

After a few months Paul Dirac (1902–85), in England, realized that matrices are not essential to the theory. They are only one possible representation of algebraic quantities like x and p, whose product depends on the order in which they are written down. Dirac named such quantities q-numbers, calling ordinary numbers c-numbers. He saw that one need not reduce q-numbers to arrays of c-numbers in order to calculate with them—they have their own mathematical rules. Dirac's words about them in those early days express something like awe:

> At present one can form no picture of what a q-number is like. One cannot say that one q-number is greater or less than another. All one knows about q-numbers is that if z_1 and z_2 are two q-numbers, or one q-number and one c-number, there exist the numbers $z_1 + z_2$, $z_1 z_2$, $z_2 z_1$, which will in general be q-numbers but may be c-numbers. One knows nothing of the way in which q-numbers are formed. (Dirac 1926)

Of course, measurements yield only c-numbers and never q-numbers, and so the theory must contain rules for extracting c-numbers from q-numbers when one wishes to calculate measurable quantities. We shall see that these rules contain a great mystery.

The physicists in Germany and Denmark were intrigued and felt that something of immense importance had been created, but their enthusiasm was modulated by the fact that, lacking any way to picture the results in their minds, they had no idea how to think about them, and much of the surviving correspondence from this period is in tones of despair (Pauli

1979, Beller 1983, Hendry 1984). One of the reasons for despair was that except for trivial examples it was very hard to calculate with matrix mechanics. Heisenberg gave up on the hydrogen atom, but in 1926 Pauli managed to do it and the result was the same as that given by Bohr's theory. Anything more difficult, even the kind of helium calculation that had contributed to the demise of Bohr's theory, seemed out of the question, but without it how could one be sure that the new theory was any better than the old one? Further, the new theory might very well dispense with electronic orbits in an atom, but it must not dispense with the electron itself. An electron may not be a little round ball, or even an ordinary kind of thing, but it is some kind of thing: its passage causes counters to fire and lights to flash, it leaves a trail behind it as it traverses a cloud chamber or a bubble chamber, and nobody would have any luck with a theory that assumed that there are $3\frac{1}{2}$ electrons in an atom. The Newtonian laws of motion that Heisenberg had transcribed into his new mathematical language referred to something at a certain place moving in a certain way. Where was it in the new theory? And the new theory had nothing to say about the old dilemma—how can light be both a wave and a particle?—for it still treated light as a wave while it did not even contain a symbol for an electron. De Broglie had at least tried to treat electrons and photons as if they were the same kind of thing, but the experiments supporting his ideas were not yet published and nobody but Einstein was yet paying any attention to de Broglie.

Wave Mechanics

When Thomas Young proved by experiment that light travels through space as a wave he did not by any means answer the question, "what is light?". It was to be another sixty years before Maxwell began to answer that question with a quantitative dynamical theory that represented light as a wave motion of electric and magnetic fields and showed how the fields are generated and how they interact with matter. The subsequent discoveries of Planck and Einstein showed that even that is not the end of the story. De Broglie's theory of matter waves was not a theory of matter any more than Young's was a theory of light. Such was the speed of scientific work in those days that a proper dynamical theory of de Broglie's waves was published only three years later.

Erwin Schrödinger (1887–1961), Viennese by birth and professor at Zurich from 1921, was outside what one might call the Bohr orbit. He tended to work alone and his interests were extremely wide, in physics and beyond. He translated Greek and Provençal lyrics and published a volume of his own poems. He wrote on physics and philosophy and human culture, and his book *What is Life?* suggesting that molecules carry the information of heredity, started a generation of scientists on the quest that led to the explanation of heredity in terms of the molecular structure of DNA. At the beginning of 1926 Schrödinger set out to find the dynamical law that controls the motion of de Broglie's waves as Maxwell's equations do for electromagnetic waves. Within five months, in a series of beautiful and eloquent papers that almost fill a book (1928), he proposed and developed a theory known as wave mechanics.

Just as Heisenberg had formed his quantum mechanics by judiciously tampering with Bohr's theory, Schrödinger's published derivation drags a wave theory out of classical dynamics. Never mind the derivation, which amounts to an inspired guess; it led to a wave equation that he at once applied to the hydrogen atom. Just as de Broglie had explained the integers n in Bohr's theory by appeal to the simple picture of a wave traveling along a curved path closed in on itself, Schrödinger also derived it from a requirement of closure, but now the waves are harder to visualize since they occupy three-dimensional space.

Schrödinger's subsequent papers show how to get approximate solutions to problems too hard to solve exactly, and one of them proves to give a very surprising result: that between wave mechanics and matrix mechanics there is complete mathematical equivalence. Though the kinds of verbiage accompanying the two seem to have nothing in common, any calculation of a physically measurable quantity must give the same answer in both. There is one great difference. For anyone familiar with the techniques of differential equations, many calculations in wave mechanics are not especially difficult. Heisenberg at once calculated enough states of the helium atom to show (1926) that the now united theory, thenceforward called quantum mechanics, gives a good approximation to the observed spectrum, and with modern calculations the agreement is almost perfect. There remained, as before, the question of interpretation, especially the question of the wave–particle duality, for the particle was talked about but had not yet, as a particle, appeared on the scene.

In Maxwell's theory light waves had turned out to have a simple if surprising physical interpretation in terms of electric and magnetic fields, but Schrödinger's wave describing any kind of particle is a wave of that kind of particle and nothing else. As soon as attention turned to systems more complicated than the hydrogen atom a remarkable feature of it was revealed: it is not a wave in the space we inhabit. Our space has three dimensions. If one assumes that the wave describing two electrons exists in this three-dimensional space one cannot account even for so simple a fact as the conservation of momentum when they collide. The wave describing them must be a wave in six dimensions, three for each, and similarly for more complicated systems. And what kind of wave is it? What does it represent? Nobody could say.

Apart from the question of interpretation, Schrödinger was sure what he had done: he had eliminated from physics the random, unvisualizable, and unanalyzable quantum jumps, for he supposed that his entire theory was expressed in a differential equation that told how an electron wave changes continuously and causally in time. His wave didn't just represent the electron; it *was* the electron, in the same sense that the electromagnetic wave is physical light. There remained, however, the question of how to relate the wave to the particular path by which, in the laboratory, a particular electron travels from one point to another and causes a particular counter to register its arrival at a particular moment. Not all particles are hidden away inside an atom and immune from observation; what about particles moving where they can be detected? The question was answered almost at once by Heisenberg's collaborator Max Born, in a paper that for many was an unhappy regression to ideas that were being abandoned: the wave does not describe the exact behavior of any particle but tells only the *probability* of finding it at a given point at a given instant of time. Born recalled that Einstein had once interpreted the electromagnetic field as a "ghost field" in exactly the same sense. But such an interpretation does not deny the fully causal character of Schrödinger's equation: "One can state it a bit paradoxically: the motion of the particle obeys laws of probability, while the probability itself develops in accordance with the causal law" (Born 1926, p. 804). To Heisenberg also, Born's proposal amounted to giving up territory that had just been won, for when he first constructed quantum mechanics he had no intention that probability should play a part in it.

The word "probability" suggests that someone lacks information. If we knew enough about horses and riders we could make a fortune at the race track; if our nervous systems were a little more finely tuned we could control the throw of dice or see how a coin is going to fall. We speak of good or bad luck only because we are clumsy and ignorant. Information exists in principle but is not available to us. But according to Born's interpretation it is the theory that, so to speak, lacks information, for the q-numbers that express it are not numbers in the ordinary sense and do not denote quantities as numbers do. Whether this is because the theory is incomplete, or simply that the wrong question has been asked and the information does not exist, is a question that borders on metaphysics. It provoked passionate discussion without producing agreement, and when it was finally settled (understanding, of course, that few questions in science are ever *really* settled), it was, surprisingly, not by theoretical or epistemological arguments, but by experiment, as we shall see in the next chapter. At the end of 1926 everyone realized that there must be a better understanding in words of what the formulas of quantum mechanics were talking about.

In spite of its obscurity the new theory began to attract followers. It demanded strange mathematics, and the few calculations that had been carried through tended to be difficult. Far from illuminating the dark areas where the quantum dwelt, it had removed the entire subject further from the realm of understanding with its destruction of visualizable images. And what image to put in its place? None, yet.

Nobody claimed to understand what was going on but in two years results began to arrive. Heisenberg explained the spectrum of helium. The general features of molecular spectra were accounted for and perhaps most important of all, Walter Heitler and Fritz London (1927) solved the ancient problem of chemical bonding; they showed that the force that holds atoms together is electric in nature and is a necessary consequence of the quantum-mechanical formalism. All this was done by people who could not have explained clearly what they were doing. The situation at this time provides a good example of a principle that by now is taken for granted: conceptual clarity and explanatory power are not the first criteria for judging a scientific theory. The first one is that it should work. The new theory accounted for a variety of experimental facts. Did it explain them? That depended, and still depends, on what one means by "explain."

Indeterminacy

There are several laboratory situations in which a particle travels through space and a counter signals its arrival. The language of quantum mechanics ought to be able to say something about where it is and how it is moving. In the mathematical formalism of the theory that cannot be done exactly. After the translation from q-numbers to c-numbers the theory can only tell where the electron will probably be found and what its momentum will probably be, with some necessary statistical uncertainty in each, but the magnitudes of these uncertainties are related in a simple way: if the system has been prepared so that one knows accurately where the electron will be found, then the momentum is uncertain and vice versa. Quantitatively, let Δx be the statistical uncertainty in the particle's position x (essentially, the width of the statistical spread found in a number of repetitions of the same measurement), and Δp be the corresponding uncertainty in the momentum found in the same experiment. It then follows from the theory that the reciprocal relation between the two is expressed by

$$\Delta x \, \Delta p \geq h/4\pi.$$

(The sign $\geq$ means "is at least," and it must be there because there is no practical limit on how large an error one can make.) This relation was first derived by Heisenberg in 1927, in a paper that has profoundly influenced people's thinking as to what the theory means. Note that in stating the relation Heisenberg has abandoned his former goal of a theory without reference to particles, their positions, their motions, or probability. The particle is here, it is somewhere in space and it is moving somehow, and the theory can be interpreted as talking about these things with limited precision.

Having derived the formula, Heisenberg writes that before we can decide exactly what it means we must first decide what the terms position and momentum of a particle mean, and for this purpose he adopts the operational point of view that Einstein had used in clarifying the equations of relativity. "If one wants to say clearly what is meant by the 'position of a thing,' say an electron (relative to a particular system of reference), one must specify experiments by which one thinks to measure the 'position of the electron'; otherwise the term has no meaning." Similarly for the momentum. Heisenberg stresses that either quantity can in principle be measured and known, and hence might reasonably be incorporated in the formulation of a theory, but the question is, can they both be measured and known at the same time?

There follows a discussion of a conceptual experiment that is a little confused, but when straightened out later goes like this. Suppose I want to know the position and momentum of a particle. First, I might prepare a beam of particles whose momentum is determined as accurately as possible (there is essentially no limit to the precision possible here). Then I might mount a powerful microscope above the beam and look at an electron as it goes by (Figure 15.4). A microscope requires light, so the picture shows a light source. Since, as we shall see, the light introduces uncertainty into the result, let us minimize the uncertainty by using only a single photon—it would of course be impossible to do this in practice, but one can at least imagine the photon bouncing off the electron into the lens and then being registered somehow at the eyepiece of the microscope.

The photon has an initial momentum that can be accurately controlled, and as it bounces off the electron it conveys some of this momentum to the electron in recoil. The electron's momentum after the observation will be composed of its original momentum plus the amount of the recoil. How much is that? If we knew the photon's path subsequent to the collision we could answer the question, for all that is involved is the Compton process already mentioned. But we do not know it, for the microscope lens has a width and we have no way of knowing through

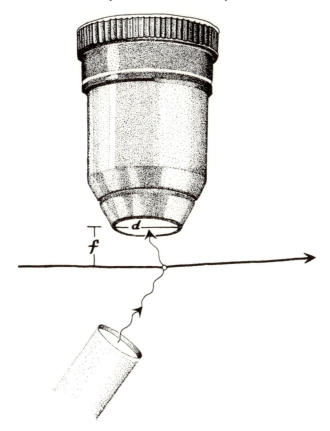

FIGURE 15.4. Observation of an electron in Heisenberg's imaginary microscope. The photon bounces off the electron into some part of the lens, leaving the electron with its motion slightly changed by the recoil.

what part of the lens the photon passes. Therefore the observation has introduced an uncertainty, which we shall call Δp, into the electron's momentum. If p_p represents the photon's momentum, Δp is roughly

$$\Delta p = p_p\, d/f,$$

where d is the diameter of the objective lens and f is its focal length.

Now what about the accuracy with which the electron's position is determined? It is a known property of microscopes that the images they give are not perfectly sharp, and this is because the rays of light are diffracted by the lenses, especially by the bottom one. Even after looking through the microscope and seeing the photon, one still knows the electron's position only with an error of about

$$\Delta x = f\lambda_p/d,$$

where λ_p is the wavelength of the light. Multiply the two uncertainties together; the product is

$$\Delta x\, \Delta p = \lambda_p\, p_p,$$

and this, by de Broglie's relation, is just h. That is the very best that can be done. Real measurements are more uncertain, especially because nobody can see anything with just one photon. The "equals" should read "is at least." Thus we recover Heisenberg's formula (except for the 4π, which depends on the quantitative definition of uncertainty and is unimportant), but in an entirely new context: it now appears as a limitation of the precision with which we know about the electron after the measurement has been made. The uncertainty that at first appeared as a defect of the theory appears in this discussion as inherent in the process of measurement itself. Or, if we base our claim to know the values of x and p on our ability to measure them, it is an uncertainty inherent in what we can know. Still, the basic question remains undecided. The theory, rightly or wrongly, does not represent the particle as having a definite position or momentum. The argument by which we try to clarify matters seems to assume that in some sense the particle has these attributes, but that we cannot find out what they are. These are two different assumptions. Which is right? We shall return to this question in Chapter 16.

Before we try to draw far-reaching conclusions from an argument like this we must be convinced that we are not dealing with a special case. Is there perhaps some other procedure, some other instrument we could use to reduce the uncertainty? Einstein, and others who did not like the direction in which the quantum theory was moving, tried very hard to find one but without success. Failure to find something does not necessarily mean it isn't there, but ultimately even Einstein stopped looking, and by now the matter is pretty well settled.

At the end of his paper Heisenberg adds a general remark: the idea of strict causality in nature is dead, for "in the strict form of the principle of causality, 'If we know the present exactly we can calculate what will happen in the future' it is not the conclusion but the premise which is false. We cannot, even in principle, know every detail of the present" (1927, p. 197). It is to emphasize this aspect of Heisenberg's argument that one often refers to indeterminacy rather than uncertainty.

We are left with an urgent question in our minds: is it just that by some curious conspiracy of nature we are prevented from knowing the world exactly and hence from predicting exactly what will happen next, or is it that the future is truly not determined by the present? Ever since the rise of modern science people have realized that if one faces facts without blinking one must reckon with the possibility that there may some day be scientific proof that determinism operates in every aspect of nature, even within the sanctuary of the human brain—that *res extensa* is all, that the universe, including ourselves, is a vast machine, and that free will is an illusion. Just as it was important for man's view of himself in the universe to know whether his planet was the center of all the spheres, it is important to answer the question of determinism. Heisenberg's principle is a factor in the answer, but it is not the whole answer. I believe that the answer lies in studies of the brain and the relation of consciousness to its function, but that we cannot now anticipate even in the roughest way what the outcome will be, since before the work is done we will have to make such radical changes in our vocabulary that words like "knowledge," "fact," "reality," "determinism," "consciousness," "symbol," and "meaning" will either be used in entirely new ways we cannot now anticipate, or they will have dropped from sight altogether. I write with the same optimism that Heisenberg expressed in an essay written many years ago: "The exact sciences also start from the assumption that in the end it will always be possible to understand nature, even in every new field of experience, but that we may make no *a priori* assumption about the meaning of the word 'understand' " (1958b, p. 28).

Complementarity

I think that during the 1920s and 1930s people were hoping that with new developments the wave–particle dilemma would just go away. From about 1927 until the end of his life in 1962, Niels Bohr, more than anyone else, brought light into the quantum world. During these years he developed a consistent doctrine designed to accommodate the discordant experimental facts that had puzzled people for so long, and to interpret the abstract mathematical theories that agreed so well with experiment. This doctrine is generally known as the Copenhagen interpretation, and the fact that people are not at all unanimous as to what it is shows that the central questions are still alive. Most teachers explaining the theory to their students, as well as most researchers trying to puzzle their way out of conceptual dilemmas, start from this interpretation as they understand it, even if they are not entirely comfortable with it. It never satisfied everybody and it does not today. Careful thinkers like Einstein and Schrödinger and de Broglie were not convinced. It is easy, and true, to say that this was because their metaphysical presuppositions disagreed with Bohr's, but there is a good deal more to it than that. In this section I shall sketch the Copenhagen line of thought, not as Bohr first formulated it in

1927, but as it slowly matured over the years. In the next chapter I shall point out some of the questions that were not answered in Copenhagen.

How can light (or matter, but let us talk about light) be a wave and a particle at the same time? Bohr points out that the question is improperly put, for although there are some experiments in which light exhibits wave properties and others in which it exhibits particle properties, there is none in which it exhibits both at the same time. Let us start again. An experiment asks nature a question. Young's experiment asks "Wave?" Nature answers "Yes." The Compton experiment asks "Particle?" Again, "Yes." What we would like is an experiment that asks "Wave or particle?" Try to design such an experiment, in which the result will affirm one while denying the other. It does not seem possible, and this is the center of Bohr's argument. Light shows different properties under different and mutually exclusive modes of questioning. If I look at a sculpture from the front I cannot see the back, but nothing prevents me from walking around to the back where I can no longer see the front. We piece together our knowledge from the results of various observations. One is not enough; the different ones complement each other in forming a picture in our minds. This is the fundamental idea, and since 1927 it has been known as complementarity.

Now let us refine the idea a little. We must be very careful that the "picture in our minds" matches what we actually know, for the picture, if we are naive, portrays things as we think they *are* or would like them to *be*, whereas what we actually know is what *happens* in certain experiments. Experiments on invisible particles do not tell us what they are. The temptation to talk about things as they are arises as soon as we try to define and describe a thing in isolation from its surroundings, even though it is these surroundings that determine how we perceive it. Instead of imagining light by itself as a wave or a particle we should be recalling what it does in a certain experimental situation. Bohr's term for what we can actually observe and describe to others is *phenomenon*. In common language the word refers to something that is observed. In Bohr's language the phenomenon includes the experimental circumstances—apparatus and procedures, as well as what is observed. Thus we may speak of the wave and particle phenomena associated with light and not encounter any contradiction, since they are found in different situations.

The proposer of a new theory must specify the phenomena that it predicts. If in actual trials something else occurs, the theory is wrong. In order that there may be no ambiguities in the comparison of theory with experiment Bohr insists that phenomena must be described in plain, ordinary language. A theory may carry out all its calculations in q-numbers but it must describe the expected phenomena in the ordinary c-numbers that result from measurement. We cannot translate directly from q-numbers to *c*-numbers. There is necessarily a mismatch at this point, and it is here, in Bohr's opinion, that statistical considerations enter physics. A calculation in *q*-numbers can be exactly causal, so that the *q*-number account of a state at the end of a process follows exactly and unambiguously from the state at the beginning, and yet *q*-numbers do not furnish a pictorial account of what has happened. After the translation into *c*-numbers, when the phenomenon can finally be described in plain language, the language is usually statistical. The theory must furnish mathematical recipes to translate *q*-numbers into statements about statistical distributions.[3]

The wave–particle duality is the classic example of this line of thought. Figure 15.5 illustrates a two-slit device, essentially that used by Young in 1802, designed to exhibit the wave character of light and explained in Note G. If light from a single source strikes the two slits the screen will show the regular alternation of light and dark bands that is taken as evidence that the wave, passing through both slits at once, recombines at the screen to give light or darkness according to the relative phases of the two waves.

But sometimes light acts as if the beam were composed of particles. Suppose we want to exhibit its particle character. Like "wave," "particle" is a word that refers to our mental picture of what happens in a certain experiment. Nobody claims to see the particle or follow its motion through space, but if in talking about this experiment we say "particle," we are talking about something that can go through only one slit, and an experiment to validate our use of the particle picture must show that it does. Figure 15.6 shows, in the same purely conceptual style as Figure 15.4, how this might be done. Mount the upper slit on such delicate springs that it will respond to the tiny recoil impulse it receives when it changes the direction of a particle that goes through it. The recoil must be large enough to register on the little dial, but there is a difficulty, for by the uncertainty principle we cannot exactly know the momentum of the slit either before or after the interaction unless its location is completely unspecified, and of course that would ruin the experiment, since the separation of the bands on the screen depends on the separation of the slits. To make the observation we have to compromise: not too much

---

[3] I have hedged a little, saying that the plain language is generally statistical. Sometimes the results are definite and not statistically distributed, but in such cases one usually sees on closer examination that this is because some convenient idealization was made in order to simplify thinking and calculation. The development of Bohr's thoughts can be followed in Bohr's three collections of essays (1934, 1958, 1963), and in Petersen's excellent commentary (1968). By now the literature is enormous, but for a single short exposition of Bohr's mature thought I recommend his essay in Klibansky (1958, reprinted in Bohr 1963, pp. 1–7).

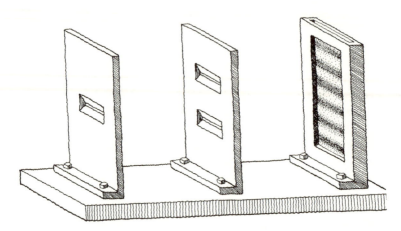

FIGURE 15.5. Bohr's representation of a device for observing the interference
tern produced when light from a single source passes through two fixed slits.

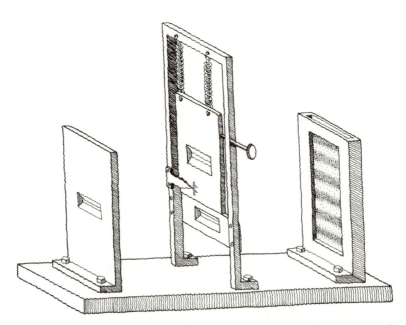

FIGURE 15.6. Two-slit apparatus. If the pin is removed one slit is now movab
so that the passage of a particle through it can be detected by noting how the sl
recoils.

uncertainty in the momentum of the movable slit, not too much uncertainty in its position. We have seen in Chapter 13 (cf. Note G) that if D is the distance on the screen to the first dark band, the relation between the wavelength of the light and the various distances is

$$D/L = \lambda/2d \quad \text{or} \quad Dd/L = \lambda/2.$$

If light bands are not to get mixed up with dark bands on the screen, which would destroy the pattern, the uncertainty Δd in the position of the slit must be substantially less than d:

$$\Delta d < d.$$

If p represents the momentum of a photon arriving at the slit, a simple calculation shows that the transverse momentum it receives when the slit recoils is of the order of Dp/L, and if the pattern is to survive the uncertainty in the slit's momentum must be less than this:

$$\Delta p < Dp/L.$$

Multiplying the last two relations gives

$$\Delta d \, \Delta p < d \, Dp/L = \lambda p/2 = h/2$$

(the last step following from de Broglie's relation $p = h/\lambda$), but this is impossible, for according to the uncertainty principle $\Delta d \, \Delta p > h$. Therefore the instrument cannot perform its function of telling which slit the particle went through without obliterating the interference pattern.

In the drawing there is a pin. Push it in and the upper slit is locked in place; pull it out and it can move. With the pin in, the situation is as in Figure 15.5; the pattern is seen but we cannot verify that a particle goes through one slit and not the other. With the pin out, the recoil can be observed and we can indeed say which slit a particle went through, but in performing the experiment we have obliterated the pattern and with it all evidence of a wave. Simply by moving the pin we can change the experiment from one that asks "Wave?" to one that asks "Particle?" and each time the answer is "Yes." What can we conclude as to which light really is? Nothing. When we ask such a question we have forgotten about the pin and imagine that light *is* something in itself, without regard for the situation in which it is observed. The experiment gives us a choice of phenomena, nothing more. Remember Newton's advice: make all the hypotheses you want to, if they help you to think, but do not build them into your theory. No choice between waves and particles is built into the quantum theory.

The foregoing argument is not definitive because it is not quantum-mechanical. Instead, it treats the slit mechanism using classical physics modified by the uncertainty formula. Note J gives a proper calculation and reaches the same conclusion: the installation of a detector destroys the pattern.

This much of the doctrine of complementarity flowed in a fairly straightforward way from the conceptual and mathematical structure of quantum mechanics and the limits that nature sets on the possibilities of experiment. Complementarity focuses on how, not why. When we say we understand the effects produced by electrons in an interference experiment it is not because we can explain *why* by saying that electrons are waves and waves do such things, but because we have begun to understand analytically *how* electrons produce their effects without needing to say what they are.

Bohr was always convinced that the doctrine could be applied in the wider realm where one does not have mathematics and experiment to control one's speculations. For example, he saw in complementarity a possible way to resolve the problem of free will that does not rest on any assumed atomic mechanisms:

> Just as the freedom of the will is an experiential category of our psychic life, causality may be considered as a mode of perception by which we reduce our sense impressions to order. At the same time, however, we are concerned in both cases with idealizations whose natural limitations are open to investigation and which depend upon one another in the sense that the feeling of volition and the demand for causality are equally indispensable elements in the relation between subject and object which forms the core of the problem of knowledge. (Bohr 1934, p. 116)

Note that Bohr is not trying to give us a *theory* that explains how freedom and causality can coexist; he is simply showing that, given that each of these apparently opposed concepts is solidly based in our knowledge and experience of the world, there is a way of thinking about them that does not require us to accept one and reject the other.

Quantum mechanics has proved its value in so many fields of science that it is surely the best theory we have. Its predictions are, as I have said, generally statistical. Later we shall have to ask what a statistical statement actually tells us about the world, but from a practical point of view we know what the results mean. I have discussed indeterminacy and complementarity here so as to show something of a remarkable episode in the history of ideas in which one tested and self-consistent way of looking at the world suddenly gave way to another, and to give some idea of the puzzles and confusions that can arise in interpreting a theory even when one is sure that it is correct.

As time went on most physicists learned to get along with the paradoxes of the quantum theory. Einstein was not one of them. He believed to the end of his life that the doctrines of Niels Bohr provided only superficial resolutions and that the truth lay deeper. This conviction extended to the entire complex of arguments with which people tried to understand the quantum theory and even to the mathematical theory itself, and the last years of Einstein's life and work were passed in comparative isolation from the main currents of his profession. Though he always said he enjoyed the solitude, his friends knew that he was lonely.

The Universality of Quanta

In the late 1920s there occurred a development that cleared up a number of conceptual problems. It was found that not only dynamical variables like energy and momentum, but also the quantities that give the strength of an electromagnetic field or a matter field must be represented by *q*-numbers. It follows that when these fields are observed they exhibit typical quantum effects: for instance if we measure the number of their quanta we will find 0, 1, 2, . . . , but not $1\frac{1}{2}$.[4] This helps to clarify the picture. The atoms of Leucippus are quanta of matter, as are electrons and photons, and if their energy and momentum are measured we find they carry quanta of energy and momentum. It turns out that even sound waves passing through a solid are quantized. With the quantization of fields we begin to appreciate the universality of quantum phenomena. We see the microscopic part of nature being consonant to itself: for every wave there is a particle and for every particle there is a wave. The instinct of the first atomists was the right one. Atoms are more than matter; they are light and sound. Whether it is advantageous to consider that soul also is atomic must, I think, be decided when we know very much more about soul than we know now.

[4] I am ignoring some recent calculations that suggest bizarre exceptions to this rule.

CHAPTER 16

Does It Make Sense?

I always said that truth is a relation between the proposition and the way things are, but I could never exhibit such a relation.

—WITTGENSTEIN

THE last chapter told about theories dealing with things on the atomic scale of size. It turned out that if we try to talk about microphysics in plain language we must introduce some new ideas, notably indeterminacy and complementarity, that help to turn our thoughts away from the old obsession with states of being ("what *is* an electron: wave or particle?") to more fruitful descriptions of nature in terms of what happens when we perform certain actions. This chapter continues the account. It shows how the presence of apparatus for detecting a particle changes the character of what is observed; it shows that when we try to understand what is actually there, we may be led to think about unphysical interactions between distant objects; and it offers an answer to the question raised by the principle of least action: "How does a particle know what path between two points minimizes the action unless it tries them all?"

Probability and the Individual Event

In microphysics most experiments involve large numbers of trials, and the predictions of probability that issue from quantum-mechanical theory translate easily and naturally into statistical predictions as to what will be

334

experimentally observed. Even after sixty years there are no signs that anything is wrong with the basic theory when it is applied to situations we think we understand. The statistical predictions always agree with the statistics of what is observed. Why then does the rhetoric that accompanies the mathematics refer to probabilities rather than to statistics? It is because we believe that a calculation ought to indicate what we should expect to find, even in a single trial. Probability belongs to a different category of thought from statistics. My insurance company gathers data and makes its decisions on statistical grounds, but when I buy insurance I think about probability and don't care about statistics. I have only one house to burn and one life to lose.

Probability pertains to the individual case, and it has also an element of subjectivity. Suppose I am standing with you on a street corner and we watch a little man walking on the other side of the street. I ask you, "What is the probability that he goes into house #211?" You mention some small number. Then I whisper into your ear, "But he lives in #211." Nothing about the house or the little man has changed, but your number changes. The same thing happens in quantum mechanics. If I have no idea where a particle is I assign it a wave function that is broadly spread in space, Figure 16.1. If I think I know, I give it a wave function that is localized. What happens to the wave function when I suddenly learn where a particle is? Apparently the wave function collapses from broad to narrow, but is this a *physical* change? Of course, all of physics ultimately filters through the human psyche, but need the psyche invade the mathematical theory? We can begin to understand what happens only by careful consideration of the way in which information is obtained. Here are two examples.

The first example is the formation of a track in an old-fashioned instrument called a cloud chamber, used to detect the passage of individual charged particles. Inside the chamber is a vapor, and when a particle passes through it forms a trail of droplets just as an airplane leaves a trail of cloud when it moves through cold moist air. If the particle is imagined as a particle it is easy enough to understand what happens: it bumps its way through the gas, hitting one molecule of vapor after another and changing their electric charges. As a result the molecules move together and in a moment droplets form. A treatment of the same process by quantum mechanics arrives at a completely different explanation of the same phenomenon (Mott 1929, Park 1974, p. 343). If initially there is no reason to predict where the particle will enter the chamber its wave function will be spread out evenly across it. Suppose the particle interacts with one of the molecules so as to give it a little energy. We then find by calculation that the wave function collapses into a narrow jet extending behind the struck molecule, so that the next molecule to be struck will lie

FIGURE 16.1. Wave functions of a particle whose position is (*top*) uncertain and (*bottom*) more accurately known.

directly behind the first, and so on in sequence. There is no prediction as to which if any molecule will be struck first, but the calculation shows that *if* a molecule is struck then the next one will almost certainly lie behind it. In this sense it does not take an intelligence to collapse a wave; a molecule will do.

For a second example of the way in which observation changes the objective description of a system let us return to the two-slit apparatus illustrated in Figure 15.5. It is designed to point up a paradox: In order to form an interference pattern on the screen a "particle" must go through both slits at once. This sounds crazy, but if I try to show that it goes through only one slit, the pattern is not formed. With the pin out and the detector functioning, the mechanical action of the detector on the ray passing through it destroys the interference pattern on the screen and therefore destroys the paradox at the same time. When the process is analyzed in Note J, a remarkable result emerges. The effect on the interference pattern is the same as before: if the detector interacts efficiently with particles going past, the pattern disappears; but the language of the explanation is changed. It is the mere presence of the detector that destroys the pattern, even if (as is shown there) the detector is a magical one that does not disturb the particles in any way. I do not know how to explain this result in terms of the kinds of causes and effects we ordinarily think about. It is a characteristically quantum-mechanical effect, and it shows that when we try to understand events in terms of visualizable processes we may find ourselves stumbling further from real insight.

In the first of the experiments just described the detector is a single molecule, chosen at random. In the second it is some kind of oscillator installed near one slit. In each, the detector greatly changes the probabilities of the various possible outcomes, and this is the quantum-mechanical analogue of the way your assignment of probabilities changed when

I whispered into your ear. The difference is that the probabilities calculated in quantum mechanics should not be understood subjectively.

What Is an Individual Event?

My discussion of the two-slit experiment may leave sensitive readers with a faint feeling of disappointment. The task of science is to explain the world in terms that, as far as possible, relate to our experience of it. Especially in the physics of the very large and the very small we would like intuitively plausible models that help us to understand what happens without resorting at every stage to calculation. The discussion of the two-slit experiment shows both the possibilities and the limitations of such models. When the pin in the apparatus is moved we can explain in words what happens, but our explanation is justified only by calculation along somewhat different lines.

The central mathematical quantity in most calculations in quantum mechanics is the wave function, usually referred to as ψ, the Greek letter psi. This is the unknown quantity in the equations of the theory, and once known it can be used to find the values, or the probable values, of observable quantities. The wave function itself is not observable; this is perhaps strange, and it has sometimes been taken as a sign that the present theory is only provisional. Of course, many quantities of physical theory are not directly observed—energy, momentum, electric charge, for example—but the unobservability of ψ is of a different kind. Consider de Broglie's formula that gives the wavelength of a matter wave as h/p (where h is Planck's constant and p is the momentum of the corresponding particle). If I take photographs of some ocean waves from a point on the shore and then find the wavelength by measuring them, and if you in a moving boat do the same thing, we expect (neglecting the tiny effects of relativistic contraction) to get the same number, since the wave is objectively there in the water. But for the wave of an electron the situation is different. If you walk along beside the electron so that for you its momentum is almost zero then you must say that the wavelength is very large, while for me it may be very small. If the wavelength depends strongly on the observer's state of motion we must take it as reflecting the observer's view of things—and that is fine, since it is natural for a description to reflect an observer's view of things—but not as reflecting an observer-independent aspect of nature such as we imagine a water wave to be.

The other peculiar fact about quantum mechanics I have already mentioned: its results are essentially probabilistic, and therefore the language into which it is interpreted is essentially indeterminate. But when we imagine the world for ourselves and talk about it with each other, we use language that does not represent it as either observer-dependent or probabilistic. The Washington Monument is *there*; the pin in Figure 15.6 is either in or out. There is a disjunction between the mathematical description of the world and its description in ordinary language. This rarely causes trouble, since in specific physical situations the mathematical variables are related to experimental numbers in a definite way. The realm of phenomena to which quantum mechanics is normally applied is a little part of the realm of our experience, and there we have no trouble; it is when we leave that little part and try to apply the theory to phenomena outside it, especially phenomena of everyday life, that we come across questions that are hard to answer.

The traditional way of exhibiting such difficulties is to find a paradox, that is, an experiment (real or imaginary) in which our normal modes of thought and language lead us to expect one thing, while a proper use of the theory leads us to expect something that is quite different, and experiment must decide. The experiment with the two slits is an example; each of us is able to imagine the situation in terms of waves or of particles, but neither representation is enough to explain everything that happens. Another is the type of conceptual experiment that was proposed more than fifty years ago by Einstein, Podolsky, and Rosen (1935; Wheeler and Zurek 1983, p. 138). The original context of the proposal was a debate, now almost forgotten, between Einstein and Bohr over the nature of physical reality. In order to point out the singular features of quantum mechanics EPR, as they are called, invented a kind of conceptual experiment that contains surprises for anyone eager to find that quantum mechanics can be understood in intuitive terms.

Simple intuition imagines a particle as traveling through space carrying with it certain properties. We have already seen that this model gets into trouble when the particle is asked to pass through two slits: the particle seems to follow two different paths. An imaginary EPR experiment, to be described in a moment, shows the same difficulty in a different way.

First, a short digression on polarized light, phrased in the language of the market place. Figure 16.2 shows a beam of unpolarized light, such as is emitted from any ordinary source. PA is a polarizer with its axis oriented vertically. This means that the light that passes through it will have its electric field vibrating in the vertical plane as in Figure 13.7 (in unpolarized light the electric field vibrates in every direction perpendicular to the direction of travel). To verify that the light from PA is polarized, pass it through a second polarizer, PB. If its axis is also vertical, $\theta = 0$ in the figure (and if its optical properties are of supernatural perfection), then all the light from PA will pass through it. If its axis is horizontal none will pass, and if θ is some intermediate angle, then the fraction that passes is given by the nineteenth-century law of Malus as $\cos^2\theta$. Talking photon

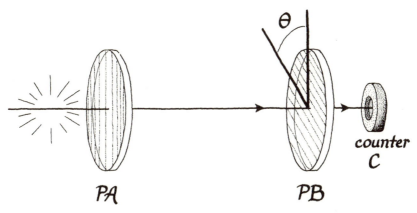

FIGURE 16.2. Light from a source at left encounters two polarizers on its way to a counter.

language, we say that $\cos^2\theta$ is the probability that a photon that passes PA will also pass PB.

What is the probability that a photon from the unpolarized source in the figure will pass through PA? The natural answer to the question is to say that in an unpolarized beam all values of θ are equally probable. The average value of $\cos^2\theta$ is $1/2$, and so one expects that half the light will go through. That turns out to be the case. After this introduction let us look at the imaginary experiment.

A neutral pion, as we have seen in Chapter 14, is a particle that decays after a very short time into two photons. If it is initially at rest or nearly so, the photons go off in opposite directions. The pion is a spinless particle; it has no angular momentum, and so the two photons must carry no angular momentum either. This imposes a connection between their states of polarization. It is known (Note K) that if one of the emitted photons passes through a polarizer with axis vertical, the other will pass through one oriented horizontally. In the imaginary experiment a succession of pions decay between two polarizers of Figure 16.3. Pairs of photons will fly off in opposite directions, and some of them will travel along the axis of the experiment and pass through the polarizers. Orient the left-hand polarizer PF with its axis of polarization vertical. Since the pions decay at random, half the photons arriving at PF will pass through. Those that pass through are vertically polarized. Now set the other polarizer PG with its axis horizontal. We have already noted that if one photon passes through PF the other will pass through PG. If counter CF counts, CG will also count, and therefore when pions disintegrate so as to send photons along the axis, half the time both counters will count.

Try to explain this according to the following argument: two photons set off with polarizations perpendicular to each other, and that of the left-

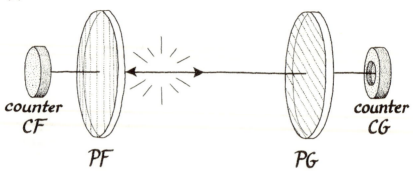

FIGURE 16.3. A neutral pion decays (*center*) into two oppositely directed photons, which encounter polarizers on their way to counters.

hand photon makes an arbitrary angle θ with the vertical. When it arrives at PF the probability that it will pass through is $\cos^2\theta$. The second photon makes the same angle θ with the horizontal axis of PG, and so the probability that it will pass through is the same. Since there is no possible causal connection between what happens at PF and what happens at PG, which may be far away, these probabilities must be absolutely independent of each other, and the probability that both photons pass through their respective polarizers is the product of these two probabilities: $\cos^4\theta$. Since θ is undetermined and there is no way in which the setting of the polarizers can influence the decay of a pion, all values of θ are equally probable, and the probability that of a pair of photons selected at random both will pass through is the average of $\cos^4\theta$ over all angles, which is 3/8. The value calculated from quantum mechanics was 1/2. Since the probability that a photon passes through PF is 1/2 we can express the new result in another way. *If* one photon passes through PF, the probability that the other one will pass through PG is 3/4. In quantum mechanics, and in reality, it is 1. The assumptions and the calculation were so simple that it is hard to see where a mistake could have been made. I assumed that photons started off with their polarizations at right angles to each other, that each had the well-attested probability of $\cos^2\theta$ that it would pass through its polarizer, and that the passage of one is *statistically independent* of that of the other.

Historically there have been two reactions to this situation. One is to say that the photons have no polarization properties as they approach the polarizer and that the polarizer creates the property when a photon passes through or perhaps later, when it is detected. The second is to say that statistical independence is violated—this would be as if you were to toss a coin in one room and I in another and there were some unanalyzable causal connection at work between the two rooms so that the probability of our both throwing heads was no longer 1/4 but somewhat greater.

There is a third point of view, which seems to be held by most physicists today, that quantum processes are so far removed from the world we know from our experience that we should consider ourselves lucky when intuition works at all, but that when, as here, it gives a discrepant answer we should check the experiments and carefully verify the quantum calculation and then believe them, since they always seem to agree.

Strongly intuitive physicists are the ones who are the least satisfied by this situation. Einstein was so offended by a theory that predicted results of this kind (the experiments had not yet been done) that he was sure quantum mechanics could not be right. In 1947 he wrote to his old friend Max Born:

> I cannot make a case for my attitude in physics which you would consider at all reasonable. I admit, of course, that there is a considerable amount of validity in the statistical approach which you were the first to recognize clearly as necessary given the framework of the existing formalism. I cannot seriously believe in it because the theory cannot be reconciled with the idea that physics should represent a reality in time and space, free from spooky actions at a distance. (Born 1971, p. 158)

Much more recently Richard Feynman expressed his uneasiness this way:

> We have always had a great deal of difficulty in understanding the world view that quantum mechanics represents. At least I do, because I'm an old enough man so that I haven't got to the point that this stuff is obvious to me. . . . It has not yet become obvious to me that there is no real problem. I cannot define the real problem, therefore I suspect there's no real problem, but I'm not sure there's no real problem. (Feynman 1982)

Feynman trusts the theory and his tone is resigned, but Einstein and others tried to do something about what they thought was an intolerable situation. One way out of it is to assume that only our ignorance makes processes seem to occur at random, that there are causal mechanisms at work that we simply do not know about, and that what appear to be "spooky actions at a distance" are really the result of these same causal mechanisms. Most physicists, including myself, regard these efforts as something of a dead end, but it is not impossible that they might succeed. The next section will describe the kind of thinking that goes into them and will show why most of us think as we do.

Hidden Variables

Especially in the early days of quantum mechanics, critics thought that a fundamental theory ought to do better than just make statistical predictions. Every observation we make is an individual case, so why has quantum mechanics nothing to say about individual cases? An obvious answer was to say that causality reigns, that what happens is determined by the values of certain dynamical variables, and that the reason things seem to us to be happening at random is that we are ignorant of these hidden variables. Hidden variables have already occurred several times in this book. When Newton sought to explain why some light corpuscles striking a glass plate pass through while others are reflected, he reasoned that if they all reach the glass traveling in the same way they ought either all to be transmitted or all reflected. To explain what happens he invoked the undetectable ether: light excites pressure waves in it, like those of sound, which alternately aid and hinder light corpuscles when they reach the surface. The ones that are aided pass through. Today most physicists would say that it is purely a matter of probability whether a given photon passes through. Newton wanted it to be physically determined, and the numerical properties of the vibrating ether would have been his hidden variables had he tried to work out the theory in detail.

Another hidden-variable theory is the kinetic theory of gases. In order to explain various thermal properties of a gas theorists in the nineteenth century imagined it to be composed of molecules moving at immense speeds through otherwise empty space. Only in this century have there been techniques for verifying the truth of this assumption by measuring molecular speeds and masses; before that the numerical properties of moving molecules were hidden variables.

Einstein, Podolsky, and Rosen rejected quantum mechanics as an incomplete theory. They did not say so explicitly, but their conclusions assumed hidden variables, and several talented people have tried to invent some (Belinfante 1973). The results have this in common: the assumptions and formulas look complicated and artificial to everyone except their proposers. This does not mean that there could not be a satisfactory hidden-variable theory if someone were ingenious enough to find it. Remarkably though, it turned out that in some important cases it is possible to show that there cannot be hidden variables, or at least most physicists think so.

In 1964 appeared a paper by John Stuart Bell, a Irish physicist who works at the European high-energy laboratory called CERN, in Geneva. In this paper he investigated the experimental predictions of hidden-variable theories that are arbitrary except that they obey two reasonable restrictions called causality and locality. Causality is the assumption that motivates the theory: that the result of a measurement depends uniquely and causally on the values of the relevant variables, both hidden and knowable. In the kind of two-particle system considered by EPR there are two variables or sets of variables, one for each particle. The assumption of locality is that the result of measuring a certain property of one particle does not depend on what kind of measurement is being made on the other particle as long as the particles are reasonably separated in space and there is no identifiable causal link between the two measurements. (This assumption turns out to be the stronger of the two; the assumption of strict causality can be weakened without changing the results.) Bell analyzed an idealized experiment, and later he and others (see Clauser and Shimony 1978) analyzed more realistic ones to see how their possible results were restricted by the two assumptions, regardless of what the details of the hidden-variable theory might be. Details are given in Bell's paper and the one just quoted; I omit them all and pass to the result.

A correlation function is a quantity that tells how the value of one variable A is correlated with that of another variable B. It varies between 0 and 1: if it is 0, knowledge of A tells us nothing about B; if it is 1, knowledge of A determines B exactly. It turns out that there are experiments in which the possible results of measuring a correlation function are distributed along the line between 0 and 1 so that the range of values possible in a causal and local hidden-variable theory is that marked HV, while quantum mechanics predicts a definite value Q that lies outside this range (see Figure 16.4). If experiment yields a value in the range HV then quantum mechanics is wrong and there exists at least one hidden-variable theory (if it can be found) that agrees with experiment. If the experiment yields Q, then quantum mechanics is supported and no hidden-variable theory that satisfies the assumptions can be found. The remaining possibility is too horrible to contemplate but apparently we need not do so, since a number of different experiments, mostly during the 1970s, have yielded values corresponding to Q.

Our naive explanation of the EPR experiment of Figure 16.3, which gave the wrong answer, is a simple example of a hidden-variable theory. The polarizations of the two photons are the hidden variables—we may have thought that in the experiment we were measuring these polarizations, but actually we were only observing counts on two counters; the rest was interpretation. The causality assumption was the assumption that a photon possesses a certain polarization that determines the probability that it will pass through a polarizer with a given orientation. The locality assumption is the assumption that the result of a measurement at PA can have no effect on what happens at PB. The correlation coefficient corresponds to the probability that if one photon passes through PA the other one will pass through PB. The hidden-variable theory gives this as 3/4. Quantum mechanics and experiment give it as 1.

Bell's theorem was a great discovery because it showed that an important question that had previously been considered as a philosophical one could be decided by experiment. The spooky actions at a distance are not

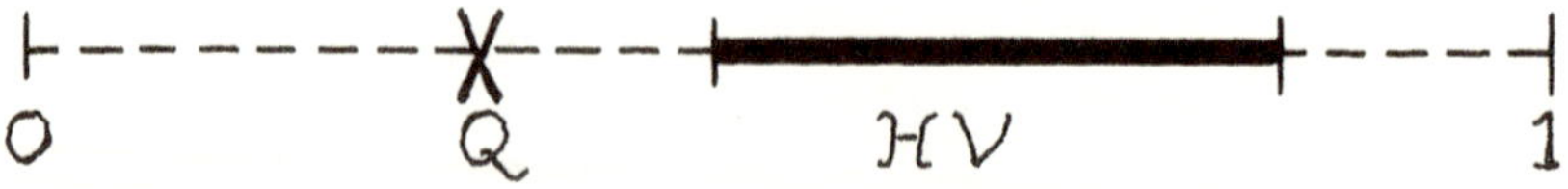

FIGURE 16.4. To illustrate Bell's theorem. The value Q of a correlation coefficient as calculated by quantum mechanics lies outside the range allowed by any hidden-variable theory.

pathological results of an erroneous theory; they seem actually to occur and they cannot be explained by hidden variables. Of course a situation like this is never solved to everyone's satisfaction. There have been murmurs from people willing to relax the assumption of locality, but what urgency the question of hidden variables ever had, and in the community at large it was never great, has by now evaporated. This is not to say that there is unanimity on how to explain the theory to one's next-door neighbor, but most physicists have more to do than argue about the deeper significance of a theory that for them is as natural as breathing and has in addition passed every experimental test. The questions of principle are of historical importance, but we can calculate observable effects without raising them any more. They have to do with intuition, and perhaps they will be important once again in the transition to a new and better theory.

Superposition

Suppose we try to make a theoretical description of a system S as observed by a set of instruments I. If the instruments are doing their job, anything that happens in S is accompanied by corresponding changes in I. Before the 1920s we would have said that the function of the instruments is to enable us to know, objectively, what is going on in S. Since then we have learned that the clean conceptual split between observer and observed has no counterpart in the theoretical description and that the word "phenomenon," for example, is best used to describe what happens to the entire arrangement, S + I, rather than the result of an attempt to disentangle them. In this description it is assumed that I contains whatever devices are necessary in order to amplify the tiny impulses reaching us from the microscopic world so that we can observe them with our unaided senses. Experience has shown that at that point there is no further need for scientific subtleties. If one human observer thinks that the needle of an instrument points to the number 6 on the dial, or that a hole has been punched in a paper tape, it appears that everyone else with a clear view of things will agree. The act of amplification ends the process.

The quantum-mechanical wave function ψ is a mathematical quantity that keeps track of the system S + I. The equations of the theory tell how ψ develops as time goes on. If at some initial moment ψ describes S together with what is essential about I, then the theory is supposed to tell what will be found after the experiment has run its course and the result has been noted and recorded. One calculates with ψ, and its mathematical properties are important. Of these perhaps the simplest is that it can be added and subtracted. We have already seen an example in the explanation of the interference pattern formed when a beam of particles splits into two parts that, after following different paths, recombine at a screen. The ψ at the screen is formed by adding the ψs of the two parts. Since ψ represents a wave, each part is alternately positive and negative in sign. If the ψs superposed have opposite signs their sum is small, perhaps zero; if they add, it is larger than either. The intensity of the beam—the number of particles arriving per second—is found by squaring this quantity. Note G shows how it alternates along the screen between positive values and zero, and this alternation matches what is actually seen. But in Newtonian mechanics, and in our daily mechanics of pushing and pulling, there is no idea of adding two objects together. One match plus one match equals two matches, not a single larger match or no match at all. We can expect that the consequences that follow from applying on the large scale this principle of superposition, as it is called, may be paradoxical. Let us look at the matter more closely.

If first a violinist and then a horn-player sounds an A the pitches are the same but they do not sound the same. We understand that this is because they are not pure tones; rather, the characteristic sound of each instrument can be resolved into several pure tones with appropriate amplitudes. We can say that the tones are composite. This is a commonplace in acoustics, but in 1925 it was not a commonplace that mechanical systems could exist in states that are in exactly the same sense composite: just as a horn produces several tones at the same time, an atom can occupy several of its possible states, with various amplitudes, at the same time—this is how, as we have seen, it can even be in two or more of its possible places at the same time.

Consider now a radioactive atomic nucleus which at some initial instant we know to be of a certain kind. We put it into a counting device and a few minutes later it clicks to tell us that the nucleus has emitted a particle and become a nucleus of a different kind. The exact time at which we receive the signal cannot be predicted in advance, but it is possible to characterize any radioactive species by its half-life: the probability that the signal will occur during that period of time is one-half.

At first sight the theory of such a process seems very strange. The nucleus, originally in a certain state, at once begins to make a transition into another state in which it exists simultaneously with the first state. As the amplitude of the second state increases, that of the first state goes down. At the end of a period equal to the half-life the amplitudes are exactly equal. At any moment the square of the amplitude of the second state represents the probability that the nucleus has decayed by that time, while that of the first state represents the probability that it has not decayed. The two probabilities, of course, add up to 1. The theory says nothing about the sudden signal from the counter; it cannot, since it deals only with probabilities, and the probabilities vary continuously with time. Since a radioactive sample ordinarily consists of a very large number of nuclei the probabilistic description translates readily into a statistical one. We do not feel cheated if we are told that when the half-life has elapsed about half the sample will have decayed, but if we think about any individual nucleus we tend to think of it as being either in its initial or its final state and not in both at once.

Such considerations may be a little less outlandish when we are talking of an electron or an atomic nucleus, which no one has ever seen anyhow, but consider that we ourselves and everything around us are made of electrons and nuclei, and that if quantum mechanics is a good theory it ought to apply to large groups of atoms—counters for example—as well as small ones. Since we are supposed to think not just about S but about S + I, let us think about the state of the counter with the single radioactive nucleus in front of it. Initially the counter has not fired and the nucleus is in its undivided state. As time progresses the probability that there has been a transition increases. If we assume that quantum mechanics governs large systems as well as small ones, the state of the nucleus plus counter develops the same composite character as the nucleus did previously. The counter and nucleus now exist in two states at once. Perhaps there is a dial on which a pointer would point in two different directions at the same time. This seems to be a predicted consequence of the strictly causal nature of quantum theory, but it is a thing that no one has ever seen.

The law of contradiction, in logic, states that everything is either A or not-A. It is based on experience. By now we are pretty sure that the kind of argument through which I have just tried to violate it is wrong and that the law survives without damage. Where is it wrong? The reason physics has made such great progress since Galileo is that physicists have learned to simplify their thinking by concentrating on what is important and ignoring what is unimportant. Physics deals with idealized situations in which vacuums are perfect, particles move perfectly freely through space, observations are instantaneous, and so on. The arguments that seem to violate the law of contradiction seem to carry that principle too far. The counter was assumed to be alone in the world, connected only to the decaying nucleus, whereas real counters sit on a laboratory table in a lighted room, exchanging messages with their environment all the time. Apparently, although there is no general theory, it is these exchanges that eliminate the paradoxical superposed states. Note L gives a fanciful illustration of how it might happen.

There is, of course, an enormous literature about all this, going back fifty years, but it still confronts me with a choice. The Copenhagen interpretation stops with the observed fact and has no answer for questions that wander into the regions beyond. In that region I find myself confronting an apparent contradiction between what quantum mechanics says about the atoms and molecules of my body and brain and what I say about my mental state, which I know better than I know anything else. I say it is an *apparent* contradiction because I do not know. There is a great conceptual gap between talking about atoms and molecules of the brain and talking about a mental state. I do not require or expect a physical theory of consciousness. It is a question of nonoverlapping categories of discourse. But still, there might be some physicochemical scheme of mental function not amounting to a theory, some pattern of connections between the different categories that would serve as a guide for physicists who feel the need to incorporate observers into their theories and perhaps also for those at the other side of the gap, who think about thought.

One would expect that a medieval nominalist, facing quantum mechanics, would not find it hard to understand. The theory expresses the state of a system using the mathematical quantity ψ to calculate the statistical distributions of measurable physical quantities. The symbol ψ appears in the calculation but never in the final result, since it is just the name for something that is not observable. If the calculation agrees with experience nothing else matters. The realist would demand more. He might say that the facts statistics deals with are composed of individual events and the quantity ψ tells all that can be said, in advance, about an individual event. In this sense ψ stands as an ideal and precisely knowable quantity whose examples in the world of experience reflect the uncertainties of that world. Viewed in this way, ψ is not just a quantity that occurs in the course of a calculation; it expresses all that can be truly and exactly known about the state of a system in advance of an actual trial. Both are right, of course, but that need not prevent them from having a fine argument that might last for centuries.

Quantum Mechanics and the Principle of Least Action

I mentioned in Chapter 13 that physicists find it both natural and useful to express physical laws in the form of minimum principles: of all imaginable ways to accomplish something, nature chooses the one that takes least time (Hero, Fermat) or least action (Maupertuis, Lagrange). I mentioned that in optics, nature sometimes prefers to take the longest possible time rather than the shortest, and in 1833 the Irish mathematician William Rowan Hamilton pointed out that in mechanics the action is usually only a local minimum or a local maximum (see Hamilton 1931, vol. 1, p. 311). The essential property, he found, was not that of being greater or less but that of being stationary under small variations of the motion. The principle should be called the principle of stationary action. It is in the nature of smooth maxima and minima that if the actual motion of a particle or a beam of light is such as to make a certain quantity stationary a motion close to the actual one will make only a very small difference in the value of this quantity, whereas other motions further from the actual one may make a much larger difference. This innocent remark will be important in the argument that follows.

When we turned away from Maupertuis in Chapter 13 it was with a question. If I throw a ball from P to Q and the law of motion says that the ball will follow the path that minimizes a quantity called the action, how does the ball know what path to take unless it tries them all? To begin to answer this question I point out a case in which we have already seen that it tries more than one. This is the two-slit experiment, which, if we think about it often enough, seems to ask almost all the questions and tell almost all we can know about the meaning of quantum mechanics. In that experiment we offer the particle a choice of two straight paths, and the interference pattern on the screen shows that it has indeed, in some sense, tried them both. A similar experiment passes the particles through a lattice with a very large number of slits (Figure 16.5). Note that the function of the bars between the slits is to stop some of the particles and let the rest through. They do not *make* the particles go through the slits. Take the lattice away gradually, bar by bar. Now where do the particles go? We may say that they move directly to the screen, but why would their paths stop fanning out just because the bars are taken away? We can also imagine an arrangement with many lattices in which a particle takes many possible paths as it diffracts from one to another, and then gradually take them away. Figure 16.6 shows the result of carrying the argument to its logical extreme: the particle tries every path. For enlightenment look once more at the apparatus with two slits. When a particle arrives at point P on the screen by the two routes offered, what is seen on the screen is determined by the sum of the two wave functions at that point and the interference pattern shows that both slits were traversed.

What has this to do with the principle of least action, other than to suggest that in some sense a particle does try every path? To see the relevance, recall from the two-slit experiment that when we ask what hap-

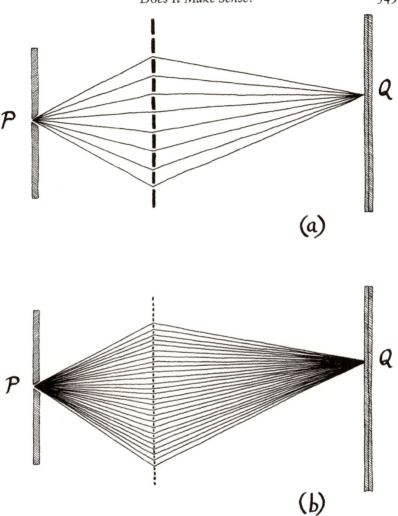

FIGURE 16.5. *Top*: A photon that we know to have started at P and ended at Q tries many paths as it passes through a grating of many slits on its way to Q. *Bottom*: the grating is taken away and its position is marked by a dotted line. The photon still tries many paths.

pens at point Q on the screen the wave function that contains the answer is formed by adding contributions from the different beams arriving there. To add these contributions we must know the relative phases with which they arrive: if they are in phase they add; if they are out of phase they subtract. This means that we must focus on the phases of the wave functions corresponding to the uncountable infinity of possible paths

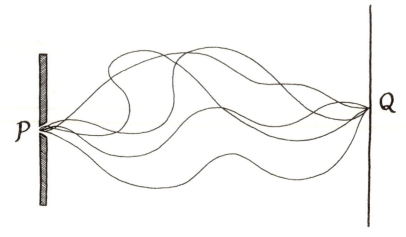

FIGURE 16.6. A photon traveling from P to Q tries every path.

leading from P to Q and occupying a given amount of time. A simple argument (Park 1979, p. 219) shows that the wave along any path changes its phase from start to finish by an amount equal to $2\pi/h$ times the action as Lagrange calculated it (the form used by Maupertuis is less useful here). If all the component wave functions start out from the source with the same phase, the difference in phase between component wave functions arriving at Q is given by the difference in actions evaluated along the paths.

Imagine a ball passing from P to Q along the actual parabolic path at which the action is stationary. It arrives at Q with a certain phase in its wave function. Imagine a path very close to this one. Because the action along the actual path is stationary, the wave arrives with almost exactly the same phase; the two reinforce each other. But now start adding wave functions corresponding to paths that are more widely different. The values of the phases will be very different, and the more of them are added the more surely they will tend to cancel each other. Thus, finally, almost the entire contribution to the wave function at Q will be furnished by a narrow sheaf of paths very close to the actual one. The resultant wave function will be large along this path and very small elsewhere. In this way we can answer those who ask how the particle knows which path is extremal if it has not tried the rest. It tries them in the quantum-mechanical sense, not that of physical presence. Physical presence implies detectability. If you install an instrument far from the actual path in the hope of detecting an adventurous ball you will be disappointed, since the various paths leading from P to the instrument will be so far from the actual one that their phases will be almost randomly distributed and the amplitude of the resultant wave at the instrument will be near zero.

That there is a connection between the action evaluated along a classical path and the phase of the wave function was pointed out by de Broglie at the very beginning. Later Feynman showed with brilliant ingenuity that he could actually derive Schrödinger's equation from this connection, and today this is considered the most direct route for the derivation of quantum-mechanical formulas starting from what we know of physics without quanta. The most difficult problems in this area are generally approached using Feynman's method.

Nothing enters the understanding except through the senses, said Saint Thomas Aquinas, and our sensory impressions are of happenings on too coarse a scale for quanta to be perceptible. When you think about it, with quanta so small and conceptually so far away, it is remarkable that we know anything about them at all. In the last two chapters I have tried to show how the quantum code was broken, as far as it has been broken, mostly in very small steps. In the next chapter we move further down the scale, and in the one after that we shall look at things that are large and far away.

CHAPTER 17

Moving Down the Scale

Let proportion be found not only in numbers and measures but also in sounds, weights, times, positions, and whatever force there is.

—LEONARDO DA VINCI

THE last three chapters celebrated some great advances in thinking about the world, but they did not talk much about the models that explain what the world is made of. At first the quantum theorists measured their success by their ability to explain what we observe by using the atomic model, and even in that limited program they encountered enough conceptual problems to keep them busy for some time. As they gained skill they began to study the nucleus, experimentally by looking at the radiations it gives out when bombarded with energetic particles, and theoretically by inventing models based on the assumption that it is a nearly homogeneous spherical mass of neutrons and protons. The model is hard to analyze because the explosive electrical force with which the protons in a nucleus repel each other must be contained by an even stronger attractive force that turns out to be of very short range and so entirely inaccessible to the kind of direct study that is possible for electric forces. Our quantitative knowledge of this force is still not very exact. Continuing, what are neutrons and protons, and what is the origin of the short-range force? Each question demanded a model, which was then analyzed according to the principles of quantum mechanics. Each answer led to new questions.

This chapter mixes models with ideas that are model-independent. The question "What are neutrons and protons?" is answered with a model in which they are composed of smaller particles called quarks. The question "What is the force that holds quarks together?" is answered with a necessarily superficial account of how fields of force can be explained by invoking the symmetries of the systems in which they act. This notion has been very fertile for new ideas and has led to new and just possibly correct modes of explanation that go beyond particles to more abstract notions that can be modeled as tiny bits of string. The next chapter will show how one can represent the whole universe in a simplified model that allows us to trace its history back to very early moments. At that point it becomes clear why the physics of fundamental fields and particles has become bound up with theories of the origin of the universe. It is essential to understand each one if we are to understand the other.

Symmetry

When Newton set out to write *The Mathematical Principles of Natural Philosophy* what emerged was an encyclopedia of special methods, some rigorous and accurate, and others merely sketched. Even Euler, replacing the Euclidean mathematics of the *Principia* with the neat and powerful methods of calculus, did not yet aim at a general formulation. For certain parts of mechanics that was the achievement of Lagrange, but it was only the beginning. In the years since 1788 it has been possible to formulate almost every new extension of physical theory in Lagrangian form. For a simple calculation like finding the trajectory of a planet, I have already described the procedure. You choose a set of mathematical variables that is convenient for locating the planet and with them you build a mathematical expression called the Lagrangian, given by the difference between the planet's kinetic energy and its potential energy. From this, for any imagined trajectory, you can derive a number called the action. The planet's actual trajectory is the one for which the size of this number is smaller (or occasionally larger) than it would have been for any other trajectory. Initially Lagrange deduced this procedure from Newton's laws of motion, but as time went on it became clear that it could be more generally applied, and the method was applied to things such as fields that are not governed by Newton's laws.

Suppose the system is a mechanical one that recognizes the difference between up and down. It might consist simply of a stone that is allowed to fall. If you describe the situation as Euler learned to do, with a coordinate system, the up-and-down coordinate plays a special role in the description. But the fundamental laws of nature do not refer to up and down. They refer to the intrinsic tendencies of things, to the way they work if they are sheltered from outside influences. For them the Lagrangian does not change when the coordinates are rotated, for if it did one orientation of the coordinates would make the action smaller than any other. The thing that is not acted upon from outside, that cannot distinguish between up and down, would insist on being described in a way that makes some such distinction where there is no real difference. Physical theory should not be like that, and so the Lagrangian for an isolated system is expressed in a formula that takes no account of directions. The pertinent word here is symmetry. If a physical system has a symmetry the Lagrangian describing it has the same symmetry expressed mathematically. An example would be the solar system. It is fairly flat, like a phonograph record; the galactic gravitational field has very little effect on planetary motions and can ordinarily be ignored. If the solar system is isolated in space we suppose that it would act just the same when placed in some other orientation—upside down, for example. In the Lagrangian this would not make any difference, and planetary motions would be calculated just as before.

There is a result of Lagrangian theory that I have already mentioned: wherever there is a symmetry there is some dynamical variable whose numerical value remains constant. In the case of spherical symmetry, as with the solar system, it is the angular momentum of the sun and all the planets.

At the beginning of this century Joseph Larmor (1900) showed that, like Newtonian mechanics, Maxwell's theory of the electromagnetic field can be derived from a Lagrangian. Because the field is spread out in space the Lagrangian is given as a density, so much of it per unit volume. The formula is remarkably simple: write down the square of the magnetic field strength and subtract from it the square of the electric field. In ordinary notation,

$$L = H^2 - E^2$$

when there is no interaction with anything else. The constants that come from the rotational symmetry of this expression give the angular momentum of the field. It is also true that this Lagrangian does not depend on where one chooses to set the origin of coordinates or when one chooses to start the clock. From these additional symmetries it follows that the field's linear momentum and its energy are conserved.

Until 1900 the logical development went this way: experiments led to a law of motion, an appropriate Lagrangian was written down, symmetries were noted, and the laws of conservation were deduced. In about 1908 Einstein and Minkowski realized that the procedure can be turned around. Suppose you believe in the principle of relativity so strongly that you wish to make it a postulate of your theory. This means that not only is the Lagrangian of the isolated system spherically symmetrical; it has also the same numerical value when viewed by a moving observer. Masses increase, clocks move more slowly, lengths shorten, but when all is calculated the numerical value of the Lagrangian must not change. To see why this is necessary consider the alternative: the Lagrangian is assigned different values by different observers. The value of the action integral will then differ from one observer to another and minimizing it will pick out some particular one for whom it is smallest. But the laws of nature are not supposed to define any preferred state of motion. . . . This is a more restrictive condition than symmetry under spatial rotation. For example, in Maxwell's theory, H^2 and E^2 are each symmetrical under rotations, but the only combination of these quantities that gives a reasonably simple law of motion and is the same when viewed by observers in different states of (uniform) motion is Larmor's $H^2 - E^2$. Another word: transformations to a moving coordinate system are called Lorentz transformations and the quality of not changing under this or some other transformation is called invariance; that is, invariance is the mathematical expression of symmetry. Larmor's Lagrangian is the simplest Lorentz-invariant expression involving E and H.[1] After Einstein's first paper appeared Minkowski discovered a way of writing an expression such as Larmor's L so that its Lorentz-invariance is immediately obvious. Einstein realized that one could now start to invent physical laws by starting with invariance principles and using simplicity as a guide. As you can see, this leads at once to Maxwell's equations for noninteracting fields.

The foregoing argument rests on two profound scientific discoveries. The first is the Lorentz transformations belonging to special relativity. The second is simplicity, no less a discovery than the first. It is not just a matter of introducing compact notations, though that helps; it is the discovery that when those laws of nature that stand the tests of experiment are compared with other possibilities that fail, the successful laws tend to be the ones that are conceptually simpler (though when written out in full they may not look very simple) and to have forms that appear beautiful, or at least neat, to adepts in mathematics. Mathematical beauty is just as hard to define as any other kind, and I shall not try, but users of mathematics tend to agree as to what is beautiful and what is not. Apparently the world is such that the laws governing it tend to be both simple and beautiful. Clearly this reflects some property of human thinking, but also some property of the natural world as well. What exactly is that prop-

---

[1] There is another invariant combination, $H{\cdot}E$, but that cannot function as a Lagrangian; see Landau and Lifschitz (1971, pp. 63, 68).

erty? The question is discussed in private but very little in print, for none of the discussions go very deep.

In the decade that followed the special theory, as Einstein came to understand that his principle of equivalence (another symmetry) had to be expressed in extremely general and unfamiliar mathematical language in order to make a theory of gravity, he came to rely more and more on the two principles of invariance and simplicity, and when he had completed his theory and sought to express it in Lagrangian form, he used the two principles explicitly (1916b). If it had not turned out that the correct equations for the gravitational field are the simplest of all that can be derived in this way, I do not see how anybody could have succeeded in finding them for a long time.

It will be evident that symmetry principles are not merely simple; they are very general. See what this means. Someone proposes a physical principle based on some limited set of observations. The principle embodies a symmetry. The proposer may not know it but perhaps the symmetry is universal, like Lorentz invariance or invariance under spatial rotations. The principle may then have applications far beyond the observations it was based on; it may even (provisionally of course) be right. In 1889 the German physicist Heinrich Hertz spoke about Maxwell's theory before a science congress in Heidelberg:

> It is impossible to study this wonderful theory without feeling as if the mathematical equations had an independent life and intelligence of their own, as if they were wiser than ourselves, indeed wiser than their discoverer, as if they gave forth more than he had put into them. (Hertz 1896, p. 313)

Both Maxwell's electrodynamics and Einstein's special relativity are descriptions that are invariant under Lorentz transformations, though neither man knew it at the beginning.

More Symmetries

There is no way of following the crests and troughs of a de Broglie wave. In the two-slit experiment we can compare the phases of the two parts of the same wave function and find the difference between them, but the phase of either part separately is beyond the reach of experiment. Let us assume it is arbitrary, that if we assigned a different phase to the entire wave function of an electron (or other particle) the description of nature would be unchanged. That is an invariance property. We ask, can it be treated within the Lagrangian framework, and if it can, what consequences follow?

I arrive at the answer in two steps. First, I pretend that I have under my control the wave function of the vast, interlocking system of all the electrons in the universe. I change the phase a little and assume that nothing physical changes. As always, out of this invariance arises a conservation law. If the description of nature is invariant under change of the phase of electrons there exists a quantity that can neither be created nor destroyed, but it turns out to be nothing new. One can tell from the formula describing it that it is a number proportional to the total electric charge carried by all the particles of the field.

Now I press on with a much more radical assumption. It seems a bit presumptuous to pretend that I can by fiat change the phase of every electron in the universe. What happens if I assume that nothing changes when I alter the phase only in some small but arbitrary region of spacetime such as my own? The change should be continuous, no sudden jumps from point to point. The mathematical conclusion is inexorable: as long as there are only electrons in the world there is no such invariance. Such a change alters the value of the Lagrangian. If I insist on invariance, some new element must be introduced into the theory that changes it back. On closer inspection it turns out that the element is not new, it is the electromagnetic field. If the field is present it can absorb and compensate the effects of a change in phase. Provided such a field exists (and we know it does!) the description of electrons is indifferent to the choice of phase, locally as well as globally. Imagine some crazy world in which scientists had discovered electrons but not the electromagnetic field. If someone in that world became convinced that Lagrangians do not depend on the local phase of ψ it would lead to the invention of electromagnetic fields. The conclusions go further. Electric charge is conserved as before, but now we see how it functions in the universe. The new theory requires that electrons interact with the electromagnetic field through their charge, and it specifies exactly how. Does it tell how the electromagnetic field behaves when no electrons are present? No, for the field is required only when there are electrons to interact with. But the mathematical nature of the field variables is specified, and we have only to choose the piece of the Lagrangian that describes the noninteracting field. I have already said how to do that: choose the simplest possible Lorentz-invariant expression, Larmor's. Now we have all of Maxwell's theory, derived from a few simple *qualitative* assumptions added to the elementary theory of de Broglie waves (see, for example, Park 1966, p. 186).

We have already encountered a different example of this same kind of argument. We ride in Einstein's train at constant speed on a straight and level track and proclaim that uniform motion is absolutely undetectable; then the train slows down and our coffee spills and we realize that departures from uniform motion are not undetectable. Acceleration seems to have an absolute meaning, which is why Newton introduced absolute space, but according to Einstein there is no need for it, for if we imagine the physics of a civilization inhabiting space capsules with no knowledge of gravity and then somebody begins to move the capsules around, the inhabitants can explain all the observed phenomena by inventing a gravitational field with suitable properties. The argument is an exact parallel of the preceding one: the principle of equivalence, if assumed to hold locally, together with an assumption of simplicity, leads to the general theory of relativity (Weinberg 1972, ch. 7).

The Form of a Theory

A mathematical theory of anything in nature must have a comprehensible mathematical structure and must also involve some built-in constants. The requirement of structure is obvious enough. The requirement of constants arises because most of the quantities we observe are not simply numerical. Five centimeters are not the same as five seconds or five grams—we perceive them differently and so a theory must involve units or dimensions that connect the numbers with our different sensory modes. We have already seen that a theory of spacetime involves the number c, with units of speed, which according to Maxwell's theory happens to be the speed of light. It is required because we perceive space and time differently and so have come to measure them in different units. Two other constants are probably enough. When conventions have been adopted concerning them, all others can be expressed in terms of them. One is usually taken to be $h/2\pi$, and the last one is sometimes taken to be the gravitational constant G. To simplify calculation all three can be set equal to 1. Further constants are put in by hand, although this ought not to be necessary. We should not have to tell a fundamental theory the value of the fundamental electric charge e or the masses of the various particles; it ought to tell us. People are beginning to perform rough calculations of some of the particle masses, but why

$$\sqrt{(hc)}/e = 29.343,$$

rather than some other number, is not yet known.

Quantum mechanics is a theory of matter, and matter touches our senses in a variety of ways. What is a water molecule? First of all, it consists of thirteen particles: one nucleus of oxygen, two of hydrogen, and ten electrons (that the nuclei are themselves composite does not matter for this discussion). We set up Schrödinger's equation for the ten-particle system and with tricks and computers and simplifying assumptions we manage to solve it. A few constants occur in the process: the masses of the nuclei and electrons, the magnitudes of the electrical charges whose interactions bind the molecule together, and Planck's constant; nothing more. When the equation is solved we understand the molecule's size and shape and begin to understand some of water's special properties, such as why solids dissolve in it more easily than in other liquids. The equation and its solution form a mathematical structure that represents the molecule in an abstract but very precise way, and it is hard to imagine that we shall ever know it much better. Then is the molecule a mathematical form? Aristotle's contemptuous dismissal comes to mind: a mathematical form has no mass. Schrödinger's molecule is not wet and it does not quench thirst, but it has a kind of existence all the same. I do not want to labor the analogy with Plato's Ideas, since the Ideas come to us fragrant with beauty and goodness and appropriateness to a theory of society and how life should be lived, but nevertheless an analogy is there. The mathematical heaven contains a formal structure to which all water molecules conform. What I am suggesting is that the mathematical form is more than just our own description of it. I suspect that if there are other planets whose inhabitants organize their thoughts about nature somewhat as we do, their mathematical water molecule is in some sense the same as ours. What goes in is a general principle applicable to all molecules alike; what comes out are numbers that are demonstrably correct. Has the mathematical form an existence independent of our own? I suppose it depends on whom you ask.

QED

We have encountered the electromagnetic field so far in three different aspects: as a force, as a wave, and as a particle. How are they related? The discussion in Chapters 15 and 16 has shown, I hope, that it is hard to relate waves and particles—they are two possible metaphors with which to talk about the thing that is at the bottom of a variety of phenomena, but we do not actually see the waves or the particles and it is not a good idea to take either of them literally. How are the forces of electricity and magnetism exerted by these waves and these particles? In Maxwell's theory there are no particles of light and the field varies continuously in space. It fills the region between two electrically charged objects, exerting a force that pulls them together or pushes them apart. But this was before the quantum theory. In that theory the field, which as we have seen is a dynamical system that changes in time and transports energy, is subject to the laws of quantum mechanics; it is quantized. A field is spread out in space, and we must associate a ψ with every point. Further, nature fills space with several fields, most of which interact everywhere, but in the entire history of the theory of interacting quantized fields, which goes back to 1929, it has not been possible to perform a single perfectly exact calculation of anything outside of artificially simplified cases. As it has evolved in the last half-century the theory amounts to pictures and recipes for calculation, some quite rough and some very accurate, but even the accurate calculations involve mathematical manipulations that do not belong to rigorous mathematics and are justified only by their results.

The method that works well when it works at all is called perturbation theory, and it approximates the actual situation of interacting fields (nobody knows what that is) by a description in terms of processes in noninteracting fields. Suppose someone wants to calculate what will happen when a beam of fast electrons is directed onto a hydrogen target. Let us consider only what happens when an electron interacts with a proton, since the electrons of the incident beam will simply push the electrons of the target out of the way. The standard way of representing what happens is with diagrams introduced by Richard Feynman (1918–88); Figure 17.1 shows a few (1949b). The charged particles exchange photons, yet they are not real, observable photons but only concrete representations of steps in the mathematical calculation. They are called virtual photons. The interaction takes place *as if* the virtual photons were there. The diagrams are to be read upward. The vertical axis represents time as measured on the laboratory clock and the horizontal one indicates all the spatial dimensions. The first diagram says that when the electron emits a virtual photon and the proton absorbs it, both particles change their directions of motion. The subsequent diagrams show a few of the infinite number of other virtual processes. The last one deserves special attention. A virtual photon hits the electron so as to reverse its direction in time. As it proceeds along its line the time as read on the laboratory clock gets earlier for a while instead of later. At the moment denoted by t_1 on the diagram there are four particles in existence, not two. The interpretation is that the particle traveling backward in time is an antiparticle—the antiparticle of an electron has the same mass and spin as an electron but its electric charge is positive, not negative (Feynman 1949a). It is called a positron and is known to exist. The diagram is therefore interpreted as follows. At A the incoming photon produces an electron and a virtual positron. The positron proceeds to B where it annihilates the incoming electron, while the electron produced at A goes on its way. Nobody thinks these virtual processes actually occur. They are pictorial and verbal representations of mathematical formulas that arise in the course of certain approximate calculations. They have. however, counterparts in the real

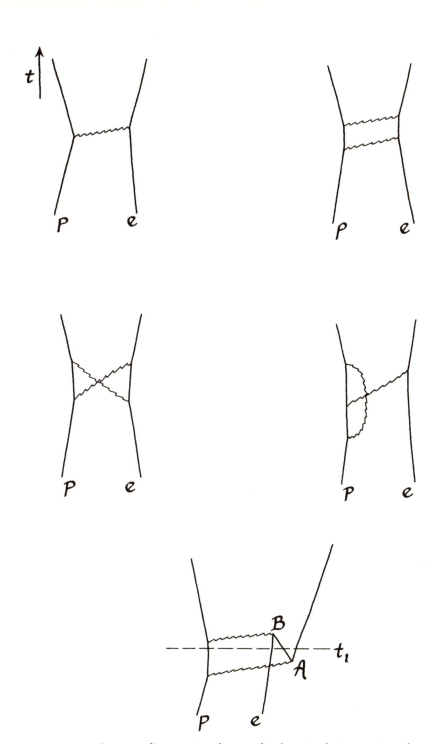

FIGURE 17.1. Feynman diagrams (read upwards) showing the interaction of an electron and a proton through processes involving one or two photons. In the last one, at time t_1, the electron line is seen as three particles. Section BA is interpreted as a virtual positron.

world: real positrons can be produced, which then annihilate real electrons.

To each diagram of Figure 17.1 corresponds a number; the diagram
tells how to find it. The number is part of the quantity called a scattering
amplitude, which tells what will happen as the electron flies past. Most
numbers calculated in this way come out at first to be infinitely large, but
there is an artful procedure called renormalization for throwing out the
infinite parts and keeping a finite residue. The figure shows only a few of
the simplest diagrams that can be drawn. The reason why the diagram-
matic approach yields useful results is that the more virtual lines a diagram contains the smaller is its contribution to the scattering amplitude
after renormalization, and so one can simply stop calculating diagrams
after a certain point and count one's winnings.

When the interaction being considered is electromagnetic the calculations, though laborious, give accurate answers. The theory is called
quantum electrodynamics, abbreviated QED, and it is one of the greatest
successes of physical theory so far. There is a general feeling that in some
way, in spite of dubious mathematics, it may even be right. You put into
the formulas the masses of electron and proton, the value of the fundamental electric charge, the speed of light, and Planck's constant, and after
a good deal of computation an answer comes out. As an example of what
theorists and experimentalists can do, here are recent calculated and
measured values (in suitable units) of the strength of the little magnet that
an electron carries:

$$\text{calculated:} \quad 1.0115965246 \pm 20$$
$$\text{measured:} \quad 1.0115965221 \pm 04$$

where the appended numbers give the uncertainties in the last two digits.[2]
As Feynman remarks, the precision of the measurement is the same as
though one were to measure the distance from Los Angeles to New York
to within the width of a human hair.

But what the theory does not tell us is as important as what it does, for
within limits we can put into the computation any values for those fundamental constants, right or wrong, and an answer will come out. In the
theory of our dreams only the right combination of numbers could be
used and any others would lead to a mathematical inconsistency. Perhaps
this would happen if we could calculate more exactly. After all, nature is
solving these equations, or perhaps some others, exactly, all the time, and

---

[2] Taken from Feynman's *QED* (1985), a book that explains nonmathematically and in
detail, for those willing to take the trouble, exactly what goes on in QED.

it always gets the same value for the mass of an electron. Too bad we cannot.

Quarks

In the most simple-minded view a nucleus is a roughly spherical mass, a few times 10^{-12}cm in diameter, composed of neutrons and protons. Most nuclei are quite stable in spite of the electric force between protons that tends to blow them apart (it is easy to calculate that a proton on the surface of a nucleus of uranium is pushed outward with a force of 20 kilograms). There must be a stronger force that keeps them together, and in 1935 the Japanese physicist Hideki Yukawa (1907-81) constructed a theory in which a nucleus is held together by the exchange of some virtual particles now called pions. These differ from photons in several respects. Photons have no rest-mass, but Yukawa calculated that his pion was about three hundred times more massive than an electron. Unlike photons, at least some pions would carry electric charge, and they would not spin as photons do. This last is important, since the algebraic sign of the force is correlated with the spin of the virtual particles: if they have no spin they produce an attractive force between similar particles, and that is necessary if they are to account for the stability of the nucleus. Yukawa's pions were discovered and studied experimentally in the 1940s. There are three kinds, bearing positive, negative, and neutral electric charges, and their masses are about 270 times that of an electron. They are denoted by the letter π. Figure 17.2 shows a neutron and a proton interacting in two possible ways by the exchange of a pion. Note that a charged pion changes a neutron into a proton or vice versa.

Yukawa's theory, though based on a very clever idea, was only a sketch. During the 1950s and 1960s immense efforts were made to make it into a satisfactory theory of nuclear forces but nothing worked, and ultimately it turned out that this was because the underlying model was wrong. Neutrons and protons and pions are not fundamental particles (for the discovery of particles see Ne'eman and Kirsh 1986), because they are made of more elementary things called quarks.

I shall not go into details, for they are not simple and it takes a book (such as Fritzsch 1983 or Davies 1984) to simplify them. The basic idea is that there are several types of quark named, in order of increasing mass, up, down, strange, charm, beauty, and truth. (the last two are also commonly called bottom and top). These are the six flavors of quarks, and each comes in three colors, usually called red, blue, and green. (None of these whimsical terms is to be understood in a literal sense. They are used because they express no assumptions about the nature of the fundamental

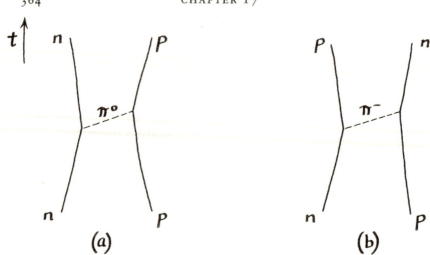

FIGURE 17.2. A neutron and a proton interact by exchanging a pion. A charged pion interchanges the particles.

properties they represent, for we do not know what those properties are. The alternative would be to use symbols, which would be hard to remember, or else sounds that have no prior meaning at all.) Though quarks of different flavors have characteristic masses and electric charges, different-colored quarks of the same flavor are thought to be exactly alike except for the color. Protons and neutrons are composed of up and down quarks, abbreviated u and d; pions consist of u, d, and their antiparticles $\bar{u}$ and $\bar{d}$. Photons, gravitons, and particles of the electron family are not supposed to be made of quarks. Pions exist in three different states of electric charge. Table 17.1 shows some relevant properties of these quarks and how to make some observed particles out of them. Note that the mass of a particle is much greater than that of the quarks composing it; this is because virtual particles occur along with the quarks.

Figure 17.3 shows how Figure 17.2(b) looks when interpreted according to the quark model. Note that for a short part of its path a u-quark travels backward in time, serving as the antiquark in the structure of the $\pi-$. Note also that four of the six quarks in the picture serve merely as spectators.

Quarks were invented independently by Murray Gell-Mann (b. 1929) and George Zweig (b. 1937). These were not the first to invent particles; Dirac, Yukawa, and others had already done so, but the other successful inventions are patently successful because the particles have been observed in the laboratory. This does not mean they have been seen or

TABLE 17.1
Quarks, Nuclear Particles, and Pions

Particle	Composition	Electric Charge	Relative Mass[a]
u	u	2/3	0.0054 ± 0.0016
d	d	$-1/3$	0.095 ± 0.003
$\bar{u}$	$\bar{u}$	$-2/3$	0.005
$\bar{d}$	$\bar{d}$	1/3	0.01
p	uud	1	1.000
n	udd	0	1.001
π^+	u$\bar{d}$	1	0.015
π°	u$\bar{u}$, d$\bar{d}$, s$\bar{s}$	0	0.015
π^-	d$\bar{u}$	-1	0.015

[a] The quark masses are estimates (Gasser and Leutwyler 1982). The wave function of a neutral pion is a superposition, in the quantum-mechanical sense used earlier, of the three particle–antiparticle states shown.

heard, but only that the particles have been produced so that experiments can be performed on them to study their properties. Quarks are different. For some years they were sought in the ground, in seawater, at the particle accelerators, but while these efforts were failing the theory slowly revealed that they must fail, for the force that binds quarks together is so strong at large distances that they cannot be separated. Figure 17.4 shows what happens if someone tries to knock a quark out of a proton by hitting it with a fast particle, here labeled an electron. The quarks act somewhat as if they were confined in an elastic bag. As the struck quark moves off the bag stretches until it breaks, and at the break a quark-antiquark pair is produced. The result of the experiment is not an isolated quark, but just some particles that are already familiar.

I have said nothing about the force that binds quarks together into pions and nuclear particles. Its general nature is the same as we have seen before; the quanta of a new field, called gluons, jump back and forth between the quarks, and what determines their interaction with the quarks is color. Gell-Mann has named this theory quantum chromodynamics (QCD). The properties of the color-bearing field are unfamiliar, however, and we must pause to see where they come from.

Gauge Fields

In 1954 appeared a short but weighty paper by Chen-Ning Yang (b. 1922) and Robert L. Mills (b. 1927) suggesting a new approach to the problem of nuclear forces. It did not become clear until later, when

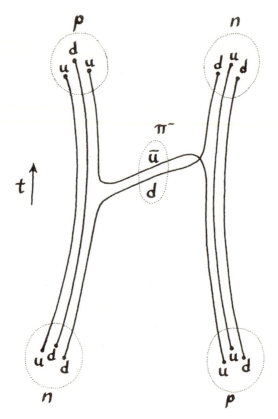

FIGURE 17.3. Interpretation of the second diagram of Figure 17.2 as the exchange of two quarks.

quarks had been invented, how their idea should be applied, so I shall sketch it as it is used in the theory of quarks.

I have mentioned that quarks occur in different colors—the word denotes some distinction that we do not understand. In every physical property we do understand, quarks of the same flavor and different colors are identical. A proton, for example, consists of three quarks, one of each color, but which quark has which color seems to make no difference and we imagine each quark to be in a color state that is a superposition of the basic colors. What follows if we assume that in a proton the choice as to the proportions of red, green, and blue superposed in each quark can be made independently at each point of spacetime? The conclusion is like the one that comes out of the argument leading to electromagnetism: a new field of force, the gluon field, is required, coupled to the quarks by property of color just as the electromagnetic field is coupled by the property

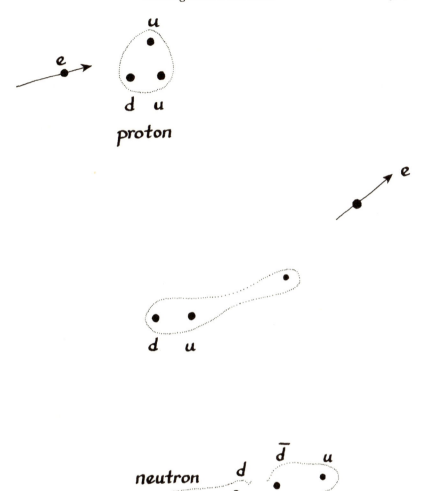

FIGURE 17.4. The attempt to knock a quark out of a proton by means of a fast electron produces a neutron and a pion.

of electric charge. There is an important difference, though. Whereas the electromagnetic field is itself electrically neutral, so that photons do not interact with each other, that is not true of the gluon field: gluons themselves are colored. This makes the Feynman diagrams much more numerous and complicated, but the present imperfect methods of calculation seem to support the claim made earlier that the force between two quarks does not decrease as the distance between them increases, so that ultimately something snaps, while other forces like those of gravity and electricity get smaller.

That is a qualitative result of the theory. Quantitative results are hard to get because of another difference from electrodynamics: if one calculates something using Feynman diagrams, the contributions from complicated diagrams in QCD are not ordinarily smaller than they are from simple ones, and therefore calculations using conventional perturbation theory would have no end. Other methods of approximation must be found, and this has turned out to be difficult. As yet there are no really accurate calculations of anything.

Yang-Mills theories are also called gauge theories. QED is a gauge theory. I have mentioned that calculations in these theories usually give answers that are infinite, and therefore absurd until they are made finite by the mathematical process of renormalization. What makes people hope that the future holds some form of gauge theory is that of all forms of field theory, only gauge theories are known to be renormalizable.

The qualitative successes of QCD would hardly be enough to establish it as a strong argument for the validity of gauge theories, but there is another theory constructed from the same principle for which calculation is easier. This is the theory of Steven Weinberg (b. 1933) and Abdus Salam (b. 1926), which extends quantum electrodynamics so as to unite it with the theory of weak interactions. Weak interactions are manifest in some transformations of unstable particles, and they are so very much weaker than electromagnetic interactions that for a long time no one imagined that there was any connection between them. The Weinberg-Salam theory describes a single interaction that manifests itself either weakly or strongly according to whether the virtual particles exerting the force (see Figure 17.1) are massive or not. That was the first step in the ambitious program that physicists have set for themselves, to find a single theory that accounts for all the interactive forces in nature: gravity, QCD, and the weak-electromagnetic force. Such grand unified theories (GUTs) are at present a distant goal. The best that has been done is what is called the standard model, which includes QCD and the Weinberg-Salam theory. There is a general feeling that this theory is largely correct. When approximate calculations are possible they agree with experiment and a consistent and reasonable picture is emerging. But the situation is not as good as it sounds for the standard model is several theories stuck together, not unified. It contains a variety of independent assumptions and requires at least seventeen constants to be put in by hand. Nobody thinks it is a complete theory; the more general mathematical structure that will bind it together and fix the arbitrary choices has not yet been found.

The standard model is like Ptolemy's astronomy: given many numbers relating to the various spheres of a planet, one could predict its path through the sky with reasonable accuracy but without understanding, in the modern sense, what was going on. After Newton, each isolated planetary orbit is completely specified by just five numbers. Two, which might be the length and the width of its ellipse, determine not only the orbit's shape but also the planet's speed at every point. The other three merely tell how the ellipse is oriented with respect to the stars, and the reason it is all so much simpler is that now we know what is going on.

The standard model is not the only approach to the problem of unification. There is a theory called supergravity, which is burdened with the prediction of a large number of unobserved particles, and finally there is a theory containing supergravity in which the fundamental units are not particles but little strings. These objects were not introduced arbitrarily into physics; rather they came out of trying to make sense of semi-empirical formulas that describe some high-energy phenomena. Superstrings, as they are called, are imagined to be very short, about 10^{-33} cm in length. This is 10^{24} times smaller than an atom: measuring an atom in superstrings would give about the same number as measuring the diameter of a galaxy in terms of the thickness of a human hair. In most versions of the theory the strings are closed into loops, and it is not these loops that represent particles but the various ways in which the loops can oscillate. The energies of these oscillations, expressed as masses by the Einstein equivalence, are the masses of the particles we know. The four dimensions of spacetime are not enough for superstrings; they can live only in a multidimensional space. Even those words, though, are misleading, for this is a mathematical space. The kind of space we know does not contain superstrings; it consists of them, and the traces of matter and force that exist everywhere are manifestations of their activity. We have seen that Newton did not believe empty space could exist. I wonder if he would have liked superstrings.

How can this multidimensional space be squared with the four dimensions of the world we know? Remember that four dimensions do not occur in holy writ. It has been found from long experience that they allow us to describe everything we see happening around us, but is this everything that does or can happen? It turns out that in the new theories there is a range of new phenomena that occur at very high energies, by which I mean energies so high that a single particle would move with the energy of a speeding locomotive. Compared with this kind of thing, whatever we experience or are likely to experience takes place at ultra-low energies, at the very bottom of the scale. The question then is: are four dimensions enough to represent in some reasonable approximation the events observed in this little range? If so, four are enough for us, and as in the

Kaluza-Klein theory of the 1920s, the extra dimensions may be rolled up in some way.

The mathematics of superstrings, involving the theory of geometrical structures embedded in spaces of many dimensions, is very difficult and still strange to most physicists. No exact calculations have been done, but what the theory says qualitatively has a number of contacts with nature. In ordinary quantum field theory, as I have mentioned, there is no room for gravity; it just doesn't seem to fit. If the superstrings are loops, it not only fits; it is compulsory. If there were no gravity the theory could not possibly be true. The standard model in particle physics is composed of various elements arbitrarily selected so as to fit the facts, and most of the numbers that go into it must be supplied by experiment. Once a superstring theory has been selected from the handful of forms that are possible, there are only a few numbers to specify and most of them only serve to adjust the scale of quantities in the theory to the scales (grams, centimeters, seconds) in which physical quantities are ordinarily measured. Once the strength of a single interaction is specified, the other interactions, "whatever force there is," and whatever kinds of particles the world contains, follow necessarily from this one choice. Right or wrong, each version of the theory stands as an unalterable whole that cannot be tinkered with so as to make it agree with experiment.

But, as I have mentioned, supersymmetry theories have a little difficulty, for they specify a vast number of particles that have not been observed. There may be a way out of this. . . . Some physicists regard superstrings as an intellectual fad something like structuralism, full of new words and fancy arguments but finally making a contribution to human knowledge that is in the neighborhood of zero—perhaps a little more, perhaps a little less.

Finally, what are we to make of proposals that assume extra dimensions of spacetime? Perhaps superstring theory contains a kernel of mathematical truth, perhaps not; but one idea is probably here to stay. Until recently the dimensions of spacetime were regarded as stage-setting for the drama of the physical world. Now they belong to the cast of characters.

Let us take a moment to compare the three theories discussed in this chapter: QED, the standard model, and superstrings. What Thomas Kuhn (1962) calls normal science proceeds by a process of patching things up. An idea looks good; one tries to find inconsistencies buried in it or facts that it cannot explain. Once this is done, perhaps the idea can be stated a little better so as to get around the difficulties. QED, while it is the result of normal science, is hard to see as a subject for further normal science; its glassy perfection has no obvious flaw that one could seek to mend. Superstrings are an immense conceptual leap, an attempt to guess the answer to all the riddles of fundamental physics at once. If it succeeds it might be classed as what Kuhn calls a scientific revolution. Compared with those two fortresses in the sky, the standard model is the ideal candidate for normal science, but human beings are strange. For them, mountain climbing and parachute dropping are sports and not just ways to escape from an enemy. As this is written, many of the finest theoretical intellects are concentrating on superstrings and many of the most imaginative experimentalists spend their time designing experiments for a future particle accelerator so vast that it can only be built by a great government. In this area at least, it would be hard to say that normal science is getting the attention it deserves.

One of the difficulties in knowing whether superstrings and other exotica are there or not is that the experiments that would decide the question lie beyond anything we can do, and some of them beyond anything we can imagine doing. They would involve particles with locomotive energies. But the road is not entirely blocked, for at the moment the universe began there may have been particles with such energies, and though they existed only briefly their properties must have had some detectable effects on the world we know. It is time to put the microscope away and take out a telescope.

CHAPTER 18

And Now the Universe

Philosophical astronomers, not concerned with saving the appearances, seek to investigate that great and admirable problem, the actual structure of the universe, because that structure exists—unique, true, real, and impossible to be otherwise; by its grandeur and dignity this problem is worthy to be set in front of all other questions of the speculative intellect.

—GALILEO

NOTHING was ever announced so modestly as Edwin Hubble's discovery of the universe. For several years there had been increasingly excited debate as to whether the nebulae that could be seen in great numbers through the biggest telescopes are "island universes," galaxies like our own and extremely far away, or whether they are only pale gaseous clouds within our own stellar system. In 1924, using the new 100-inch telescope on Mount Wilson, Hubble (1889-1953) was able to study individual stars of the type known as Cepheid variables that he found in two of the nearest nebulae. The light from such stars varies periodically, and for Cepheids in our own galaxy the period is correlated with their absolute brightness. If one assumes that the Cepheids in these nebulae are like the ones near us, then by measuring their periods one can know how bright they really are. One can also measure how bright they appear to be, and comparison of the two numbers gives the distance. At an astronomical meeting in 1924 Hubble told the audience that he had measured the distance to the Andromeda nebula as 930,000 light-years (Hubble

1927). Later, as people were getting used to such a number, Walter Baade corrected it to over 2 million light-years, which is about 2×10^{19} kilometers.

What can we do to grasp intellectually the numbers of microphysics and astronomy? The notation "10^{19}" helps, for the 19 counts the zeros and one is not left floundering in fancy names like "quadrillion" (whatever that may mean), but no one visualizes 10^{19}. It might be something like the number of grains of sand in all the beaches and coastal waters of the United States; it is also the number of water molecules in a barely visible droplet of water. Or consider this proportion:

$$\frac{\text{distance to the furthest known quasar}}{\text{New York to Boston}} = \frac{\text{New York to Boston}}{\text{size of a proton}}$$

For some years before Hubble's measurement people had been studying the spectra of the nebulae and noticing a tendency for wavelengths in the ones that look smallest and dimmest, and were therefore presumably furthest away, to be shifted toward the red end of the spectrum. After Hubble learned how to measure distances it was found that as a general thing the size of the redshift was proportional to the distance (Hubble 1929). The only physical explanation of the redshift that anybody knew was that it is a Doppler shift: the nebulae are moving away from us, and the further they are the faster they move. The relation is usually written

$$v = Hd,$$

where v is the velocity of recession, d is the distance, and H is a constant named after Hubble ("constant" means that H is independent of which galaxy one looks at; it is not thought to be constant in time). If v is measured in kilometers per second and d in millions of light-years, a rough value for H is

$$H = 20 \, \frac{\text{km/sec}}{\text{Mly}}$$

(Rowan-Robertson 1985), but the correct value may be 50 percent different, perhaps more, either way.

At first thought, to place our own galaxy at the center and assume we are so unpopular that all the rest of the universe is trying to get away from us is a grotesque parody of the ancient error that placed us at the center with the universe revolving around us, but a glance at Figure 18.1 shows

FIGURE 18.1. If a configuration of galaxies expands according to Hubble's law around a central point, it becomes larger while retaining the same shape. Therefore, as seen from any of the galaxies, the expansion follows Hubble's law.

that Hubble's law of expansion has a property that any other dependence of velocity on distance lacks: the second picture, representing the situation after a certain lapse of time, is just an enlargement of the first with all the relative directions unchanged. Thus if v is really proportional to d as viewed from the earth, the cosmic expansion will look exactly the same to an observer on any other galaxy and we have no reason to think of ourselves as specially repellent.

Cosmologists base most of their discussions of the universe on five reasonably well-founded facts of observation:

1. The universe is expanding.
2. Although galaxies tend to occur in clumps and strings with vast empty spaces between them, the average distribution, considered on the largest scale, is fairly uniform. Actually, the more the universe is explored the larger are the departures from uniformity that are discovered, but it may still be that these departures can be averaged away without destroying the picture. The smoothed-out density of known matter amounts to about 5 protons per cubic meter.
3. From where we stand there is no sign of an edge.
4. The universe is full of radiation at a wavelength in the range roughly 2–5 millimeters, radiation that would be emitted from an object at a temperature of about 3 K (i.e., 3° Celsius above absolute zero). Every cubic centimeter of space contains about five hundred photons of this background radiation, as it is called.
5. Essentially all the known matter in the universe is inside the galaxies. About 75 percent of it is hydrogen, about 24 percent helium, and the rest contains traces of all the stable atoms.

Note the reference to the *known* matter. One can measure the mass of a galaxy by using optical instruments to see what is there, and also by indirect means involving the dynamics of galactic clusters. (Essentially, the more massive the galaxies in a cluster are the faster they move.) The results of these two measurements are grossly different, and one concludes that perhaps as much as 90 percent of the mass of a galaxy is in some form or forms we cannot see. Such invisible mass might also inhabit intergalactic space. This uncertainty clouds every cosmological speculation.

How old is the universe? If everything seems to be moving away from us, how long ago did it start from here? Assuming that everything we see has always moved at the same speed, we divide distance by velocity, and using Hubble's law we find for the time of travel

$$t = d/v = 1/H$$

Taking account of the uncertainty of H, t falls somewhere between 8 and 22 billion years. The remarkable thing is that the ratio d/v is about the same for every galaxy, as if they had all started from the same point at the same time. That stupendous event, the beginning of everything we know about, is irreverently known as the Big Bang.

In order to explain why the matter spreading out from the place of origin stays so uniformly distributed one must make a very drastic assumption as to the speeds at which everything began to move. If too much of the mass started at low speeds there would be a dense mass close to the point of departure; if too much of it started at high speeds there would be a thick spherical shell far away. Viewed like this the even distribution of matter seems like a miracle, but there is another way to look at it.

The Dynamics of the Universe

According to the general theory of relativity the presence of matter curves spacetime. The mass of the sun curves it very little, but even that small curvature binds the planets to it. What would be the effect of the uniformly distributed mass of the whole universe? This is a soluble problem and Einstein solved it in 1919, but he solved it too soon, before Hubble had discovered the universe or its expansion. For Einstein and most other people at that time, the universe ended somewhere beyond the disk of the Milky Way. Einstein's calculation seemed to show that the universe expands or contracts, but believing it to be at rest he wanted his equation to have a static solution. It has none, so he tinkered with the theory (1917). He introduced an additional term containing a number selected ad hoc so as to fit what he thought were the facts, an arbitrary adjustment that marred the purity of his mathematics, rather like giving the Venus de Milo a straw hat.

In 1922 appeared a paper by a Russian meteorological physicist named Alexandr Friedmann (1888-1925), of whom no one had ever heard, in which Einstein's original equations were solved under the assumption of uniform mass density. He showed that the universe either expands or collapses; it cannot stay still for the same reason that if we glimpse a ball a few feet above the ground we know that it is either on the way up or on the way down and not just staying there. But Hubble had not yet discovered the universe and so Friedmann also wrote too early; his paper explained observations that had yet to be made. It was ignored for a decade, until the universe was known to be large and expanding. By that time Friedmann had died young and others had rediscovered most of his conclusions.

The immense new idea in Einstein's incorrect cosmology and Friedmann's (which was more nearly correct) was that the space of the entire universe may be closed on itself just as a line can close upon itself to become a loop or a two-dimensional surface can close to become a sphere. In this way the universe can be finite but unbounded. Friedmann's expanding cosmos needs no miraculous distribution of velocities at the moment of the explosion to explain why we see it uniformly filled with matter. He offers the possibility that the universe has always, from the beginning, been uniformly full, and that as space expands the astronomical bodies in it move further apart. At least on a local scale, if the distances are not so great that effects of spatial curvature become visible, every observer will see these bodies as moving away with velocities proportional to their distances. How can we say that space expands if space is emptiness, the absence of everything? Please remember that in Chapter 14 I warned that in general relativity space has thing-like properties. This is one of them.

In the excitement over the expanding universe and what Einstein's theory had to say about it, some years passed before people realized that the expanding universe is not a just question of relativity theory and that Newtonian physics also has something to say about it. Only in 1934 did W. H. McCrea and E. A. Milne work out a Newtonian version of Friedmann's calculation and show that it leads to essentially the same conclusions. There is of course no curved space in it and one has to assume the miracle of a world full of matter distributed just right, but calculations based on well-known principles clarified the situation and persuaded everyone that just as a stone does not hover above the ground, galaxies cannot hover in space. A simple version of the calculation is given in Note M.

Ancient History

The early history of the universe is the focus of a great amount of study just now, both because beginnings are interesting in themselves and because for a moment things were so hot then that the energies of the particles flying around were far greater than we shall ever be able to achieve using particle accelerators. Particle physicists therefore look at the present state of the universe in search of evidence of high-energy processes that have survived from early epochs, while cosmologists study particle physics to get some insight as to what was going on then. In the 1970s these two scientific disciplines, which had previously had little to do with each other, became closely allied. The story that emerges is complicated and has some obscure areas, and I shall summarize it very briefly because there are excellent popularizations that do it in proper detail (Weinberg 1977, Barrow and Silk 1983, Reeves 1984).

During about the first million years after the Big Bang the mass of the universe was principally in the form of radiation—some of it electromagnetic and some in the form of chargeless and possibly massless particles called neutrinos. The temperature, using a convenient unit from nuclear physics, varied approximately as

$$T = 1 \text{ MeV} / \sqrt{t_{\text{sec}}}$$

where 1 MeV (million electron-volts) is about equal to the energy that creates an electron and a positron. In temperature units it is about 10^{10} K. At one second, then, the radiation is hot enough to produce a vast density of electrons and positrons, the electrons being slightly more numerous, but a few seconds later it has cooled down and positrons start to combine with electrons, emitting more radiation. A few seconds later the positrons are gone and simple nuclear reactions are starting up, which end with the universe inhabited by photons, neutrinos, electrons, protons, and some helium nuclei. As the universe expands, gravity and other actions sweep the hydrogen and helium together into masses of galactic size in which stars begin to form. When a star becomes dense enough, nuclear reactions turn on, producing heat and, as byproducts, the nuclei of elements heavier than the hydrogen and helium with which the star started. After some billions of years these first-generation stars exhaust the sources that produce an even flow of energy. They collapse and then explode, scattering their contents through the surrounding space. If some hydrogen clouds remain there, new stars may form containing some of the old heavy materials. Our sun must be of the second generation, perhaps the third, and the materials that form the planets and ourselves seem to be left over from the process. The heavier atoms of our own bodies and of everything we see around us are not primeval; they were all formed in an instant at the core of some exploding star.

If we look backward from the first second we see a picture of increasing confusion. The temperature increases, and more and more different kinds of particles are flying around. At first the picture is fairly clear. Assuming that the formula for temperature is about right we can make some fairly positive statements. Accelerator physics has now been pushed to energies of about 2×10^6 MeV, and when new machines are built the energy will go higher. Up to this energy, which according to the formula occurred when the universe was 2×10^{-13} seconds old, we have some definite knowledge of what happens when particles interact. Further back is more conjectural. It is the domain of theories that are untestable in today's laboratories, but are required by one or another of the logical and self-consistent schemes that attempt to unify the particles and forces of nature.

What we can expect to happen in the distant future of our universe is discussed in Note M. It turns out that there are two possibilities, just as there are when Superman throws a stone into the air. If he throws it gently it will fall to earth again, but if he throws it really hard it will disappear into space and never come back. It is the same with the Big Bang. If its violence was less than a certain value, gravitation will eventually overcome the outward motion and all matter will come together again in what is called the big crunch. Otherwise expansion will continue forever. In principle it is easy to tell which is to happen. Let ρ be the smoothed-out density of matter in the universe; G is Newton's gravitational constant and H is Hubble's constant, both encountered earlier. Then if the quantity

$$\Omega = \frac{8\pi G\rho}{3H^2}$$

is greater than 1 the big crunch is certain, while if it is equal to or less than 1 the universe will continue to expand forever. Measurements of ρ and H are not yet good enough to tell which future we have to face. Ω is in the vicinity of 1 and some astrophysicists think it is exactly 1. If so, then as the universe ages it faces a long, cold future, though it seems unlikely that we shall be involved in it.

Is It Just Luck?

Almost anybody looking at the photographs of the surfaces of Mars and Venus and the moon has a feeling that we are very lucky to be where we are in the solar system. This is the only place where we can lie on the beach in sunlight and watch the clouds and the seabirds float by. To creatures from another world, used to what we might think of as extreme heat or cold or dryness, that beach might seem like a very dangerous place to be. The earth, or at least some of it, is a good place for *us* to live. In surveying the universe as a home for life we should not suppose that everyone is like us or would like to live where we do. If there are immense thinking organizations of cosmic material like Fred Hoyle's *Black Cloud* (1957), if there are beings made entirely of neutrons who live at the rate of many lifetimes per second on the periphery of a neutron star, I have nothing to say about their probable likes and dislikes, but for creatures like ourselves who build their bodies out of molecules, not just any universe will do.

Let us think about what is necessary if the universe is to accommodate the kinds of living creatures we are considering, and then consider how it got that way. First, I will suppose that their tissues involve fairly large molecules made of a variety of chemical constituents. I will also suppose that however life first starts, complex forms that can think and dance and make music develop by an evolutionary process that takes billions of years. Third, I suppose that the environment in which life develops is much more favorable if it remains fairly constant. If the earth's average temperature were suddenly to change through the relatively few degrees on the absolute scale that would put it above the boiling point or below the freezing point of water, few of our present species would survive. Finally, I suppose there must be some place for a creature to sit down. (Admittedly fish do not sit down, but neither, it seems, do they think conceptually. Perhaps there is some connection between sitting and thinking.) There are a number of special circumstances that might help to meet these requirements.

1. It seems that a star is not a good place to live. Its heat would disrupt complex molecules and in the violent currents of gas in the surface regions there would be no place to sit. A planet is best, circling a star whose output of energy is fairly constant over long ages of organic evolution. No earth-like planets have ever been observed outside the solar system, but our telescopes are not strong enough to see them if they do exist. There is no reason to think there are not lots of other planets.

2. The universe must evolve so that there are not too many stars close to a solar system, since if a star wanders through the system it is likely to change the orbit of a life-bearing planet and subject it to alternations of heat and cold if it does not fling it out of the system altogether. Luckily for us, stars throughout the galactic disk are spread quite uniformly a few light-years apart so that such intrusions are unlikely, but if they were much closer we might very well not be here.

3. Stars are made mostly of primeval hydrogen and helium with a few heavier elements mixed in, while planets are made largely of heavier elements. Stars burn slowly and evenly. Let us think about the conditions necessary for that. Stars produce their energy by a complicated series of nuclear reactions, but in lighter stars like the sun the rate of these reactions is controlled by the hydrogen reaction. In this reaction two protons come together and one of them turns into a neutron. It does this by emitting a neutrino and a positron. Then the neutron and proton stick together to form a particle called a deuteron, while the energy released in the process is carried off by the two emitted particles and the positron's share ultimately heats the star. In a simple notation,

$$p + p \rightarrow e^+ + \nu_e + \textcircled{np} \, ,$$

where e^+ is the positron and ν_e is the neutrino. In a star like the sun this reaction will go on slowly for billions of years until all the hydrogen is used up; then the star undergoes dramatic changes in size and new reactions take over. The hydrogen reaction is slow because it involves a weak interaction—all neutrino processes are weak. Basically, it is very difficult for a proton to turn into a neutron in the very short instant while the other proton is near, and there must be many collisions before that finally happens. A typical proton in the burning region of a star has to wait about 10^{10} years.

Why don't two protons stick together the way a neutron and a proton do? Because their electric charges repel each other, and the attractive nuclear force which binds a deuteron is not quite strong enough to bind two protons. This is a lucky thing for us, since if the nuclear force were only a few percent stronger, producing a bound pp particle, the reaction would go this way:

$$p + p \rightarrow \gamma + \textcircled{pp} \, ,$$

where γ represents a gamma-ray, an energetic photon (Dyson 1971). This reaction involves no neutrino and it would progress much faster. How much faster would depend on how firmly the pp was bound, but a factor of something like 10^{13} is reasonable. Forming out of gravitational contraction in a cloud of hydrogen, a star would explode as soon as it was hot enough and dense enough for the reaction to begin.

4. On the other hand, if the nuclear interaction were only a few percent weaker there would be no deuterons and then again we would be in trouble, for the process of forming the heavy elements of which we and our earth are composed begins with deuterons.

5. Given the deuterons, how were the heavier elements formed? When two deuterons come together inside a star they easily combine to form an alpha-particle, ^4He, consisting of two neutrons and two protons. Then, if two alphas collide they can form ^8Be, an isotope of beryllium. But this isotope is unstable and immediately breaks up into two alphas again unless another alpha arrives and sticks to the beryllium to form ^{12}C, a stable isotope of carbon. The third alpha has only an instant to do this, and the event would rarely occur except that by a very happy coincidence the numerical relations that determine the structure of ^{12}C produce what is called a resonance in the nucleus at the energy that most of the alphas have when they arrive. This resonance makes the capture process enormously more probable than it would otherwise be, and the carbon produced in this way is an essential step in the formation of the elements that compose us and our earth. There is a different and independent coincidence of numbers that makes it possible for carbon to be plentiful in planets, but I won't go into details (Davies 1982, ch. 3; 1983).

6. What kind of star is good to live near? Obviously it must be a kind that can possess planets, and it must be able to burn slowly and steadily for a long time. Stars like the sun (and there are many of them in the sky) are ideal. A star is held together by a delicate balance between the gravitational force tending to collapse it and the pressure of radiation tending to blow it up. Brandon Carter (in Longair 1974) has studied the equations of stellar structure to see what would happen if the numbers of nature were a little different from what they are. He finds that if the elementary electric charge were a very few percent less than it is and the other constants were the same, typical stars would exist in a more compact state that would turn them into blue giants, whereas if it were a little greater most stars would be red dwarfs. A blue giant gives a lovely light but it burns too quickly to provide a planet with eons of time for species to evolve; Carter also argues that conditions during the formation of red dwarfs would be unfavorable to the formation of planets. Thus once more, the laws of nature and the numbers that go into them smile at us.

7. The lucky coincidences are not just in the numbers of astrophysics. If the fundamental electric charge e were three times smaller than it is, elec-

trons would be very loosely bound in atoms and there would be no neutral atoms at the surface of any planet, since they would all be ionized by radiation. Molecules would be correspondingly unstable and would stay together only at very low temperatures. On the other hand if e were three times greater, nuclei would be disrupted by the mutual repulsion of the protons in them. Alpha-particles would be unstable and if any heavy nuclei were formed somehow in stellar interiors they would quickly decay by spontaneous fission (Rozental' 1980).

8. If the heavier elements are all cooked up inside stars, how do they get out to form a planet? As I have mentioned, the answer seems to be that the star has to live its life and then explode. Thus the age of the universe must be the time required for stars to form plus the entire lifespan of at least one star plus (judging from our own history) four or five billion years for intelligent life to evolve on a planet formed from the cosmic debris. This is why a universe that supports the kind of life I am talking about must be something like fifteen billion years old. Listening to astronomers and physicists rattle on about distances of millions and billions of light-years, members of the audience have been known to wonder audibly if the whole arrangement is not a bit excessive, if a smaller universe might not have done just as well. But here again the universe seems to be playing our game. We do not know whether it is finite or infinite, but if it is finite it must not be too small. According to Equation (M.9) of Note M, the total lifetime of a finite universe constructed according to the canonical model is determined by its maximum size, the size at the moment it starts to contract again. It has taken about fifteen billion years for us to arrive at the point where we can talk about the universe. Therefore it is just as well that there was a little extravagance at the creation.

These were some examples of coincidences, both in the laws of nature and in the structure of the universe we live in, that seem to be lucky accidents that enable us to live and think. A single lucky coincidence would be worth mentioning but it would hardly be enough to suggest any portentous questions. I have given eight and more have been noted.[1] Of course we have to remember that when we look out at the universe we necessarily see that it has all the special qualities it must have if we are to be here. The question is, how much should they surprise us? Lacking any standard of comparison it is impossible to say, but a number of savants have felt that perhaps it is odd that everything worked out so well. And if we decide that it is indeed odd, do we attribute any special meaning to the fact or do we just say we are lucky?

[1] For further details see the works already cited, as well as Carr and Rees (1979), Gale (1981), Barrow and Tipler (1986).

Brandon Carter has submitted for discussion two propositions which he calls the weak and strong anthropic principles (in Longair 1974). The weak principle is hardly discussible and I have already stated it:

As we look out on the universe we must expect to find all the conditions necessary for our existence.

It is innocuous enough but it imposes some severe restrictions on what we will learn as we study the universe—for example, that it must inevitably turn out to be billions of years old. It establishes certain facts but does not say why they are true. The strong principle, which Carter does not uphold but which one might uphold if one decided that so many coincidences could not have occurred by chance, states:

The universe must *be such that life can evolve in it.*

Exactly why this should be the case is left open for discussion.

Of course, the religious faith of almost every scientist who lived before the nineteenth century and of many who live today provides an explanation for the apparent coincidences, but there are other thinkers, some of whom are also believers, for whom that is not enough. Various explanations have been suggested. One is that if there are enough different universes, then one of them is bound to have the right arrangements, and of course that is the one we inhabit. The only thing I can see wrong with this is that parsimony is usually counted a virtue in argument, and it seems a bit extravagant to postulate millions, perhaps billions of barren universes in order to deal with a single scientific puzzle. John Wheeler proposes that in some sense the benign intelligence that seems to have shaped the world may be our own, or be shared with billions of other consciousnesses scattered throughout the universe as well (in Toraldo di Francia 1979). Davies (1982) and Barrow and Tipler (1986) discuss these and other possibilities of explaining the tortuous thread that leads from the initial quantum chaos imagined in our model to life as we know it. There may be billions of other habitable planets in the world but we know nothing about any of them, and this simple fact clothes most of the learned discussions in fantasy. But now we are looking for our golf ball (if indeed there is a golf ball) in very long grass, and perhaps it is best to move on.

How Good Are the Laws?

The theory of the cosmos is constructed mostly out of materials picked up around the house. Newton's and Einstein's laws were verified first in the laboratory and then by observations on the solar system. None of the quantitative information goes back more than a century or so. I have already mentioned some of the extrapolations made when these laws are applied to large systems, and how they fail when applied to very small ones. What justifies our assumption that the laws of nature as we know them apply at remote times and places? This is a question that has received much attention, and I can only sketch a little of the answer here.

First, situations in the remote past are not unavailable for study. The geological record yields its message to physics and so do astronomical observations, for when we look outward into space we look backward in time, and both can tell us something of how things were a billion years ago. There are three ways in which the laws of physics could fail to apply over long intervals of time:

1. They could contain slight errors that are magnified when big numbers are put in.
2. The laws themselves could change their form over time.
3. The laws could have the right form but the so-called constants of nature, the gravitational constant, Planck's constant, the speed of light, the charge and mass of the electron, etc., could slowly change.

These three ways are not strictly independent, but it will be convenient to discuss them separately.

1. As I tried to explain in the last chapter, the laws of nature when we know them well turn out to be embodied in simple mathematical structures. One can tinker very little with structures like these. They are either right or wrong. It is hard to imagine that they could be just a little bit wrong.
2. This is an attractive idea. Why should the laws of nature be prior to the universe? Is it not reasonable to think of them evolving as the universe does? On the other hand, how can a system of equations evolve? It certainly cannot evolve continuously. If it evolves at all it would have to be in jumps, and the records show no sign of that. Also there is observational evidence. If we look at the light from distant galaxies and even more distant quasars we can learn from the spectra whether atoms back then were governed by the same dynamical laws as they are now. A study a few years ago showed that a relationship involving both the electronic and the nuclear structure of atoms has undergone no change larger than a few parts in 10,000 during the last four-fifths of the history of the universe (Tubbs and Wolfe 1980; Dyson, in Lannutti and Williams 1978). Both astronomical and geological evidence can be explained by assuming that the laws of nature have not changed appreciably in a long time.
3. To find whether the constants of nature are constant, the work just cited is relevant and there is also evidence from a strange disaster that

took place two billion years ago in Gabon, West Africa, when a natural deposit of uranium went critical and for a while became a nuclear reactor. From materials excavated at the site it has been possible to prove that the constants of nuclear physics were very accurately the same then as they are now (Shlyakhter 1976).

I have mentioned only a little of the evidence, but it leads to a tentative conclusion there was no reason to anticipate: the laws of nature as we know them can be trusted to explain events that happened billions of years ago. It is harder to know whether they are also valid when extrapolated back to the first fraction of a second. There are some arguments designed to explain why there are about 3×10^9 photons for each heavy particle (neutron or proton) in the universe that refer to times as early as 10^{-36} seconds; I mention them not because they are convincing but to show that it is at least possible that events occurring so close to time zero might have consequences observable today.

Modeling

Do I detect a movement of impatience? Am I expecting people to believe that the beginning of the universe occurred at an instant so precisely defined that I can talk about something that happened 10^{-36} seconds later? Did the universe really plan itself so that we could be here to talk about it? Do I really believe any of this stuff? It is time to think again about models.

What is the universe, anyhow? We look up at night and see white points of light and call them stars. It is reasonable to assume that the sun is also a star and that the only reason the other stars are less bright is that they are further away. Galaxies: we measure their distances assuming that their stars are of the same kind as our stars. One assumption piles on another, assumptions as to what astronomers are looking at through the instruments they use and as to the meaning of the information they get from these instruments. What validates this tower of hypotheses? Only its self-consistency and the observations it explains. To the eye, Mars is a point of reddish light. In a telescope it looks like a round and barren world. We land a space probe on it and find it is exactly that. Take another example, the interpretation of the reddening of the light from distant objects as an indication that they are moving away from us. This does not have to be interpreted as a Doppler effect. The light with which we experiment in the laboratory travels from one place to the other in a fraction of a second while that from a distant galaxy is (by our reckoning) millions of years old. What makes us sure it has not undergone some change during that time that might redden it? Perhaps so, but there is evidence that the universe had a beginning some billions of years ago and the simple theory of an expanding model explained in Note M gives a figure for its age that agrees at least roughly with figures derived in other ways. Also, calculations along the lines of those in Note M show that this model easily explains the presence, amount, and frequency of the background radiation. And if we believe Maxwell's equations—simple, beautiful, carved in marble—then light cannot change its frequency no matter how long it spends in transit.

That is the kind of reasoning that supports a model. The universe is a model. The time 10^{-36} seconds belongs to the model. The anthropic arguments belong to the model. We make observations and interpret them in terms of the model. If it furnishes numbers that disagree seriously with the results of observations we will lose interest in it. It assumes the truth of certain natural laws. If they turn out to be incorrect, out it goes. If by assuming that it makes sense to talk about the time when the universe was 10^{-36} seconds old (and there are people who talk about times much earlier) we can explain some feature of the observations, the model is supported, but it would be very naive to say that these things happened in a perfectly realistic sense. Nature is complex; models simplify. They idealize; they include what appears to be relevant and ignore the rest. They explain the world perceived by our five senses to the minds we possess in languages we have invented. They are meant to be taken seriously but not literally, they are meant to instruct and delight and make connections between diverse experiences and thereby stir the emotions. The human race erects monuments to those who do this successfully; some are called artists, some scientists; some have other names. Scientists are less noticed than artists, perhaps because they are more numerous and they work together. As this book draws toward its close we can look back at the long labor of those who have made possible these speculations on the expanding universe, slowly creating the concepts of space and time and causality and physical law, finding out how to reason objectively while making the best use of irrational prejudice, how to explain what has been done. These are the makers of science. They made it for their own pleasure and yours. One does not step into the middle of a new art form or scientific structure and profit from it at once. One has to know something, to see it against the background of history. I thought a book like this might help with the scientific part.

CHAPTER 19

Order and Law

Order, proportion, and fitness pervade the universe. Around us, we see; within us, we feel; above us, we admire a rule, from which deviation cannot, or should not, or will not be made.
—JUSTICE JAMES WILSON

THAT the world is lawful has been said again and again: by Anaximander,

> The source from which existing things derive their existence is also that to which they return at their destruction, for they pay penalty and retribution to each other for their injustice according to the assessment of Time.

By Heraclitus,

> The sun will not transgress his measures; otherwise the Furies, ministers of justice, will find him out. (fr. 94)

By King Solomon,

> Thou has created all things in measure and number and weight. (Wisdom 11:20)

By the second Isaiah,

> Who hath measured the waters in the hollow of his hand, and meted out heaven with the span, and comprehended the dust of the earth in a measure, and weighed the mountains in scales, and the hills in a balance? (Isaiah 40:12)

By Saint Augustine,

> Behold the heaven, the earth, the sea; all that is bright in them or above them; all that creep or fly or swim; all have forms because all have number. Take away number and they will be nothing. . . . Ask what delights you in dancing and number will reply: "Lo, here am I." Examine the beauty of bodily form, and you will find that everything is in its place by number. Examine the beauty of bodily motion and you will find everything in its due time by number. (*On Free Will*, xx.16, in Augustine 1953)

And by Dante,

> "The elements
> of all things," she began, "whatever their mode,
>   observe an inner order
> that makes the universe resemble God."
>
> (*Paradiso*, I.102)

People who use their hands know that the world embodies an order expressed in form and quantity. The worker in clay or metals or dyes knows that these materials will submit only if the right procedures are followed; the sculptor controls imagination with measurement according to rule; cooks are bound by the materials they work with; the equal-arm balance serves as a symbol of impartiality because we know the natural order is lawful. Nature may not balance reward and punishment with perfect fairness from moment to moment, but over the long period inequities cancel and the truth is revealed: you will reap what you sow. The universe is just.

Science began when people discovered the regularities of nature, and in our century it has produced quantities of specific knowledge as well as the theoretical structures I have been writing about. This last chapter will comment on some aspects of fundamental theory: where it comes from, the shapes it takes, and why people seem to enjoy it. I want to show that, as I said earlier, human thought, trying to solve the riddles of nature, encounters again and again the same situations, and that modern science instead of transcending them merely provides clearer examples of them.

What Has Been Accomplished

During the last three hundred years two systems of dynamical law have been developed: Newtonian, for large objects, and quantum mechanics, for small ones. Relativity theory modifies Newton's model without totally transforming it. Newton's laws were guesses and so, much later, were those of quantum mechanics. Since large objects are made up of small objects it ought to be possible to derive Newtonian physics from quantum mechanics and it is, though to do it right involves rather delicate arguments. With these tools we can understand almost anything in the physical world, as long as we know what is going on. We know what is going on in atoms and to some extent in nuclei. The fundamental properties of solids, liquids, and gases are quite well understood, though many puzzles remain. For example, nobody knows how to calculate numbers such as the boiling and freezing temperatures of water. We understand the molecules and the forces, but the changes that occur at those two temperatures are so complicated that we do not know how to encode them into equations and probably could not solve them if we did. As we go further from the subject matter of conventional physics, to ask questions about the behavior of large molecules or the living cell, our lack of detailed physical knowledge becomes acute, and the special techniques of chemists and biologists normally work better than those of conventional physics. Physicists' knowledge is based on principles that have been invented over the centuries and have been incorporated into an increasingly consistent and interrelated structure of ideas. The ideal, of course, is an intellectual structure so perfect and so complete that not one element of it could be changed without changing the whole. The ideal may well be unattainable, but I think that every scientist believes that even if that is so, some progress toward it will always be possible.

Look around you at the devices in your neighborhood that make music, show pictures, keep records, make war, cover paper with symbols, transport you, and help keep you healthy. You may not like them all, but most of them are visible evidence of the enormous advances in understanding the physical world that have been made in the last ten or twenty years. The great organism of ideas rises slowly, puts out branches, intertwines and becomes so complicated that nobody can master more than a small part of it. There is no plan. Yet the entire structure is based on a comparatively small number of fundamental principles whose meaning and historical development have concerned us in these pages. Not all these principles are independent of each other, and occasionally, as with de Broglie's comparison of the principle of least action with that of least time, two of them may turn out to be saying the same thing. Rather, the principles represent particularly general and understandable statements that can be made about nature. It is time to see what can be said about them.

Simplicity

As we compare Kepler's planets with Ptolemy's, or the modern explanation of comets with Aristotle's, we are struck by how many fewer hypotheses are needed, how much simpler it all seems now than it did in the past. Not only the explanatory models but, when we finally find them, the basic ideas tend to be simple. I do not mean that they necessarily look simple when expressed in mathematical notation—although they are often simple in that way too—but rather that there is simplicity in the mathematical and physical concepts they express. The interlocked non-linear differential equations that emerge when we try to solve the equations of general relativity are not especially simple, but what the equations say in this clumsy language is a simple and natural analogue, expressed in terms of geometry, to the equations of electromagnetic theory (Misner et al. 1973, ch. 15).

The second principle that seems to govern the laws of physics is that they can often be stated so as to express some kind of balance involving a maximum or a minimum. The principles of least time and stationary action govern the trajectories of light rays and material objects. In general relativity the basic law of motion is that the path of a body moving between two given points in spacetime is the one that takes more time than any other as measured by the body's own watch. Phrased in this way the laws are simple, at least in concept, though not always in details.

Modern physics is a physics of fields. Fields are spread out in space, but similar principles apply. We have seen that Maxwell's equations can be derived from a Lagrangian density that is formed in an extremely simple way from the field quantities, and the same sort of thing happens with general relativity and the various fields used in quantum mechanics to represent particles. In every case a Lagrangian can be defined to describe the system; it is comparatively simple in form and the equations of motion that tell what the system will do follow from the simple requirement that the action should be stationary under small variations of the fields.

The idea that nature seems to allow simple descriptions has been slowly formed from long experience; only lately has it tended to be used as an a priori requirement in the construction of new theories. I referred earlier to the desirability of being able to derive fundamental laws from qualitative, not quantitative principles: that the world shows tendencies toward economy at the most fundamental level, even though it may in the long run not turn out to be as true as it seems to be at present, is that kind of principle. Symmetry is another.

Symmetry

I have already said quite a lot about the role of symmetry in physical theory but perhaps a recital of basic notions will be useful. The primitive idea of symmetry arises when we notice that a certain geometric figure is unchanged after it has been reoriented: the appearance of a square piece of blank paper rotated through 90 or 180 degrees or flipped over in various ways does not change. A circle remains the same when it is flipped or rotated. Such reorientations are examples of transformations, and symmetry is defined as the property of invariance under a transformation.

There are more interesting spatial symmetries than these; in two dimensions they define classes of repeating ornamental patterns and in three dimensions the classes of crystals, but even squares and circles are useful to think about (Weyl 1952). In dynamics, it seems that in the formulation of natural laws no position or direction in space should be picked out to play a special role. That is, if one were floating in a closed box in interstellar space it should not be possible to perform any experiment to show where the box is or how it is oriented with respect to distant constellations. The Lagrangian that describes something in the box must be invariant under all possible spatial displacements and rotations. These are very restrictive requirements. Thus, for example, the Lagrangian density of the electromagnetic field can depend only on the magnitude and relative directions of the electric and magnetic fields and not on their absolute directions with respect to some fixed system of coordinates. If uniform motion of the box is undetectable the Lagrangian must also be Lorentz-invariant. Then, as we have seen in Chapter 17, the simplest Lagrangian is the combination $E^2 - H^2$ and it turns out to be the right one. It should also be impossible, from inside the box, to tell what time it is; that is, any experiment should give the same result whenever it is performed. The effect of this requirement is that the Lagrangian should depend only on the spatial coordinates describing the system's interior arrangements and not at all on the time coordinate.

Recall that whenever a Lagrangian is invariant under some transformation the system it describes has a dynamical property that remains constant. It follows from the invariances just mentioned that linear momentum, angular momentum, and energy are conserved. In quantum mechanics the same conclusions follow, and if we require that the Lagrangian be independent of the phase of matter waves, then, as we have seen in Chapter 17, the conservation of the total electric charge is an immediate consequence. If we suppose that the phase can be chosen arbitrarily in every place at every time, this requires the existence of a field not otherwise necessary, which turns out to be the electromagnetic field. And finally, the requirement of other local invariances of similar kinds leads to the necessity of gauge fields not previously known. Arguments of this kind helped in constructing the Weinberg–Salam theory, combining weak and electromagnetic interactions, as well as quantum chromodynamics, and both seem to be largely correct.

How are creative physicists led to formulate invariance principles like these? Obviously they are a mixture of economy, common sense, and experimental knowledge. On the whole they are simple to state, understand, and test; even the more recondite ones involving the phase of wave functions and the indistinguishability of particles are firmly based on experiment.

What else are they based on? Given that the universe seems to be pretty much the same everywhere and in all directions we are not terribly surprised to learn that Lagrangians are invariant under spatial transformations. But the universe changes over time; it gets larger every second, and yet Lagrangians are independent of time. It seems that these symmetries may be autonomous expressions of natural law, not merely reflecting symmetries of the universe. Remember that superstring theory and other new speculations are suggesting that spacetime, far from being the stage on which nature presents its juggling act, may be a derived and approximate concept, good enough to describe the performance as seen from our seats in the back row. If this is true, the spacetime symmetries of which I have made so much cannot be fundamental at all, but must emerge from more basic ideas when simplifications are made. It may be so also for other symmetries as well; the Yang–Mills scheme for explaining fundamental interactions, now a postulate, may well end as a consequence.

A scientific program has a purpose: that those who work on it may some day be able to say they understand. Understanding does not mean that some complex system of equations is found that saves all the phenomena; rather it means, as it always has meant, that some contact has been made with principles so simple and so natural that one is ready to accept them, provisionally of course, as representing some version of the way things really are. Formal simplicity, economy, and symmetry seem to be three such principles. There is a fourth, so natural to the practicing physicist that it may almost escape notice: that *known* mathematics (though it may need to be extended a bit) has always sufficed for physical theory. Earthlings surely do not know all the mathematics that can be known or made; there is no reason to assume that dwellers in some unearthly place, if they cultivate mathematics, do it the same way we do. Earthly mathematics is a human endeavor, just as physical theory is, created to satisfy the emotional needs of people who devote themselves to it. Why does it work so well? In a memorable essay, "The unreasonable effectiveness of mathematics in the natural sciences" (1967), Eugene Wigner writes, "We are in a position similar to that of a man who was provided with a bunch of keys and who, having to open several doors in succession, always hit on the right key on the first or second trial. He became skeptical concerning the uniqueness of the coordination between keys and doors." In even plainer language, this unreasonable effectiveness suggests once more that our theories are not uniquely determined by what we observe.

The four principles I have mentioned (formal simplicity, economy, symmetry, and being comprehensible in terms of known mathematics) are all man-made, and underlie the kinds of theories we are able to think up. But it should not be forgotten that the world is such that it permits these principles and these theories.

Causality and Determinism

The same structure of physical laws that prescribes the spatial symmetry of a snowflake prescribes a causal order also. But are the present and future completely determined by the past? The question is a very old one and is given weight by its connection with the problem of free will. The whole story is too long to tell here, but the part of it that belongs to physics really begins with Newtonian mechanics (or Ptolemaic, if you do not mind the inaccuracies) and the plain proof that planetary motions can be calculated far back and far ahead from our own time if we do enough work—that there is no room at all, for example, to believe that an eclipse is sent as a comment on human affairs unless one is willing to suppose that everything was planned that way from the very beginning. Of course one can argue that quantum-mechanical indeterminacy has put an end to determinism, but Planck's constant is very small and the normal context of quantum mechanics is atoms rather than objects of planetary size. Let us therefore keep our categories straight by first considering Newtonian mechanics. The classic statement of the determinism that flows from this theory is that of Laplace:

> Assume an intelligence which at a given moment knows all the forces that animate nature as well as the situations of all the bodies that compose it, and further that it is vast enough to perform a calculation based on these data. It would then include in the same formulation the numbers of the largest bodies in the universe and those of the smallest atom. For it nothing would be uncertain, and the future, like the past, would be present before its eyes. (Laplace 1843, vol. 7, p. vi)

This is simple enough and at first sight obvious, but for many years people sensitive to the nuances of physics have warned us that as an example of causality Newtonian physics contains subtleties that are not obvious if you only look at special cases like planetary theory. One of them can be seen in a simple imaginary example.

Let a hard elastic ball be dropped onto a knife edge from a position to the left of it. The ball will end up on the left side of the knife. If it is dropped to the right, the ball will end up on the right. What happens if the ball is dropped from an extremely small distance d to the left? We need not try it, for Laplace's *daimon* can calculate what will happen: the ball will bounce up and down on the knife edge, perhaps for years, but it will finally fall to the left. If d is made still smaller it will bounce still longer. Now suppose we say that not even the *daimon* can perform an infinite calculation, and we set an upper limit on the length of time during which we will allow him to keep on figuring. There will then be a small range of values of d on each side of zero for which the daimon cannot calculate what will happen, and it will have to say that the motion is indeterminate. The point $d = 0$ is known as a singular point. Another example is a pendulum that turns on an axle. Push it and it swings back and forth. Push it harder and it turns round and round. Now push it so that it lingers at the top, deciding whether to go over or fall back. Here again, if you push it just right, you encounter a singular point that the *daimon* must call indeterminate.

Since the time of Henri Poincaré it has been known that analogous singular points occur in real problems, even in a system as simple as three bodies moving under the influence of their mutual gravitational attraction. The existence of these points can cause certain dynamical systems moving in more than one dimension to behave unpredictably, as if their motions were governed by chance and not by causality. The modern term for such behavior is "chaotic." It has taken mathematicians a century to learn how to talk about chaos, but for many years the lack of a theory of it has been noticeable in physics. For example, in calculating the properties of a gas one does not think of following the motion of every molecule à la Laplace, instead one assumes that the molecules move according to laws of chance. This leads to feasible calculations and correct answers, but for a long time it was not clear why, if all molecular motions are causally determined, we are allowed to calculate them as if they occurred at random. We are beginning to understand: under some circumstances the chaotic behavior of a system whose parts obey determinate laws will mimic ideal random behavior so closely that it can be studied using the laws of probability. What occurs at random is by definition unpredictable. The unpredictable, in the sense I have described, exists in the Newtonian world as well as in the quantum world.

In quantum mechanics the wave function contains the most complete possible specification of the system it describes, and in simple systems Laplace's *daimon* can predict the wave function's future course. More complex systems have wave functions that seem to be chaotic, incalculable. Further, as we have seen, the wave function is not itself observable, and even if we know it, statistical uncertainties arise when we use it to calculate what will happen in a specific case. That electronic motions are indeterminate is an accepted fact. Does this imply that our own thoughts and actions, which as far as is known depend on the motions of electrons, are therefore to any appreciable degree indeterminate? Perhaps, but I have told why I think it will be a long time before anything about thoughts and perceptions is concluded from physical arguments.

The Role of Chance

In daily life, as in science, we are continually estimating probabilities and acting upon them. This is true even if it is hard to explain in objective terms exactly what we are doing. When a coin is tossed we believe in the honesty of the coin and the person who throws it; this means that we assign equal probabilities to the two possible outcomes of the throw. But what does that tell us about what will happen when the toss occurs? Nothing. The coin will come down heads or tails, that is all. Exactly the same thing would happen if the coin were crooked. Still, we want the probabilities to be equal; for most of us it seems better that things should be done that way.

What does it mean to say that the probabilities of heads and tails are equal? If we try to answer by saying something like "the toss does not favor one side or the other," we are explaining probability by invoking the concept of probability that is hidden in the word "favor." After long reflection one tends to conclude that probability, not in the abstract but as it applies to the real world, is a primitive concept that cannot be defined in terms of anything else.

These remarks about tossing a coin apply only to a single toss. One cannot judge from one toss whether it is honest or not. If it falls "heads" five times in succession we become suspicious, and if the run continues to a thousand we approach a condition of certain knowledge that something is wrong, though there is still one chance in about 10^{301} that our suspicion is unfounded. But we should not think that there has been some change in the kind of knowledge we possess as we continue the trials. Only the degree of certainty has changed. We know quite well without talking about it that in most matters absolute certainty is beyond our reach, and that what we call knowledge has only a greater or smaller probability of being true. The conclusion that if a coin falls the same way a thousand times all is not well is as close as we could ever come, in the experience of the world outside ourselves, to certainty.

In physics we do statistical calculations concerning such enormous numbers of atoms and molecules that probability amounts to certainty. Henri Poincaré put it this way:

You ask me to predict the phenomena about to happen. If, unluckily, I know the laws of these phenomena I could make the prediction only by inextricable calculations and would have to renounce attempting to answer you; but as I have the good fortune not to know them, I will answer you at once. And what is most surprising, my answer will be right. ("Chance," in Poincaré 1914)

That is, there is an overwhelming probability that his answer will be right. Furthermore, his answer will be more useful than any answer that Laplace's calculating demon would give, for its calculation, following the fortunes of every molecule as they interact, would necessarily deal with only a single case. Science deals with general statements. If I measure the pressure in a gas I am not interested in following the tiny fluctuations registered on the gauge in some particular trial; I want averages, for these are what will be the same from one experiment to the next, and they are what the statistical treatment, by its nature, is designed to give.

The science of statistical mechanics began with Maxwell and Boltzmann and Clausius over a century ago, and its proofs depended on the assumption that molecular motions are, in a sense that came to be more and more exactly defined, chaotic; that is, that in spite of being mechanically caused at every step, they can be analyzed by mathematical methods appropriate to random processes. On this assumption a splendid science has been built; for some it is the most beautiful room in the house of theoretical physics. But what of the assumption? In mathematics propositions do not usually need to be assumed, for there are such things as theorems and there is such a thing as proof. To my knowledge there is a single relevant theorem, stated and proved by the Russian mathematician Yakov Sinai, and it goes something like this. Imagine a cubical box with perfectly elastic walls containing N perfectly elastic spheres. There is no gravity, and once the spheres are set in motion their elasticity keeps them bouncing around forever. The theorem states that except for certain special initial motions that would be impossible to establish in any case, the motion of this system will have the chaotic behavior necessary if the theorems of statistics are to be applied, for any N greater than 1. The proof is enormously long and difficult (summarized in Sinai 1970), and as I understand it an important element is the existence of singular points in the motion such as I illustrated with the parables of the knife and the pendulum. The surprise is that randomness turns out not to be an affair of large numbers; it exists even when the number of interacting particles is small. As for real gases in real containers, the details remain to be filled in.

Necessity

Long ago, in Chapter 5, I quoted Averroes' complaint that the designers of the celestial spheres had never shown that those constructions are necessary to explain the phenomena. Talking with his collaborator Ernst Straus, Einstein raised another question:

> What I'm really interested in is whether God could have made the world in a different way; that is, whether the necessity of logical simplicity leaves any freedom at all. (Holton 1978, p. xii)

Averroes asked, "Given what we know about the world, could our theories be any different?" Einstein asks, "Given what we know about theories, could the world be any different?" Averroes' question is easier to answer, and an example that points it up very sharply can be found in Einstein's own experience with theories of gravitation.

At the beginning of this century Newton's theory of gravity was regarded as one of the highest achievements of human speculation. According to it every object in the universe produces a gravitational field obeying the inverse-square law in the space around it. It follows that every two bodies in the universe attract each other with a force proportional to the product of their masses and, at least at large distances, inversely proportional to the distance between them. Assuming this, together with the Newtonian law of motion, astronomers could explain even very recondite phenomena of the solar system. They also assumed that Euclidean geometry describes space exactly, but that was considered so obvious that they rarely mentioned it. There was a tiny discrepancy between theory and observation in the motion of Mercury, but the observational number was not well established and few people worried about it. It was not because of this discrepancy that Einstein created a new theory of gravity but rather because he wanted to believe that only relative motions are detectable. When he had finished, as we have seen, no trace of the Newtonian mode of explanation was left. The force of gravity is gone and its effects are explained in terms of curved spacetime.

Now suppose our astronomical observations were not very good—it is enough to suppose that the earth's climate might always be a little cloudy—so that the small discrepancy in Mercury's orbit and the other relativistic effects like the bending of starlight by the sun could not be observed. We would then, in the twentieth century, have two experimentally indistinguishable but conceptually unreconcilable theories of gravity. The fact that such a thing might have happened, even if it didn't, was obviously of enormous significance to Einstein, and it caused him to say on many occasions that scientific theories are free creations of the human imagination (e.g., in "Physics and reality" and "On the method of theoretical physics," in Einstein 1954). This does not mean that one can publish anything one likes, for theories that do not pass the test of experience must go into the wastebasket, but only that theories are not forced on us by the observational material; they originate in some other way, and this suggests that the answer to Averroes' question is that as long as we do not know everything about what happens there will probably be more than one possible theory that explains what we do know.

The question of whether the net of laws in which we seek to capture nature can ever be uniquely determined can be better thought about if we break it into smaller pieces. First, and easiest, is there any chance that if we finally talk to people on a distant planet we will find that their science is the same as ours? Of course it is possible, but it seems very unlikely. In the first place one would expect that the science of any race would depend on the observations they make; these in turn would depend on the organization of their nervous and sensory systems, what they have to look at, and what they think are interesting questions; there is no reason to think that in such matters those distant people would very much resemble ourselves. And second, our entire conception of what makes a scientific theory has changed so rapidly during the last centuries that to catch even a biologically similar race in a state of development exactly corresponding to our own would be unlikely. More interesting is the question whether we, given our own cognitive abilities and ways of thinking, might have organized our fundamental science according to different principles than we have. The story of the two theories of gravity suggests that we might, for it was not observational considerations that impelled Einstein to develop the general theory. A more careful look makes this conclusion less certain.

There is no way to combine special relativity, Newtonian gravity, and the principle of equivalence into a consistent whole, so that in the years after 1905 gravity was a sore spot in physical theory. Sooner or later someone would have to find a theory of gravitation that is in harmony with the rest of physics. Is general relativity the only way? Modern quantum field theory gives an entirely different approach to the problem. It can be shown that the most natural assumption about the carriers of the gravitational force is that they are massless particles with twice the spin of a photon. Now try to make a consistent theory of the interactions produced by the exchange of virtual particles of this kind, requiring that the principle of equivalence be satisfied. It is quite difficult (Weinberg 1964, 1965; Boulware and Deser 1975), but what results when the job is finished, as if by magic, is Einstein's theory as it applies to the interactions of large objects. Further, although the derivation is carried out in the flat spacetime of special relativity, the requirement that the principle of equivalence be satisfied assures that the resulting mathematics can be interpreted in terms of the curved spacetime that Einstein introduced ad hoc. This result suggests that the physics at which we eventually arrive is likely to form a consistent whole in which there is little room to change parts of it. As to whether there could exist a completely different consistent whole, agreeing with the same experiments (but not necessarily with all possible experiments), obviously this is a logical possibility, but nothing we know about the world gives any hint as to what it might be.

Understanding

A few pages ago, looking back at the long history of people's attempts to understand the universe or even small parts of it, I suggested that forms may be the units of understanding; that is, that when we are able to relate a certain body of concepts to a form that is familiar to us we begin to experience the easing of curiosity that culminates in "I see." That is very loose talk. What are these forms? It is easy, and true, to characterize them as cognitive structures that we develop in the process of learning to think, but some of the forms we encounter in the study of nature originate, after all, in nature itself. There is no reason to argue as to whether a flower or a snowflake or a crystal exhibits form; it is only a matter of a definition to which most people would readily assent. That the earth is nearly a sphere, that planetary orbits are nearly ellipses show us formal principles at work on a larger scale as well, and in these examples and innumerable others we recognize that the form does not sprout fortuitously out of nature but that it emerges as a consequence of fundamental principles at work. That all the leaves on a plant and all the petals on its flower are usually alike reflects the fact that if one wants to avoid errors of transcription it is a good idea to transcribe a short message, and the DNA that specifies an organism is economized if the same instructions can be used again and again. The shapes of crystals and snowflakes are determined by the properties of the atoms and molecules involved; that stars and planets tend to be round and satellite orbits tend to be ellipses follow at once from the laws of dynamics and gravity and some elementary principles of structure. Is it not striking that the same universal principles that assure the visible manifestations of form are themselves understandable in terms of laws that express symmetry and economy? Of course, these laws are manmade, but nature is such that it allows them to be made, and assuming that the snow of some distant planet also makes hexagonal patterns I would be very much surprised if science on that planet, provided it concerned itself with causes that underlie the appearances, did not create formal causes to explain the appearances of form. The creators of our own scientific tradition have done that from the beginning.

It is often said that science tells how but not why, meaning that its laws do no more than describe how nature behaves. Whether or not a description is correct can be verified by experiment. Experiment is the test of scientific truth, and since it cannot answer the question why something happens, the question lies outside science. Very well then, since we must not quarrel over definitions: though it pertains to science it is not a scientific question. Nevertheless, most scientists I have talked to admit that it interests them very much, or perhaps it would be better to say it motivates them, since answers to the question why come more in a sense of the feel of things than in a set of propositions. Suppose one were to try to answer it in propositions: how could we tell whether the answer is correct? I could ask the same thing concerning questions of goodness or morality or beauty. These questions cannot even be asked very clearly and we are used to not having verifiable answers to them, but that does not mean that they have no meaning or are unimportant. If I suggest that the why of natural law expresses itself in principles of symmetry and economy, in the existence of certain mathematical forms, I am really only describing the intellectual climate of physics in our own day. I have already suggested that the symmetries we find may be no more than observable manifestations of more abstract principles, and it may equally be that economy is not a guiding principle but a necessity imposed by some iron-clad law we do not yet know. Then what happens to the why? It is a matter of faith, I suppose, but if past experience is any guide one might expect that when we come to understand them the new principles will have some formal features that will once more encourage us to think we understand.

As to the language of explanation, that trails along behind the leading edge of scientific discovery like the camp-followers that traveled with the Crusaders. Explanation respects common opinions and common speech; its task is to make some adjustment between newly discovered facts of nature and conventional modes of representing the world. Earth-centered astronomy was the most reasonable adjustment of astronomical facts to a general belief in the absolute centrality of man. Mechanical explanations of electromagnetic phenomena trailed behind Maxwell's equations for decades, as long as people believed that nature was best understood in terms of material mechanisms. The lesson that was slowly learned was that the formal principles may provide the best explanation: that if Maxwell's equations are true it is not because the ethereal machinery is constructed that way, but because that way the mathematical form is especially clear.

But in speaking of mathematics are we speaking of anything real? That is of course a foolish question, but the point it addresses is a real one. If an electron, for example, or a water molecule were really a thing, in the sense that one could describe it and say where it is and how it is moving and what it looks like under a microscope, any explanation of atoms would emphasize these particulars, and specifying them with numbers would serve to make the picture quantitatively exact. But in the physics of the last half-century this has not been the situation. Objects of the microworld defy description as things; they cannot be assigned positions and motions; they can be in two places at the same time. Nevertheless, we have a mathematical theory that allows us to answer many other questions than pictorial ones, questions as to what electrons or water molecules will do when placed in various experimental situations. Questions beginning "What is . . . ?" cannot be answered by calculation and their answers cannot be tested experimentally—they do not live in the laboratory.

In this situation the role of calculation is to provide numbers that describe not a particle itself but the observable properties of things in whose structure the particle plays a role, to tell what these things will do in various situations, and finally to suggest new questions. But what about the particles themselves? If mathematics is our only conceptual tool, is there some sense in which we have to say that these fundamental objects in our explanations of nature are only mathematical entities? What then is their relation to physical reality? Should the word "reality" be reserved only for what is presented by sensory experience? Or should it be abandoned altogether? Opinions differ. No possible answer can be tested by experiment and I had better avoid them all.

Still, words ought to have definitions. How should we define an elementary particle? I think Heisenberg said it very clearly:

> Modern atomic theory shares with the old one the fundamental idea that we explain the external world's multiplicity by setting it into correspondence with a multiplicity of forms that can be distinguished and analyzed. To serve as forms of this kind the Greek philosophers had only geometrical configurations at their disposal, and therefore the old theory explains the qualities of matter by grouping atoms in different ways. . . . The indivisible elementary particle of modern physics does not have the property of occupying space any more than it has properties like color and solidity. Fundamentally, it is not a material structure in space and time but only a symbol that allows the laws of nature to be expressed in especially simple form. (Heisenberg 1959, pp. 79, 80)

It would be a mistake to stand back at this point and pretend that, having checked its metaphysical baggage, physical theory can finally begin to give a coherent explanation of all the natural phenomena we have so far observed. Quantum mechanics consists of recipes that have been tested a thousand times and found to work. For example, one recipe says that a quantum process comes to an end with the irreversible act of amplification that enables us to detect it at all; after that our account of what happens is to be expressed in the plain language of classical physics. But recipes have reasons behind them. To understand why this one works it would be necessary to analyze the process of observation in a consistent way, starting from the quantum phenomenon and ending, perhaps, with an account of the observer's experience. This is the so-called problem of measurement, and it is not yet solved. It involves many bridges; the probabilistic bridge between q-numbers and c-numbers is only one of them. To solve the problem may not require a dramatic reformulation of anything, but obviously as long as physical theory is unable to explain or even to accommodate the simple process of observation, we can hardly claim that it gives a complete account. During the last fifty years the mathematics of quantum theory (except at its speculative fringes) has undergone little change while our metaphysics has changed very much. Perhaps this is an indication of what will continue to happen. Perhaps the problem of observation, like so many others, will turn out to be a pseudoproblem, but at the moment it is very much alive.[1] In one way it is not very important that the problem remains unsolved, for it does not seem to slow the progress of scientific knowledge in any field, and in any case we should not be surprised if more is known about how things happen than about why they appear to us to happen as they do.

Third Pastoral

Same hill, same tree. The sun has set and the western sky shows orange and red and very dark blue. Phyllis and Corydon watch as the light changes.

PHYL. I've been thinking about what you said the other day, that when someone writes down a formula or a diagram, maybe even a number, they ought to color it. Maybe it's already colored. Maybe they only write down the colored ones.

COR. Give an example of one that wouldn't be colored.

PHYL. Let's count the leaves on this tree or find out how much water there is in the pond down there. We work very hard; then we write down the answer and there is a fact that nobody knew before.

COR. Solid gray. What's a fact?

[1] For a very careful technical discussion of the problems involved, see Joos and Zeh (1985).

PHYL. He doesn't know what a fact is. I think I know, but it's hard to say.

COR. I can listen.

PHYL. Let's suppose we already have some statements that we've provisionally labeled facts. They can be statements about what we think or just what we've seen. Then somebody comes with a new statement and asks if it states a fact. First we see whether it makes sense. Then we check to see if it fits the evidence, I mean if it doesn't contradict any of the statements already labeled as fact. If it doesn't, we can provisionally call the new one a fact.

COR. At what stage do we stop calling it provisional?

PHYL. Quite soon, but we never forget. That's the point. What did you mean by color?

COR. A theory shouldn't just be a diagram or a formula that explains how something happens. It has to do something for us. It has to make someone happy.

PHYL. It might make someone happy to know just how many pails of water are in that pond.

COR. I doubt it. But I happen to think it's just great that $3 \times 7 \times 37 \times 73 \times 137 = 7770777$. It's pink.

PHYL. How on earth did you know that?

COR. I've got all this free time. But that's not what the book was about. It was about generalizations. They seem to be colored too. I don't think people would be willing to work very hard on black and white generalizations.

PHYL. I was talking with Cleanth the other day and he said that those formulas only tell how something happens and not why, and that to talk about the how and the why is stupid.

COR. He's a little out of date. Copernicus told how; De revolutionibus is an instruction manual, but Kepler saw these immense spheres slowly floating along geometrical paths in empty space. The ellipses were part of why.

PHYL. How about quantum mechanics and the rest of those abstract theories?

COR. There was a whole generation when people told each other that quantum mechanics is only an instruction manual, but I think they are beginning to see more than that in it.

PHYL. Maybe theories don't just describe what happens.

COR. Instead—?

PHYL. Maybe they say something about what is.

COR. Of course they may be wrong.

PHYL. Of course. What does that matter? But how does it happen that there are these colors in nature?

COR. That's a funny way to say it. I thought we put in the colors.

PHYL. No, that's not possible. When you finally find out how, you find that it's bright blue.

COR. Perhaps there is a guarantee.

PHYL. What does it say?

COR. It is hereby guaranteed that if you arrive at some idea that saves the phenomena and fits in with the rest of what you know, it will be bright blue.

PHYL. Signed?

COR. Experience.

PHYL. That's not enough. I wanted you to say "Nature" or something so I could cut you down.

COR. Experience or nature.

PHYL. Perhaps they are hard to distinguish.

COR. They're hard for *us* to distinguish. What I meant by experience is that even when none of the theories were right people had the experience of thinking their ideas were beautiful.

PHYL. There never was a theory that wasn't any good. Even Aristotle's law of motion lit everything up for people. Even astrology. You could use it to explain what happened all around you and it gave you a feeling that the world is comprehensible.

COR. Isn't that the point?

PHYL. What?

COR. That the world is comprehensible. It didn't need to be.

PHYL. I don't know what you mean by that last remark, but the world is more than just comprehensible. It is comprehensible in terms of beautiful, simple forms.

COR. You mean mathematics that looks beautiful when we see how much it explains.

PHYL. No, mathematics that is beautiful anyway. Like your silly multiplication—you seem to think that's beautiful. Of course it doesn't prove anything at all. . . . Mathematicians don't need to make scientific theories to think that what they are making is beautiful. But explaining nature is more than writing down mathematics, it's beautiful strings of words that suddenly make you see the connections between things you never thought were connected. What property of nature is it that makes good theories be beautiful? It asks that somewhere in the book.

COR. I guess that's the property.

PHYL. No, but I mean is there anything you can see in nature to tell you that if you look for a correct theory you will find a beautiful theory? Those people at the beginning of that last chapter didn't know anything about natural law but they knew the world was full of beautiful numbers anyhow. I remember you said that motion can be a form. Did you notice Augustine said it too? How did they know?

COR.   It seems to me the world has a tendency to bring forth things we like. Good things.

(*There is a pause.*)

PHYL.   Is that connected with the anthropic business, that the world seems to have been designed so we can sit here and talk like this?

COR.   That's why we are here.

PHYL.   You mean because otherwise we wouldn't have been born?

COR.   Of course, but I mean we might not have evolved into you and me if thinking weren't so much fun from the beginning. We could have been two oysters, just sitting and growing.

PHYL.   And what you said about good things is the law of the universe? The law that explains all the other laws?

COR.   Yes.

PHYL.   Why is it that way? How do you explain that law?

COR.   I don't know. I'm not even sure it exists. Don't you think it's time we went to sleep?

PHYL.   Go ahead if you want to but I want to think.

(*Business with a blanket. It has grown darker. Venus shines in the western sky, Saturn is a little higher. Toward the south, Antares is suddenly visible. The air is still. They sleep.*)

Hero's Principle

(Chapter 8)

I illustrate with reflection of light from a plane surface, where the mathematics is trivial. Figure 8.1(a) shows a light ray passing from P to Q along a path whose length is $S = s_1 + s_2$,

$$S = \sqrt{(h_1^2 + x^2)} + \sqrt{[h_2^2 + (D-x)^2]}.$$

The minimum value of S occurs at the value of x for which the x-derivative of this quantity vanishes, and this very quickly gives

$$\sin i = \sin r,$$

from which it follows that $i = r$, as was to be proved.

Fermat's Principle

(Chapter 8)

Pierre de Fermat showed that the principle of least time leads to Snell's law of refraction at a plane surface. Figure 8.1(b) shows a ray of light that travels between points P and Q. Let the speed of light in the two media be v_i and v_r; then the time to go from P to Q is

$$s_i/v_i + s_r/v_r \,,$$

and this is proportional to

$$T = s_i + n\, s_r \,, \qquad \text{where } n = v_i/v_r \,.$$

The ratio n is called the relative index of refraction of the two media. From the figure,

$$T(x) = \sqrt{(x^2 + h_1^2)} + n \sqrt{[(D - x)^2 + h_2^2]} \,.$$

Minimizing this with respect to x gives

$$\sin i = n \sin r \,,$$

and this is Snell's law.

Newton's Theorem

(Chapter 10)

Suppose a body moves uniformly in a circle of radius r, always accelerating toward the center, and suppose a second body is given a constant acceleration equal in magnitude to the centripetal acceleration of the first. Then during the time when the first body travels a distance equal to r in its curved path, the second one travels a distance $r/2$.

To prove this theorem, imagine a body moving in a circular path with velocity v (Figure C). Then during a short time t it will fall a distance h below where it would have been had it moved in a straight line. To find h we use Pythagoras's theorem,

$$(h + r)^2 = (vt)^2 + r^2 ,$$

so that

$$h^2 + 2rh + r^2 = v^2t^2 + r^2 .$$

If t is very short the first term is negligible compared to the second and may be dropped. There remains

$$h = (v^2/2r)t^2 .$$

Thus, in the neighborhood of the starting point (which can be any point), h varies as t^2, which is the sign of constant acceleration. (The reason why t is taken to be small is that, as one can see from Figure C, the direction of h changes continually when the body moves, but over a short enough time it changes very little.) Suppose now that this same instantaneous acceleration persists for the longer time r/v during which the orbiting body travels a distance r along its path. The distance fallen is

$$(v^2/2r)(r/v)^2 = r/2 ,$$

which is Newton's starting point in his early calculation of orbital motion.

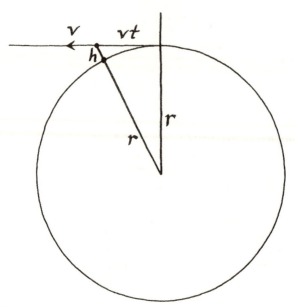

FIGURE C. To illustrate Newton's theorem.

Calculation of the Moon's Period

(Chapter 10)

To apply the theorem of Note C to determine the moon's motion, let the circulating body be the moon and the second one be a body falling near the earth. Its acceleration is 60^2 times greater than the moon's, so during the time t required for the moon to travel a distance r, it will fall a distance not $r/2$ but $60^2r/2$. Since the circumference of the moon's orbit is $2\pi r$, the time T required for the moon to make a complete circuit is equal to $2\pi t$. Suppose the second object falls during this entire period. How far does it fall?

Newton seems to have measured carefully the distance a body near the earth falls in one second and found 196 inches. From Galileo, we know that the distance increases as the square of the time, so that in t seconds it will fall a distance $196t^2$ inches. Then, by Newton's theorem,

$$60^2 r/2 = 196t^2 = 196(T/2\pi)^2 .$$

Let the radius of the earth be R; then $r = 60R$. Nobody knows what value Newton used for R. If the earth's circumference is taken as 25,000 miles then T comes out as 27.1 days. To compare this with the observed value we must remember that the time from new moon to new moon, called the synodic period, is not the same as the siderial period just calculated. This is because the new moon is an effect of illumination by the sun, and if we imagine the earth as stationary the sun does not stand still. It moves around the earth once a year in the same direction as the moon, and therefore the rate at which the moon's illuminated shape changes is given by the difference in the two rates:

$$\frac{1}{27.1} - \frac{1}{365} = \frac{1}{29.3} .$$

Thus the calculated synodic period is 29.3 days and the observed one is about 29.5. The error is reduced when more accurate data are used, but of course the real situation, with the moon revolving around the earth while the earth itself revolves around the sun, both orbits being elliptical, requires a more detailed calculation.

The Law of Areas

(Chapter 11)

To see how the methods of calculus give much simpler and more rigorous proofs of the theorems of mechanics than do those used in the *Principia*, consider Kepler's law of areas. Suppose a body moves in a field of force F always directed towards (or away from) the same point. Taking that point as origin of coordinates, write the equation of motion in component form,

$$m\ddot{x} = F(r)\, x/r\,, \qquad m\ddot{y} = F(r)\, y/r\,,$$

where each dot (Newton's notation) means differentiation with respect to time. Next, define the angular momentum of the orbiting body as

$$l = m(x\dot{y} - \dot{x}y)\,.$$

The time rate of change of this quantity is

$$\dot{l} = m(x\ddot{y} - \ddot{x}y)\,,$$

and substitution of the equations of motion shows that this is zero; therefore l is a constant.

To derive the law of areas it is convenient to express l in polar coordinates. Setting

$$x = r\sin\theta, \qquad y = r\cos\theta\,,$$

one finds easily

$$l = mr^2\dot{\theta}\,.$$

Figure E shows that the rate at which an area is swept out by the radius vector is $r^2\dot{\theta}/2$, so that the constancy of l implies the constancy of this rate. Note that the law of areas, unlike Kepler's other two laws, does not require the force to vary as the inverse square of the distance.

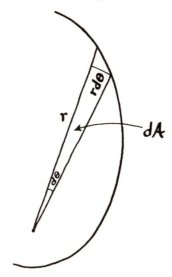

FIGURE E. For Kepler's area law: $dA = \frac{1}{2}r^2 d\theta$.

Elliptical Orbits

(Chapter 11)

Figure F reproduces Jean Bernoulli's proof of Kepler's law of ellipses (1742, vol. 1, p. 471). Jakob Hermann had derived a differential equation for the planetary orbit and integrated it incorrectly; Bernoulli does it right. Though the starting equation below is now obvious, Hermann's derivation is quite delicate (see ibid.), because it was only forty years later (see Chapter 13) that Euler showed how to resolve a force into orthogonal components. I will express the proof in modern terms. The equation for the x component of the planet's motion is

$$m\ddot{x} = -\frac{kx}{r^3}, \qquad r = \sqrt{(x^2 + y^2)}$$

where $\ddot{x} = \mathrm{d}^2x/\mathrm{d}t^2$. Write this as

$$m\ddot{x} = -\frac{km^2}{l^2}\frac{x(y\dot{x} - x\dot{y})^2}{r^3} \qquad \text{where } y\dot{x} - x\dot{y} = -l/m = \text{const.}$$

or

$$-a\ddot{x} = \frac{x(y\dot{x} - x\dot{y})^2}{r^3}, \qquad a = \frac{l^2}{km}.$$

That seems like a crazy way to write it, but now multiply by $\mathrm{d}t^2$; time variables disappear and what is left involves only the spatial variables appropriate to discussing an orbit. Time does not yet appear in the equations of dynamics, and Hermann has not used it in his derivation. I will keep the time variable so as to make the steps easier to follow. Remarkably, the last expression is a perfect derivative. It integrates to

$$-a\dot{x} = \frac{y(y\dot{x} - x\dot{y})}{r} + \text{const.}$$

This extraordinary integration can best be understood by differentiating it again. It is possible because the constant is proportional to the y-component of Laplace's

EXTRAIT DE LA REPONSE

De Monſieur BERNOULLI à Monſieur HERMAN, dattée de Basle le 7 Octobre 1710.

Dans vôtre équation differentio-differentielle — $a\,ddx$ = $(y\,dx - x\,dy) \times (x\,y\,dx - x\,x\,dy) : (x\,x + y\,y)^{\frac{3}{2}}$ je ne mets pas feulement [comme vous] — $a\,dx$ pour l'intégrale de — $a\,ddx$, mais — $a\,dx \pm$ une quantité conſtante. c'eſt-à-dire, — $a\,dx \pm e\,(y\,dx - x\,dy)$; pour le reſte je le fais comme vous; de ſorte qu'en intégrant vôtre précédente équation differentio-differentielle, je trouve — $a\,dx \pm e\,(y\,dx - x\,dy) = -y\,(y\,dx - x\,dy) : \sqrt{(x\,x + y\,y)} = (x\,y\,dy - y\,y\,dx) : \sqrt{(x\,x + y\,y)}$, ou — $a\,b\,dx : x\,x \pm e\,b\,(y\,dx - x\,dy) : x\,x = (b\,x\,y\,dy - b\,y\,y\,dx) : x\,x\sqrt{(x\,x + y\,y)}$, dont l'intégrale eſt $a\,b : x \pm e\,b\,y : x \pm c = b\,\sqrt{(x\,x + y\,y)} : x$ c'eſt-à-dire (en prenant $h = e\,b$, & en reduiſant l'équation) $a\,b \pm h\,y \pm c\,x = b\,\sqrt{(x\,x + y\,y)}$: laquelle équation, quoiqu'elle renferme $h\,y$, que la vôtre ne renfermoit pas, eſt cependant [comme elle] aux trois Sections Coniques.

FIGURE F. Jean Bernoulli's proof that the orbits of objects bound by an inverse-square force are conic sections. From *Opera omnia* (1742).

vector integral (1798, vol. 1, pp. 160–68), commonly attributed to Lenz, Pauli, and other latecomers (see Park 1979, p. 77).

Now comes another trick. Write the constant of integration as $\pm e(y\dot{x} - x\dot{y})$ and introduce the integrating factor b/x^2. Once more, as if by magic, everything can be integrated to give

$$\frac{ab}{x} \mp eb\frac{y}{x} = b\frac{r}{x} \pm \text{const. } c,$$

so that with $eb = h$,

$$br \pm hy \pm cx = ab,$$

where h and c are arbitrary. Go to polar coordinates $y = r\sin\theta$, $x = r\cos\theta$ and introduce two other arbitrary constants ε and ϕ by

$$h = \varepsilon b\cos\phi, \qquad c = \varepsilon b\sin\phi.$$

Then

$$r = \frac{a}{1 \pm \varepsilon \sin (\theta \pm \phi)}.$$

This is the equation for an ellipse ($0 < \varepsilon < 1$), a parabola ($\varepsilon = 1$), or an hyperbola ($1 < \varepsilon$), and ε is the eccentricity denoted by e in Chapter 6. All three types of orbit are (as Newton knew) possible when the force varies as $1/r^2$.

Bernoulli acknowledges that this is not the best way to calculate the orbit because it works only when the force is inverse-square. A few pages further on (1742, p. 474) he solves the general problem by reducing it to the same integral that is used to solve it today. The derivation is complicated because he lacks the concept of conserved energy and still does not use time derivatives in expressing Newton's law.

The only thing the proof given here has in common with Newton's discussion of the inverse problem (finding the law of force when the orbit is known to be an ellipse) is that each requires the talents of a tremendous virtuoso. As we shall see in Chapter 13, many years elapsed before the development of techniques that put such calculations within reach of the average mathematician.

Derivation of Young's Formula

(Chapter 13)

Figure G shows, in exaggerated scale, two light waves produced by a single source and issuing from two narrow slits. We want to find the wavelength of the light from measurements on the apparatus. As the figure is drawn, the path from Q to S is half a wavelength longer than that from P to S; therefore the wave PS arrives exactly out of phase from the wave QS and their effects cancel at S, producing a dark band on the screen. To calculate the wavelength λ from known quantities, we note that since PR is perpendicular to QS, and PQ is perpendicular to TU, triangles PQR and STU are similar. Therefore, if the distance SU is called D,

$$\frac{\lambda/2}{d} = \frac{D}{L} \qquad \text{so that} \qquad \lambda = 2d\frac{D}{L}.$$

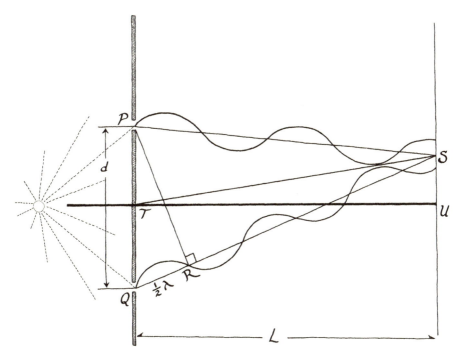

FIGURE G. Formation of an interference pattern. At S, where the two waves cancel, there will be a minimum of intensity.

Of Time and the River

(Chapter 14)

If a boat moving through the water at speed c is to move straight toward the opposite shore of a river that flows at speed v, it must head upstream so that the combined velocities of boat and stream carry it straight across (Figure H). The resultant speed is $\sqrt{(c^2 - v^2)}$, and so the time required to cross the river and back, a distance $2L$, is

$$t_{\text{across and back}} = \frac{2L}{\sqrt{(c^2 - v^2)}} = \frac{2L/c}{\sqrt{(1 - v^2/c^2)}}.$$

To find the time required to travel the same distance downstream and back, note that the river aids the boat on the way down and slows it on the way back:

$$t_{\text{down}} = \frac{L}{c + v}, \qquad t_{\text{back}} = \frac{L}{c - v}.$$

The sum of these gives

$$t_{\text{down and back}} = \frac{2Lc}{c^2 - v^2} = \frac{2L/c}{1 - v^2/c^2}.$$

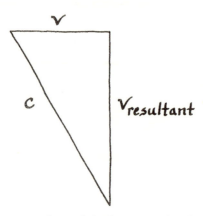

FIGURE H. The river's speed v and the boat's speed c determine the resultant motion of the boat.

NOTE I

The Mass of a Moving Object

(Chapter 14)

Assume that increasing the energy of an object by an amount E increases its mass by E/c^2, and that the increase in kinetic energy of a moving object when a force F pushes it through a distance dx is Fdx. Assume also that force increases momentum according to the law first stated in words by Newton and in symbols by Euler, $dp = Fdt$, where $p = mv$. Then

$$dE = Fdx = \frac{d(mv)}{dt}\,dx = \frac{dx}{dt}\,d(mv) = v\,d(mv).$$

But this is also equal to $dE = c^2dm$. Equating them gives

$$c^2dm = v\,d(mv) \qquad \text{or} \qquad c^2m\,dm = mv\,d(mv)\,.$$

Now integrate both sides of this relation:

$$c^2m^2 = m^2v^2 + c^2K^2\,,$$

where c^2K^2 is a constant of integration. Solving for m gives

$$m = \frac{K}{\sqrt{(1 - v^2/c^2)}}\,.$$

As we increase the object's speed and kinetic energy by doing work on it, its mass increases, and this formula gives the law of increase. Evidently K represents the mass of the object determined as it sits motionless in the laboratory. This is called the rest mass and is usually denoted m_0.

The Two-Slit Experiment in Quantum Mechanics

(Chapter 15)

This note is written for people who know some quantum mechanics. Its purpose is to study what happens to a two-slit interference pattern when someone tries to discover which slit the particle went through. In Chapter 15 we analyzed such an experiment in a rough-and-ready way, imagining an implausibly sensitive macroscopic detector, and found that if the experiment is done the interference pattern disappears and there is no longer any evidence to suggest that one particle went through both slits. The argument was crude because I wanted to treat the detector without using quantum mechanics, but really any device made to detect a single particle must itself belong to microphysics. This will be a more refined discussion, but it has a further purpose. The earlier argument concluded that the interference pattern is destroyed by the action of the detecting device on the ray that goes through it. Here I will show that the same thing happens even if the detector does not interfere at all with the ray.

Figure J shows the two-slit apparatus. The detector is indicated as a ball, near enough to the ray to interact with at least some of the particles going past. I draw it with a spring to suggest that the passage of a particle sets it into oscillation, but I don't care how it works. It is assumed to be in its ground state at the start of the measurement and to have a single excited state. Let $u_o(x)$ be the wave function of a ray that goes through the top slit without exciting the detector, and $u_1(x)$ be the changed wave function to the right of the detector if the detector has been excited. Since the lower ray never excites the detector it can be written simply as $v(x)$. Call the coordinate of the particle x and that of the detector q; let the normalized wave functions of the ground state and the excited state be $f_o(q)$ and $f_1(q)$. Finally, let s be a (real) number between 0 and 1 giving the probability amplitude that a particle in the upper ray will excite the detector; the probability is s^2. Adding the two waves together to find the pattern on the screen S gives

$$U_S(x,q) = \sqrt{(1 - s^2)}u_o(x)f_o(q) + su_1(x)f_1(q) + v(x) f_o(q) .$$

The next steps can be worked out on a scrap of paper; I omit the details. First, square the absolute magnitude of U_S. Considered as a function of x and q, this gives the probability that the particle arrives at the point P on the screen and that simultaneously the oscillator coordinate is q. But the oscillator is a thing of atomic size and we are not interested in the value of q; we want the probability that the particle arrives at the screen, no matter where the oscillator particle is. To find

420

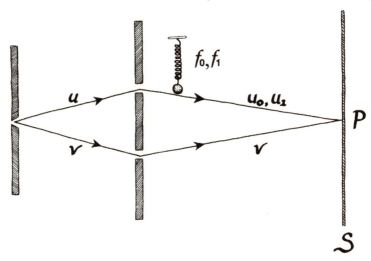

FIGURE J. A detector, here shown as a simple oscillator, hangs where it interacts with some of the particles passing the upper slit.

this, integrate over the oscillator coordinate. The integral of the squares of f_0 and f_1 are each equal to 1, and that over the product of the two is zero. (This is a fundamental rule of the theory.) The result is

$$\int |U_S(x,q)|^2 \, dq = |\sqrt{(1 - s^2)} \, u_0 + v|^2 + s^2 |u_1|^2.$$

This contains everything we need. To read it, look at the extreme cases when $s = 0$ (no coupling with the detector) or 1 (every particle interacts).

$$s = 0: \quad |u_0 + v|^2, \qquad s = 1: \quad |u_1|^2 + |v|^2.$$

In the first case the amplitudes add and we have interference; in the second, the intensities add. It is as if two different sources illuminated the two slits so that the light at P is just the sum of the intensities due to both. The intermediate cases represent an interference pattern that is partly washed out.

Thus far it is possible to conclude that the pattern is destroyed by the action of the detector on the particles it detects. The situation is more subtle than this, however. Suppose the detector is a magic detector that has no effect at all on the ray passing it. This can be represented by letting $u_1 = u_0$. The conclusion is exactly the same. This is an example of a remarkable feature of quantum mechanics that one would hardly have expected: the mere presence of something coupled to one ray can destroy the interference pattern. The something need only be a passive spectator, acted upon by the beam but not acting on it.

Quantum mechanics offers two distinct ways of describing and explaining what happens: one can talk of waves or particles. The "spectator effect" just described can be explained either way: one can say either that the spectator produces an unanalyzable change in the phase of one of the beams or that because it establishes that a particle goes through one slit or the other, not both, it prevents the formation of an interference pattern. The detailed theory provides the common basis for the two explanations.

Quantum Correlations That Suggest Action at a Distance

(Chapter 16)

If a beam of plane-polarized light passes through an analyzer whose axis is tilted at an angle θ away from the direction of polarization, the fraction of the incident intensity that gets transmitted varies as $\cos^2\theta$. When we consider the light as consisting of photons we must interpret this as the probability that a given photon is transmitted. This note will examine in more detail, for people who know some quantum mechanics, the conceptual experiment illustrated in Figure 16.3 and show that its results are hard to interpret in terms of photons passing through analyzers.

When a pion disintegrates it produces a pair of photons whose states of polarization are correlated. The pion is a pseudoscalar particle of spin zero, and one finds without difficulty (Yang 1950) the state describing the two photons into which it decays. Let e_{Fx} be the polarization vector describing the state of a photon polarized along the x-axis and traveling in the z-direction that takes it through polarizer PF, and similarly with other plane-polarized states. The two-photon function turns out to be

$$e_{Fx}e_{Gy} + e_{Fy}e_{Gx} .$$

We can use this expression to predict the result of some measurements performed on the apparatus of Figure 16.3. We set polarizer PF permanently with its axis of polarization vertical, parallel to the y-axis. Now look at the photons that travel toward PG. Since PF is oriented vertically, the state after CF has counted is $e_{Fy}e_{Gx}$. Thus what remains in the right-hand beam is e_{Gx}, and so we know that if one photon passes through PF and PG is oriented with its axis horizontal, the other photon will certainly pass through PG. Now suppose PG is at an angle θ away from the horizontal. The law of Malus again applies, and we can conclude that the probability that one photon passes PF and the other passes PG is again

$$P_Q = \tfrac{1}{2} \cos^2 \theta$$

(Q stands for quantum). This prediction has been borne out by careful experiments (Aspect et al. 1981, 1982a) that are similar in principle if not in detail.

There remains the problem of understanding what has happened. In Figure

16.2, when photons passed through PA and were about to strike PB, we said that PA had polarized them, and we could verify this by rotating either PB or PA. What has happened in the previous paragraph seems to be that in Figure 16.3, PF has polarized the photons that are about to strike PG—at least, both mathematically and experimentally, everything happens as if that were the case. The probability that both photons go through depends only on the relative orientation of PF and PG and again, we can change it by changing either PF or PG. But how can rotating PF affect what happens to the photons traveling to the right? When photons travel through PA to get to PB in Figure 16.2 we can say there is a causal connection, but how can there be in the other experiment in which two different photons are involved and nothing travels between PF and PG? To imagine what is happening using the ordinary notions of cause and effect, one has to introduce strange hypotheses such as hidden variables or hidden causal influences between PF and PG, any of which would greatly complicate a situation that mathematically is very simple. These are the "spooky interactions" Einstein referred to.

I have already mentioned a consequence of this reasoning that seems to defy common sense. Set PG at right angles to PF; then the probability that both counters count is 1/2, and that is not the same as the 3/8 calculated earlier on the basis of statistical independence and common sense. It is hard to see where the plausible argument has gone wrong as long as we assume that a photon carries a quantity known as polarization and that when photons strike two spatially separated screens their fates are determined independently of each other.

Though these and other conclusions may seem paradoxical to the mind, they do not contradict anything we actually know, and in fact they seem to be supported by the experiments that have been done (for earlier work see Clauser and Shimony 1978). What they contradict is our ideas about statistically independent events, our well-founded knowledge that if I toss a coin in one room and you toss one in another, the fall of one coin cannot possibly affect what happens to the other. It is the apparent violation of this principle—not just on the scale of photons but also on the scale of events that can be observed in the laboratory—that leads some people to claim that they do not feel they really understand the theory.

The Troublesome Question of How Things Look

(Chapter 16)

The theory of observation in quantum mechanics is complicated by the fact that we do not directly observe the tiny objects with which the theory normally deals. We think that macroscopic objects obey the same physical laws as microscopic ones, but when we try to apply these laws to them we sometimes run into trouble, mostly because the principle of superposition, a commonplace in the quantum mechanics of particles, seems to have no counterpart in our experience of things we can see and feel. This is illustrated in the following very artificial experiment.

Let us consider the spatial orientation of an object such as a large molecule M, large enough to be observed in some kind of supermicroscope. M is the kind of thing that we are accustomed to think of as having a definite orientation just as a book has. Suppose that M is held in place under the microscope by a combination of electric and magnetic fields. Let its orientation be represented by M_o, a function of three angles. At a certain moment a photon starts in the direction of a half-opaque screen that absorbs half the photons that strike it. A photon that goes through enters a counter that registers its arrival. If this happens, fields are applied so as to rotate the molecule through 90° into a new state M_1. If we apply the principles of quantum mechanics to this idealized picture the final wave function is a coherent superposition of two states, one representing the molecule in M_o and the other in M_1. What would we expect to see in the microscope? If the molecule's state is given by $(M_o + M_1)/\sqrt{2}$, then the probability distribution will be

$$\frac{1}{2}\left[|M_o|^2 + |M_1|^2 + 2\,\mathrm{Re}\,M_o^* M_1\right],$$

where Re means that the real part of what follows is to be taken. Microscopes are supposed to give sharp images (at least this one is), but the probability distribution contains a term corresponding to a molecule that is in some sense in both states at once. Will the image be a blur, or what? Anyhow, nobody has ever reported seeing such a thing in the real world.

In the real world things are never as simple as they are in idealized experiments like this one. The various parts of the world interact in so complex a fashion that if one part changes the rest are never quite the same. We have assumed here that they are. What should we put in to make the situation a little more realistic? Let us put in a fly, F, with coordinate x, that sits on the counter. As long as no photon arrives at the counter the fly does not move, but if the counter counts there is a probability given by s^2 that it is disturbed and begins to fly around. $F_o(x)$ represents the fly at rest, F_1 represents it flying around, and there is essentially no overlap between the two spatial distributions. Rewritten to include the fly, the state function is

$$\frac{1}{\sqrt{2}}\left[M_0 F_0(x) + \sqrt{(1-s^2)}M_1 F_0(x) + sM_1 F_1(x)\right].$$

The probability distribution is found by squaring this:

$$\frac{1}{2}\left\{|M_0 + \sqrt{(1-s^2)}M_1|^2|F_0(x)|^2 + s^2|M_1|^2|F_1(x)|^2 \right.$$
$$\left. + 2s\,\mathrm{Re}\,[M_0^* + \sqrt{(1-s^2)}M_1^*]M_1 F_0^*(x)F_1(x)\right\}.$$

But the fly is irrelevant to the experiment as long as we are looking through the microscope and not observing the fly. To find what the microscope shows we integrate over the coordinate x and find

$$\frac{1}{2}\left[|M_0|^2 + |M_1|^2 + 2\sqrt{(1-s^2)}\mathrm{Re}\,M_0^* M_1\right].$$

The interference term is still there but its amplitude depends on how likely the fly is to take off when the counter fires. If it is a nervous fly then $s = 1$ and the probability distribution reduces to the sum of two terms describing the possible orientations; we will see the molecule in one or the other of them with equal probability. Introducing a little bit of the world's complexity has simplified the result. Of course flies are not perfectly reliable, but real apparatus contains the equivalent of many flies, and the effect always seems to be that the interference term disappears.

The fly was a trick introduced so that I would not have to do a lot of physics. There is no question of representing a fly by a single vector in Hilbert space. And the real world is disturbed in much more subtle ways. To deal plainly with you, there was no need for the fly. The molecular transitions when a photon strikes a surface, the processes involved in the operation of the counter, the effects of the light needed for the microscope—any or all of these would have served as well as the fly. Realistic situations are much harder to analyze but it can be done (Joos and Zeh 1985), and the result is essentially the same: interference between two states on the macroscopic level is a prediction of quantum mechanics that is not intrinsically false but does not happen in the world we know. Perhaps some day it may be produced for a brief instant in the laboratory; only then will we know what it would be like to experience it.

Theory of the Expanding Universe

(Chapter 18)

Let us assume that the universe is infinite, or at least very large, and try to analyze its dynamics in purely Newtonian terms, making whatever simplifying assumptions are necessary to fill in the gaps in reasoning. Assume that the universe looks the same to every observer and that every point can therefore equally well be taken as the center. Now take our own galaxy as the center, calling it O. Assume that the space surrounding O is uniformly filled with galaxies and focus attention on one of them; call it G. Draw a sphere through G and centered on O (Figure M.1). It is a theorem of Newtonian gravity that G will be attracted toward O with a force depending only on the mass M of the matter contained within the sphere. This assumes that the rest of the universe, however far it extends, is symmetrically distributed about O. Let us assume it. Hubble's law states that as the universe expands all the galaxies lying on the sphere will move away at the same rate; those within the sphere will move more slowly and those outside more quickly, so no galaxies will ever cross the surface and the mass M will remain constant. Let m be the mass of the galaxy G. The total energy of G can be taken to be

$$E = \frac{1}{2} m \left(\frac{dR}{dt}\right)^2 - \frac{GmM}{R} = \text{const.}, \qquad (M.1)$$

where G is the Newtonian constant of gravitation. We may consider E to be positive, negative, or zero. Rewrite (M.1) as

$$\left(\frac{dR}{dt}\right)^2 = \frac{2GM}{R} + \frac{2E}{m}, \qquad (M.2)$$

and note the effect of the sign of E as expansion proceeds and R increases. Suppose $E < 0$. Then when

$$R = R_{\text{max}} = \frac{GMm}{-E},$$

the right side of equation (M.2) becomes zero and further expansion is impossible. At this point R reaches its maximum value and the universe collapses again. If $E \geq 0$ the expansion continues forever. The three cases are illustrated in Figure M.2. If $E < 0$, solution of (M.1) shows that

427

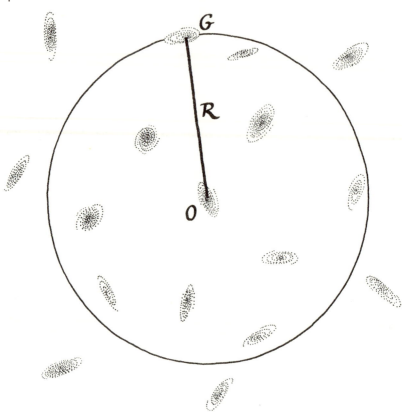

FIGURE M.1. A galaxy G, observed from O, is chosen to represent the rest.

$$t_c = \pi \left(\frac{m}{-2E} \right)^{1/2} R_{\max}.$$ (M.3)

Obviously it is of interest to know the sign of E. How can we find out? Going back to equation (M.2), note that

$$\frac{1}{2} \left(\frac{dR}{dt} \right)^2 \begin{array}{c} > \\ = \\ < \end{array} \frac{2GM}{R} \qquad \text{according as } E \begin{array}{c} > \\ = \\ < \end{array} 0 .$$ (M.4)

Make two changes in this. On the left put in Hubble's law

$$\frac{dR}{dt} = HR .$$

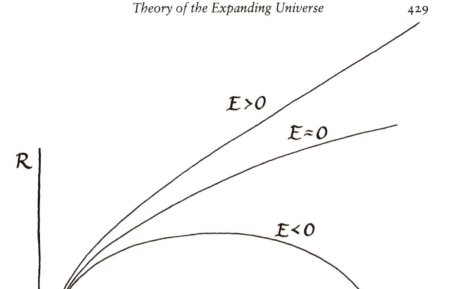

FIGURE M.2. Depending on the algebraic sign of the total energy of the model universe, the universe will either continue to expand, or it will collapse.

On the right express M in terms of the average matter density, ρ:

$$M = \frac{4}{3}\pi R^3 \rho .$$ (M.5)

Now the irrelevant R disappears, and equation (M.4) reduces to

$$\Omega = \frac{8\pi}{3}\frac{G\rho}{H^2} \begin{matrix}<\\=\\>\end{matrix} 1 \qquad \text{according as } E \begin{matrix}>\\=\\<\end{matrix} 0.$$ (M.6)

All the quantities on the left are measurable, at least in principle, though at present the values of ρ and H are not well known. I have mentioned the uncertainty in ρ, and H is uncertain because although the Doppler effect enables the velocities of recession to be found quite well the scale of cosmic distances is still in dispute. The remarkable fact is that when plausible values are introduced, Ω comes out to be in the vicinity of 1, perhaps somewhat less, which implies that $E > 0$ and the universe will expand forever.

Since E is not far from zero we can get a reasonably accurate picture by setting it equal to zero and proceeding. It is now easy to integrate equation (M.1), and when this is done we find that

$$R = Ct^{2/3} \tag{M.7}$$

where C is a constant depending on M. With this we can correct the estimate of the age of the universe given in Chapter 18. There, assuming that the galaxies have always been expanding at the same rate so that Hubble's constant really is a constant, we found $t = 1/H$. Obviously the assumption is false and t is less than this, since the effect of gravitation must be to slow the expansion as time goes on. From the last result we see how H varies with time:

$$H = \frac{dR/dt}{R} = \frac{2Ct^{-1/3}/3}{Ct^{2/3}} = \frac{2}{3t},$$

and if t is the present age of the universe and H the present value of Hubble's constant, it follows that

$$t = \frac{2}{3H} \tag{M.8}$$

which is somewhat less than before. In Chapter 18 it was mentioned that $1/H$ is measured at 8 to 22 billion years. With the correction, we arrive at about 5 to 15 billion years. This is not the only way to find the age of the universe. From the amounts of certain radioactive substances and from studies of the evolution of galaxies and star clusters, William Fowler (1987) has deduced an age of 11.0 ± 1.6 billion years, though some astronomers would assign a larger uncertainty. This may be an indication that the model is roughly correct.

If the same calculations are performed according to general relativity, equation (M.1) reappears but E need no longer be arbitrary.[1] The scale can be chosen so that if space is curved R represents its radius of curvature, while if it is flat R is a convenient scale factor. Then E is replaced by $\pm \frac{1}{2}mc^2$ or 0 instead of the continuum of values assumed above. If the negative sign is chosen, the curvature is that of a sphere and the universe has the finite volume $2\pi^2 R^3$. It is finite in extent as well as in time, and from equation (M.3) we at once find the connection: the

---

[1] Pauli (1958, note 19) gives a brief summary. For a detailed treatment of relativistic cosmology see Weinberg (1972) or Misner et al. (1973).

time between the big bang that opens the universe and the big crunch that closes
it is

$$t_c = \pi \, R_{max}/c \, .$$ (M.9)

That is, it is the time required for a light signal to travel halfway around the
universe at maximum expansion.

Whether there will be a big crunch depends on the value of Ω: is it greater than,
equal to, or less than 1? Observation does not say, but a theoretical argument
suggests that rather than taking on some adventitious value determined long ago
by the strength of the initial explosion, it is exactly 1.

It seems safe, from available data, to say that Ω_o, the present value of Ω, lies
between 0.01 and 10, so that

$$|\Omega_o - 1| < 9.$$

If this is so, then what was the value of the same quantity at some early instant t_i?
From equation (M.1), we find

$$\frac{2E}{m} = v^2 - \frac{8\pi}{3} G \rho R^2.$$

Divide this by $v^2 = H^2 R^2$ and refer to the definition of Ω:

$$|\Omega - 1| = \frac{2|E|}{mv^2}$$

and thus, comparing values at the initial and present times, we have

$$|\Omega_i - 1| = (v_o/v_i)^2 |\Omega_o - 1|.$$

In its early phase the energy of the universe was mostly in the form of radiation
and a calculation like the one above shows that the radius increased as $\sqrt{t}$. At
about 1 million years the radiation-dominated universe became matter-domi-
nated, and a simple calculation gives the ratio of the velocities in the last equation:
if t_i is given in seconds, then

$$|\Omega_i - 1| = \frac{t_i}{3 \times 10^{16}} |\Omega_o - 1|.$$ (M.10)

Choose some instant when the simple model first begins to apply, say 10^{-35} sec-
onds. Then if the present value of $|\Omega - 1|$ is less than 10, the initial value must
have been less than 3×10^{-52}. This would be very fine tuning. It seems less likely

that chance placed the initial Ω so close to 1 than that for some good reason it has been equal to 1 from the beginning. There is at present a theoretical model known as the inflationary universe (Barrow and Silk 1983, Reeves 1984) that explains why this might be so.

If $\Omega = 1$, then $E = 0$: the energy of the entire universe is zero (the mass-energy omitted from equation (M.1) is taken care of in the relativistic theory), and this has an interesting consequence.

No present theory says anything about what happened before the big bang, but if we suppose that a few basic conservation laws connect the time we know with some imagined epoch before it all started, the possibility that the total energy is zero takes on a striking significance: to make the entire universe no energy needed to be supplied. Two other laws that are firmly trusted are those that assert the conservation of angular momentum and electric charge. There is no indication that at present either the angular momentum or the total electric charge of the universe is anything other than zero. This being the case it is possible to argue (but not with equations since there are none) that to construct the universe required no space, no energy, no matter, no electric charge, and no spin. The universe may therefore be considered as an immense elaboration of nothing at all.

Bibliography

Agathias, S. *The Histories*, trans. J. D. Frendo (de Gruyter, Berlin, 1975).

Agrippa, H. C., *Three Books of Occult Philosophy or Magic*, ed. W. F. Whitehead, trans. J. F. (Hahn & Whitehead, Chicago, 1898; repr. AMS Press, New York, n.d.).

Aiton, E. J. "Celestial Spheres and Circles," *History of Science* 19: 75–114 (1981).

d'Alembert, J. L. *Traité de dynamique* (David, Paris, 1743; 2d ed., 1758).

Alexander, H. G. *The Leibniz–Clarke Correspondence* (Manchester University Press, 1956).

Alfargano (Al Fragani). *Il "Libro dell' aggregazione delle stelle,"* ed. and trans. R. Campani (Citta di Castello, Lapi, 1910).

Alväger, T., Farley, F.J.M., Kjellman, J., and Wallin, I. "Test of the second postulate of special relativity in the GeV region," *Phy Lett* 12: 260–62 (1964).

Anon. *Placides et Timéo ou Li sécrès as philosophes*, ed. C. A. Thomasset (Droz, Geneva, 1980).

——— (Henry Brougham). "The Bakerian lecture on the theory of light and colours," *Edinburgh Rev.* 1: 450–56 (1808).

——— (Isaac Newton). "An account of a book entitled, Commercium epistolicum Collinii et aliorum, de analysi promota," *Phil. Trans R. Soc. London* (1714). Repr. Baldwin, London (1809), pp. 116–53.

Aquinas, Saint Thomas. *On the Physics*, trans. R. J. Blackwell, R. J. Spaeth, and W. E. Thirlkel (Yale University Press, New Haven, 1963).

Archimedes. *The Works of Archimedes*, ed. and trans. T. L. Heath (Cambridge University Press, 1895, 1912).

Aristotle. *Complete Works*, ed. J. Barnes, 2 vols. (Princeton University Press, 1983).

Armitage, A. *Copernicus* (Allen & Unwin, London, 1938).

——— "The deviation of falling bodies," *Ann. Sci.* 5: 342–51 (1947).

——— *John Kepler* (Faber, London, 1966).

Armstrong, A. H., ed. *Cambridge History of Later Greek and Early Medieval Philosophy* (Cambridge University Press, 1967).

Aspect, A., Grangier, P., and Roger, G. "Experimental tests of realistic local theories via Bell's theorem," *Phys. Rev. Lett.* 47: 460–63 (1981).

——— "Experimental realizations of Einstein–Podolsky–Rosen–Bohm *Gedankenexperiment*: A new violation of Bell's inequalities," *Phys. Rev. Lett.* 49: 91–94 (1982a).

49: 91–94 (1982a).

Aspect, A., Dalibard, J., and Roger, G. "Experimental test of Bell's inequalities using time-varying analyzers," *Phys. Rev. Lett.* 49: 1804–807 (1982b).

Augustine, Saint. *The Writings of Saint Augustine*, 3 vols. (Cima, New York, 1948).

———— *Against the Academics*, trans. J. J. O'Meara (Newman Press, Westminster, Md., 1950).

———— *Earlier Writings*, trans. J.H.S. Burleigh (Westminster Press, Philadelphia, 1953).

———— *Later Works*, trans. J. Burnaby (Westminster Press, Philadelphia, 1955).

———— *Aurelii Augustini opera*, Pt. II, 2, ed. W. M. Green: Corpus Christianorum, Series Latina XXIX (Brepols, Turnhout, 1970).

———— *The Literal Meaning of Genesis*, trans. J. H. Taylor, 2 vols. (Newman Press, New York, 1982).

Bachelard, G. *Le Rationalisme appliqué* (Presses Universitaires de France, Paris, 1949).

Bacon, R. *The Opus Majus of Roger Bacon*, trans. R. B. Burke (University of Pennsylvania Press, Philadelphia, 1928).

Bailey, C. *Epicurus* (Clarendon Press, Oxford, 1926).

Bailey J. et al. "Measurement of relativistic time dilatation for positive and negative muons in a circular orbit," *Nature* 268: 301–305 (1977).

Barocas, V. "Jeremiah Horrocks (1619–1641)," *J. Brit. Astron. Assoc.* 79: 223–26 (1969).

Barrow, I. *The Usefulness of Mathematical Learning Explained and Demonstrated. Being Mathematical Lectures Read in the Publick Schools of the University of Cambridge*, trans. J. Kirkby (Austen, London, 1734).

Barrow, J. D. and Silk, J. *The Left Hand of Creation* (Basic Books, New York, 1983).

Barrow, J. D. and Tipler, F. J. *The Anthropic Cosmological Principle* (Oxford University Press, 1986).

Baumgardt, C. *Johannes Kepler: Life and Letters* (Philosophical Library, New York, 1951).

Bede. *Opera*, 8 vols. (Hervagius, Basel, 1563).

———— *Historical Works*, 2 vols., trans. J. E. King (Heinemann, London, 1930).

Belinfante, F. J. *A Survey of Hidden–Variable Theories* (Pergamon, Oxford, 1973).

Bell, J. S. "On the Einstein Podolsky Rosen paradox," *Physics* (NY) 1: 195–200 (1964).

Beller, M. "Matrix theory before Schrödinger," *Isis* 74: 469–91 (1983).

Benjamin, P. *The Intellectual Rise in Electricity* (Appleton, New York, 1895).

Bentley, R. *Sermons Preached at Boyle's Lecture* (Macpherson, London, 1838).

Bernardus Sylvestris. *The Cosmographia of Bernardus Sylvestris*, ed. and trans. W. Wetherbee (Columbia University Press, New York, 1973).

Bernoulli, D. "Examen principiorum mechanicae," *Commentarii Academiae Scientiarum Imperialis Petropolitanae* 1: 122–37 (1728).

———— *Hydrodynamica* (Dulsecker, Strassburg, 1738).

Bernoulli, J. *Opera omnia*, 4 vols. (Bousquet, Lausanne, 1742).

Bernstein J. and Feinberg G., eds. *Cosmological Constants* (Columbia University Press, New York, 1986).

Beyer, R. T., ed. *Foundations of Nuclear Physics* (Dover, New York, 1949).

Bochner, S. *The Role of Mathematics in the Rise of Science* (Princeton University Press, 1966).

Boethius. *De institutione arithmetica, De institutione musica,* etc., ed. G. Friedlein (Teubner, Leipzig, 1867). Parts trans. in *Boethian Number Theory,* trans. M. Masi (Rodopi, Amsterdam, 1983).

Bohr, N. "On the constitution of atoms and molecules," *Phil. Mag.* 26: 1–25, 476–502, 857–75 (1913).

—— *Atomic Theory and the Description of Nature* (Cambridge University Press, 1934).

—— "Can quantum–mechanical description of reality be considered complete?" *Phys. Rev.* 48: 696–702 (1935).

—— *Atomic Theory and Human Knowledge* (Wiley, New York, 1958).

—— *Essays 1958–1962 on Atomic Physics and Human Knowledge* (Wiley, New York, 1963).

—— *Collected Works,* ed. L. Rosenfeld et al., 9 vols. (North-Holland, Amsterdam, 1972–).

Boorse, H. A. and Motz, L., eds. *The World of the Atom* (Basic Books, New York, 1966).

Bopp, F., ed. *Werner Heisenberg und die Physik unserer Zeit* (Vieweg, Braunschweig, 1961).

Born, M. "Quantenmechanik der Stossvorgänge," *Z. Physik* 38: 803–27 (1926).

—— ed. *The Born–Einstein Letters* (Walker, New York, 1971).

Boscovich, R. J. *A Theory of Natural Philosophy,* trans. from the 2d ed. by J. M. Child (Open Court, London, 1922; repr. MIT Press, Cambridge, 1966).

Boulware, D. G. and Deser, S. "Classical general relativity derived from quantum gravity," *Ann. of Physics* (NY) 89: 193–240 (1975).

Boyer, C. B. "Early estimates of the velocity of light," *Isis* 33: 24–40 (1941).

Boyle R. *The Sceptical Chymist* (Crooke, London, 1661; repr. Dawson's, London, 1965).

Brecher, K. "Is the speed of light independent of the velocity of the source?" *Phys. Rev. Lett.* 39: 1051–54 (1977).

Brehault, E. *An Encyclopedist of the Dark Ages: Isidore of Seville* (Columbia University Press, New York, 1912).

Brewster, D. *Memoirs of the Life, Writings, and Discoveries of Isaac Newton,* 2 vols. (Constable, Edinburgh, 1855).

Brillet, A. and Hall, J. L. "Improved laser test of the isotropy of space," *Phys. Rev. Lett.* 42: 549–52 (1979).

de Broglie, L. "Ondes et quanta," *C. R. Acad Sci. Paris* 177: 507–10 (1923).

—— "Recherches sur la théorie des quanta," *Annales de Physique* 3: 22–128 (1925).

Bronowski, J. *The Ascent of Man* (Little, Brown, Boston, 1974).

Brush, S. *The Kind of Motion We Call Heat,* 2 vols. (North-Holland, Amsterdam, 1976).

Burkert, W. *Lore and Science in Ancient Pythagoreanism* (Harvard University Press, Cambridge, 1972).

Burckhardt, J. *The Civilization of the Renaissance in Italy*, trans. J. Middlemore (Phaidon, New York, 1944).

Burnet, J. *Early Greek Philosophy* (Macmillan, London, 4th ed., 1930).

Burtt, E. A. *The Metaphysical Foundations of Modern Science* (Humanities Press, New York, 2d ed. 1932; Doubleday, Garden City, 1954).

Buti G. and Bertagni, R. *Commento astronomico della Divina Commedia* (Remo Sandron, Firenze, 1966).

Campbell, L. and Garnett, W. *The Life of James Clerk Maxwell* (Macmillan, London, 1882).

Cantor, J. "The reception of the wave theory of light in Britain," *Historical Studies in the Physical Sciences* 6: 109–32 (1975).

Carnot, S. *Reflexions on the Motive Power of Fire*, ed. and trans. R. Fox (Manchester University Press, 1986).

Carpenter, N. C. *Music in the Medieval and Renaissance Universities* (University of Oklahoma Press, Norman, 1958).

Carr, B. J. and Rees, M. J. "The anthropic principle and the structure of the physical world," *Nature* 278: 605–12 (1979).

Cassiodorus. *An Introduction to Divine and Human Readings*, trans. L. W. Jones (Columbia University Press, 1946).

Charleton, W. *Physiologia Epicuro-Gassendo-Charltoniana* (Heath, London, 1654; repr. Johnson, New York, 1966).

Cicero. *De re publica & de legibus*, ed. and trans. C. W. Keyes (Loeb Classical Library, 1928).

——— *De natura deorum & academica*, trans. H. Rackham (Loeb Classical Library, 1933).

Claggett, M. *The Science of Mechanics in the Middle Ages* (University of Wisconsin Press, Madison, 1959).

——— *Archimedes in the Middle Ages* (University of Wisconsin Press, Madison, 1964–).

Clauser, J. F. and Shimony, A. "Bell's theorem: Experimental tests and implications," *Rep. Progr. Phys.* 41: 1881–927 (1978).

Clausius, R. "Über verschiedene für die Anwendung bequeme Formen der Hauptgleichungen der mechanischen Wärmetheorie," *Ann. Physik Chemie* 125: 353–400 (1865).

Clay, D. *Lucretius and Epicurus* (Cornell University Press, Ithaca, 1983).

Clement, J. "Students' perceptions in introductory mechanics," *Am. J. Phys.* 50: 66–71 (1980).

Cohen, I. B. "Roemer and the first determination of the velocity of light," *Isis* 31: 327–79 (1940).

Cohen M. R. and Drabkin, I. E. *A Source Book in Greek Science* (Harvard University Press, Cambridge, 1948).

Collins, J., ed. *Commercium epistolicum D. Johannis Collins et aliorum de analysi promota* (Royal Society, London, 1713).

Compton, A. H. "A quantum theory of the scattering of X-rays by light elements," *Phys. Rev.* 21: 483–502 (1923).

Copernicus, N. *On the Revolutions*, ed. J. Dobrzycki, trans. E. Rosen (Johns Hopkins University Press, Baltimore, 1978).

Cornford, F. M. *Plato's Cosmology* (Humanities Press, New York, 1937).

Cosmas (Indicopleustes). *The Christian Topography* (Hakluyt Society, London, 1897).

Crombie, A. C. *Robert Grosseteste and the Origins of Experimental Science, 1100–1700* (Clarendon Press, Oxford, 1953).

—— *Medieval and Early Modern Science* (Doubleday, New York, 1959), first pub. as *Augustine to Galileo: The History of Science A.D. 400–1650* (Harvard University Press, Cambridge, 1953).

Cusanus, N. *On Learned Ignorance*, trans. J. Hopkins (Banning, Minneapolis, 1981).

Darrow, K. K., ed. *Princeton Bicentennial Conference on the Future of Nuclear Science* (Princeton University Press, 1947).

Darwin, F. and Seward, A. C., eds. *More Letters of Charles Darwin*, 2 vols. (Appleton, New York, 1903).

Darwin, C. G., "The clock paradox in relativity," *Nature* 180: 976–77 (1957).

Davies, P.C.W. *The Search for Gravity Waves* (Cambridge University Press, 1980).

—— *The Accidental Universe* (Cambridge University Press, 1982).

—— "The anthropic principle," *Progr. Particle and Nuclear Phys.* 10: 1–38 (1983).

—— *Superforce* (Simon & Schuster, New York, 1984).

Delambre, J.B.J. *Histoire de l'astronomie ancienne*, 2 vols. (Courcier, Paris, 1817).

Descartes, R. *Oeuvres*, ed. C. Adam and C. Tannery, 11 vols. (Vrin, Paris, new ed. 1964–1974).

—— *Principles of Philosophy*, trans. V. R. Miller and R. P. Miller (Reidel, Dordrecht, 1983).

—— *The Philosophical Writings of Descartes*, trans. J. Cottingham, R. Stoothoff, and D. Murdoch, 2 vols. (Cambridge University Press, 1984).

Diels, H. *Die Fragmente der Vorsokratiker*, ed. W. Kranz (Weidmann, Berlin, 9th ed., 1959–60).

Dijksterhuis, E. J. *The Mechanization of the World Picture*, trans. C. Dikshoorn (Oxford University Press, 1961; repr. Princeton University Press, 1986).

Diogenes Laertius. *Lives of Eminent Philosophers*, trans. R. D. Hicks (Loeb Classical Library, 1925).

Dirac, P.A.M. "Quantum mechanics and a preliminary investigation of the hydrogen atom," *Proc. R. Soc. London* A110: 561–69 (1926).

—— "Lagrangian in quantum mechanics," *Physikalische Zeitschrift für Sowjetunion* 3: 64–72 (1933).

Dobbs, B.J.T. *The Foundations of Newton's Alchemy* (Cambridge University Press, 1975).

—— "Newton's alchemy and his theory of matter," *Isis* 73: 511–28 (1982).

Drake, S. *Galileo at Work* (University of Chicago Press, 1978).

Dreyer, J.L.E. *History of the Planetary Systems from Thales to Kepler* (Cambridge University Press, 1906).

Dugas, R. *A History of Mechanics*, trans. J. R. Maddox (Griffon, Neuchatel, 1955).

—— *Mechanics in the Seventeenth Century*, trans. F. Jaquot (Griffon, Neuchatel, 1958).

Duhem, P. *Le Système du monde*, 8 vols. (Hermann, Paris, 1913–59); one-volume condensation: *Medieval Cosmology*, ed. and trans. R. Ariew (University of Chicago Press, 1985).

—— *The Aim and Structure of Physical Theory*, trans. P. P. Wiener (Princeton University Press, 1954; repr. Athenaeum, New York, 1962).

—— *To Save the Phenomena*, trans. E. Doland and C. Maschler (University of Chicago Press, 1969).

Dyson, F. J. "Energy in the universe," *Sci. Am.* 225 (9): 50–59 (September, 1971).

Einstein, A. "Über einen die Erzeugung und Verwandlung des Lichtes betreffenden heuristisches Gesichtspunkt," *Ann. Physik* 17: 132–48 (1905a). Trans. in Boorse and Motz (1966, vol. 1, p. 544).

—— "Die von der molekularkinetischen Theorie der Wärme geforderte Bewegung von in ruhenden Flüssigkeiten suspendierten Teilchen," *Ann. Physik* 17: 549–60 (1905b). Trans. in Einstein (1926).

—— "Zur Elektrodynamik bewegter Körper," *Ann. Physik* 17: 891–921 (1905c). Trans. in Lorentz et al. (1923).

—— "Eine neue Bestimmung der Moleküldimensionen," *Ann. Physik* 19: 289–306 (1906); 34: 591–92 (1911). Trans. in Einstein (1926).

—— "Relativitätsprinzip und die aus demselben gezogenen Folgerungen," *Jahrbuch für Radioactivität und Elektronik* 4: 411–62; 5, 98–99 (1907).

—— "Einfluss der Schwerkraft auf die Ausbreitung des Lichtes," *Ann. Physik* 35: 898–908 (1911). Trans. in Lorentz et al. (1923).

—— "Lichtgeschwindigkeit und Statik des Gravitationfeldes," *Ann. Physik* 38: 355–69 (1912). Trans. in Lorentz et al. (1923).

—— "Die Grundlage der allgemeinen Relativitätstheorie," *Ann. Physik* 49: 769–822 (1916a). Trans. in Lorentz et al. (1923).

—— "Hamiltonsches Prinzip und allgemeine Relativitätstheorie," *Preussische Akad. Wiss. Sitzungsber.* 1916: 1111–16 (1916b). Trans. in 3d and later eds. of Lorentz et al. (1923).

—— "Kosmologische Betrachtungen zur allgemeinen Relativitätstheorie," *Preussische Akad. Wiss. Sitzungsber.* 1917: 142–52 (1917). Trans. in 3d and later eds. of Lorentz et al. (1923).

—— *Sidelights on Relativity*, trans. G. B. Jeffery and W. Perrett (Methuen, London, 1922).

—— *Investigations on the Theory of the Brownian Movement*, ed. R. Fürth, trans. A. D. Cowper (Methuen, London, 1926).

—— *Out of My Later Years* (Philosophical Library, New York, 1950).

—— *Ideas and Opinions* (Crown, New York, 1954).

Einstein, A., Podolsky, B., and Rosen, N. "Can quantum-mechanical description of reality be considered complete?" *Phys. Rev.* 47: 777–80 (1935).

Eliade, M. *The Forge and the Crucible* (Rider, London, 1962).

Elkana, Y. *The Discovery of the Conservation of Energy* (Harvard University Press, Cambridge, 1974).

Emerton, N. E. *The Scientific Reinterpretation of Form* (Cornell University Press, Ithaca, 1984).

Euclid. *Euclide's Elements*, ed. and trans. I. Barrow (Daniel, London, 1660).

—— "The Optics of Euclid," trans. H. E. Burton, *J. Optical Soc. America* 35: 357–72 (1945).

Euler, L. *Opera omnia*, ed. F. Rudio, A. Krazer, P. Stäckel et al. (Teubner, Leipzig, 1911–).

—— *Lettres à une princesse d'Allemagne sur diverses sujets de physique et de philosophie*, 3 vols. (Académie impériale des sciences, St. Petersbourg, 1768–74). Trans. as *Letters of Euler on Different Subjects in Natural Philosophy Addressed to a German Princess*, 2 vols. (Harper, New York, 2d ed., 1842).

Faraday, M. *Experimental Researches in Electricity*, 3 vols. (Taylor, London, 1839–1855).

—— "A speculation touching electric conduction and the nature of matter," Phil. Mag. 24: 136–43 (1844).

Feynman, R. P. "Space-time approach to non-relativistic quantum mechanics," *Rev. Modern Phys.* 20: 367–87 (1948).

—— "Space–time approach to quantum electrodynamics," *Phys. Rev.* 76: 769–89 (1949a).

—— "The theory of positrons," *Phys. Rev.* 76: 749–59 (1949b).

—— *The Character of Physical Law* (BBC, London, 1965; MIT Press, Cambridge, 1967).

—— "Simulating physics with computers," *Int. J. Theoretical Phys.* 21: 467–88 (1982).

—— *QED* (Princeton University Press, 1985).

Findlay, J. N. *Plato: The Written and Unwritten Doctrines* (Humanities Press, New York, 1974).

Fischer, I. "Another look at Eratosthenes' and Posidonius' determinations of the earth's circumference," *Q.J.R. Astron. Soc.* 16: 52–167 (1975).

Flamel, N. *Nicholas Flammel, His Exposition of the Hieroglyphicall Figures Which he Caused to bee Painted upon an Arch in St. Innocents Churchyard in Paris,* trans. "E. Orandus" (Walkley, London, 1624; recent ed. Heptangle Books, 1980).

de Fontenelle, B. *Entretiens sur la pluralité des mondes* (Blageart, Paris, 1686, and ed. A. Calame, Paris, Didier, 1966).

Forsee, A. *Albert Einstein, Theoretical Physicist* (Macmillan, New York, 1963).

Fowler, A. *Spenser and the Numbers of Time* (Routledge & Kegan Paul, London, 1964).

Fowler, W. "The age of the observable universe," *Q.J.R. Astron. Soc.* 28: 87–108 (1987).

Freeman, K. *Ancilla to the Presocratic Philosophers* (Harvard University Press, Cambridge, 1978).

Friedmann, A. "Über die Krümmung des Raumes," *Z. Physik* 10: 377–86 (1922). Trans. in Bernstein and Feinberg (1986).

Fritzsch, H. *Quarks* (Basic Books, New York, 1983).

Gale, G. "The anthropic principle," *Sci. Am.* 235 (12): 154–71 (December, 1981).

Galileo Galilei. *Discourses and Mathematical Demonstrations Concerning Two New Sciences*, trans. H. Crew and A. Favaro (Macmillan, New York, 1914).

—— *On Motion* and *On Mechanics*, ed. and trans. I. E. Drabkin and S. Drake (University of Wisconsin Press, Madison, 1960).

—— *Dialogue on the Great World Systems*, trans. T. Salusbury (University of Chigago Press, 1953).

—— *Opere*, ed. A. Favaro, 20 vols. (Barbera, Firenze, 1968).

Galileo Galilei, Grassi, H., Guiducci, M., and Kepler, J. *The Controversy on the Comet of 1618*, trans. S. Drake and C. D. O'Malley (University of Pennsylvania Press, Philadelphia, 1960).

Gaos, J. *Historia de nuestra idea del mundo* (FCE, Mexico, 1973).

Gassendi, P. *Animadversiones in decimum librum Diogenis Laertii*, 2 vols. (Barbier, Paris, 1649).

Gasser J. and Leutwyler, H. "Quark masses," *Phys. Reports* 87: 77–169 (1982).

Gershenson, D. E. and Greenberg, D. A. "The first chapter of Aristotle's *Foundations of Scientific Thought*," *Natural Philosopher* 2: 5–55 (1963).

Gilbert, W. *De magnete* (Short, London, 1600), repr. *Culture et Civilisation*, Brussels, 1967; trans. P. F. Mottelay (Quarich, London, 1893).

Gillispie, C. C. *The Edge of Objectivity* (Princeton University Press, 1960).

Gingerich, O. " 'Crisis' versus aesthetics in the Copernican revolution," *Vistas in Astron.* 17: 85–95 (1975a).

—— "The origins of Kepler's third law," *Vistas in Astron.* 18: 595–601 (1975b).

—— "Was Ptolemy a Fraud?" *Q.J.R. Astron. Soc.* 21: 253–66 (1980).

—— "Ptolemy Revisited: a reply to R. R. Newton," *Q.J.R. Astron. Soc.* 22: 40–44 (1981).

Gombrich, E. H. *Norm and Form* (Phaidon, New York, 1966).

Gregory of Nyssa, Saint, *Apologia in hexaemeron*, ed. W. Jaeger (Brill, Leiden, 1960).

Grosseteste, *see* Robert.

Grynaeus, S. *Novus orbis* (Hervagius, Basel, 1532).

Gurevich, L. and Mostepanenko, V. "On the existence of atoms in n-dimensional space," *Phys. Lett.* A35: 201–202 (1971).

Haefele, J. C. and Keating, R. E. "Around-the-world atomic clocks," *Science* 177: 166–168; 168–170 (1972).

—— "Theories of Gravitation," *Physics Today* 26 (3): 11 (March 1973).

Hall, R. "Correcting the *Principia*," *Osiris* 13: 291–326 (1958).

Hamilton, W. R. "On a general method of expressing the paths of light, and of the planets, by the coefficients of a characteristic function," *Dublin University Review* (1833), 795–826.

—— *Mathematical Papers*, ed. A. W. Conway *et al.*, 3 vols. (Cambridge University Press, 1931–1965).

Harries, K. "The infinite sphere. Comments on the history of a metaphor," *J. History of Philosophy* 13: 5–15 (1975).

Harris, J. *Lexicon Technicum Or, An Universal English Dictionary of Arts and Sciences*, 2 vols. (Brown et al., London, 1710).

Heath, T. L. *Aristarchus of Samos, the Ancient Copernicus* (Clarendon Press, Oxford, 1913).

———— *A History of Greek Mathematics*, 2 vols. (Clarendon Press, Oxford, 1921).

———— *Greek Astronomy* (Dent, London, 1932).

Heilbron, J. *Historical Studies in the Theory of Atomic Structure* (Arno, New York, 1981).

Heisenberg, W. "Über quantentheoretische Umdeutung kinematischer und mechanischer Beziehungen," *Z. Physik* 33: 879–93 (1925). Trans. in van der Waerden (1967).

———— "Über die Spektra von Atomsystemen mit zwei Elektronen," *Z. Physik* 39: 499–518 (1926).

———— "Uber den anschaulichen Inhalt der quantentheoretischen Kinematik und Mechanik," *Z. Physik* 43: 172–98 (1927). Trans. in Wheeler and Zurek (1982).

———— *The Physical Principles of the Quantum Theory* (University of Chicago Press, 1930).

———— *Philosophic Problems of Nuclear Science* (Pantheon, New York, 1952).

———— *The Physicist's Conception of Nature* (Harcourt, Brace, New York, 1958a).

———— *Physics and Philosophy* (Harper, New York, 1958b).

———— *Wandlungen in den Grundlagen der Naturwissenschaft* (Hirzel, Stuttgart, 8th ed. 1959). Trans. in part as *Philosophic Problems of Nuclear Science* (Fawcett, Greenwich, 1966).

———— *Physics and Beyond* (Harper, New York, 1971).

———— *Across the Frontiers* (Harper, New York, 1974).

Heitler, W. and London, F. "Wechselwirkung neutraler Atome und homöopolare Bindung nach der Quantenmechanik," *Z. Physik* 40: 883–92 (1927).

Helmholtz, H. von. *Über der Erhaltung der Kraft* (Reimer, Berlin, 1847).

———— *Selected Writings*, ed. R. Kahl (Wesleyan University Press, Middletown, 1971).

Hendry, J. *The Creation of Quantum Mechanics and the Bohr–Pauli Dialogue* (Reidel, Dordrecht, 1984).

Heninger, S. K. *Touches of Sweet Harmony* (Huntington Library, San Marino, 1974).

Herivel, J. *The Background to Newton's Principia* (Clarendon Press, Oxford, 1965).

Hero, *The Pneumatics of Hero of Alexandria*, ed. B. Woodcroft, trans. J. G. Greenwood (Taylor, Walton & Maberly, London, 1851).

Hermann, A. *The Genesis of Quantum Theory (1899–1913)*, trans. C. W. Nash (MIT Press, Cambridge, 1971).

Herschel, J.F.W. *A Preliminary Discourse on the Study of Natural Philosophy* (Harper, New York, 1840).

Hertz, H. *Miscellaneous Papers*, trans. G. E. Jones and G. H. Schott (Macmillan, London, 1896).

Holmyard, E. J. *Alchemy* (Penguin, London, 1957).

Holton, G. *The Scientific Imagination* (Cambridge University Press, 1978).

Hooke, R. *Posthumous Works* (Waller, London, 1705).

Hopper, V. F. *Medieval Number Symbolism* (Columbia University Press, New York, 1938).

Horrocks, J. *Jeremiae Horroccii opera posthuma*, ed. J. Wallis (Godbid, London, 1672).

Hoyer, U. "Kepler's celestial dynamics," *Vistas in Astron.* 23: 69–74 (1979).

Hoyle, F. *The Black Cloud* (Heinemann, London, 1957).

Hubble, E. P. "Cepheids in spiral nebulae," *Publ. Am. Astron. Soc.* 5: 261–64 (1927).

——— "A relation between distance and radial velocity among extra-galactic nebulae," *Proc. Nat. Acad. Sci.* 15: 168–73 (1929).

Huizinga, J. *Homo Ludens*, trans. R.F.C. Hull (Routledge & Kegan Paul, London, 1949).

Humphreys, D. *The Esoteric Structure of Bach's Clavierübung III* (University College Cardiff Press, 1983).

Hutchison, K. "What happened to occult qualities in the scientific revolution?" *Isis* 73: 233–53 (1982).

Huygens, C. *Horologium oscillatorium* (Muguet, Paris, 1673, repr. *Culture et Civilisation*, Bruxelles, 1966).

——— *Traité de la lumière* (vander Aa, Leyden, 1690; English trans. *Treatise on Light*, Macmillan, London, 1912; pap. Dover, New York, 1962).

Hyman A. and Walsh, J. J. *Philosophy in the Middle Ages* (Harper & Row, New York, 1967).

Ibn Khaldun. *The Muqaddimah*, 2 vols., trans. F. Rosenthal (Pantheon, New York, 1938).

Isidore of Seville, Saint. *Etymologiarum sive originum libri xx*, ed. W. M. Lindsay (Clarendon Press, Oxford, 1911).

Jacobi, C.G.J. *Vorlesungen über Dynamik*, ed. A. Clebsch (Reimer, Berlin, 1884).

Jaeger, W. *Aristotle* (Oxford University Press, 2d ed., 1948).

——— *The Theology of the Early Greek Philosophers* (Oxford University Press, 1947).

Jammer, M. *Concepts of Space* (Harvard University Press, Cambridge, 1954, 1969).

——— *Concepts of Force* (Harvard University Press, Cambridge, 1957).

——— *The Conceptual Development of Quantum Mechanics* (McGraw-Hill, New York, 1966).

Jerome, Saint. *S. Hieronymi presbyteri opera*, Pt. 1, 2, ed. M. Adriaen: Corpus Christianorum Series Latina LXXIII (Brepols, Turnhout, 1963).

Jones, C. W. "A Note on concepts of interior planets in the early Middle Ages," *Isis* 24: 397–99 (1936).

Joos, E. and Zeh, H. D. "The emergence of classical properties through interaction with the environment," *Z. Physik* B59: 223–43 (1985).

Joule, J. P. *Scientific Papers*, 2 vols. (Taylor & Francis, London, 1884).

Jung, C. G. and Pauli, W. *The Interpretation of Nature and the Psyche* (Pantheon, New York, 1955).

Jungnickel, C. and McCormmach, R. *Intellectual Mastery of Nature*, 2 vols. (University of Chicago Press, 1986).

Kaivola, M., Poulsen, O., Riis, E., and Lee, S. A., "Measurement of the relativistic Doppler effect in neon," *Phys. Rev. Lett.* 54: 255–58 (1985).

Kaluza, T. "Zum Unitätsproblem der Physik," *Preussische Akad. Wiss. Sitzungsber.* 1922: 966–72 (1922).

Kargon, R. H. *Atomism in England from Hariot to Newton* (Oxford University Press, 1966).

Kelvin, see Thomson, W.

Kepler, J. *Ad Vitellionem paralipomena* (Marnius, Frankfurt, 1604).

——— *Astronomia nova* (Heidelberg, 1609).

——— *Tertius interveniens* (Frankfurt, 1610).

——— *Dioptrice* (Franck, Augsburg, 1611).

——— *Harmonices mundi* (Tampach, Linz, 1619).

——— *Gesammelte Werke*, ed. M. Caspar et al., 20 vols. (Beck, München, 1937–75).

——— *The Six–Cornered Snowflake*, ed. L. L. Whyte, trans. C. Hardie (Clarendon Press, Oxford, 1966).

——— *Mysterium cosmographicum (The Secret of the Universe)*, trans. A. M. Duncan (Abaris, New York, 1981).

Keynes, J. M. *The Collected Writings of John Maynard Keynes*, 29 vols. (Macmillan, London, 1971–83).

Kimble, G.H.T. *Geography in the Middle Ages* (Methuen, London, 1938).

King-Heale, D. G. *Vistas in Astron.* 18, 497–517 (1975).

Kirk, G. S., Raven, J. E., and Schofield, M. *The Presocratic Philosophers* (Cambridge University Press, Cambridge, 2d ed., 1983).

Klein, O. "Quantentheorie und fünfdimensionale Relativitätstheorie," *Z. Physik* 37: 895–906 (1926).

——— "The atomicity of electricity as a quantum theory law," *Nature* 118: 516 (1926).

Klibansky, R. *The Continuity of the Platonic Tradition During the Middle Ages* (Warburg Institute, London, 1939).

——— ed. *Philosophy in the Mid-Century* (Nuova Italia, Firenze, 1958).

Koyré, A. *Etudes galiléennes*, 3 vols. (Hermann, Paris, 1939).

——— *From the Closed World to the Infinite Universe* (Johns Hopkins University Press, Baltimore, 1957).

——— *Newtonian Studies* (University of Chicago Press, 1965).

Kren, C. *Medieval Science and Technology* (Garland, New York, 1985).

Kuhn, T. S. *The Copernican Revolution* (Harvard University Press, Cambridge, 1957).

——— *The Structure of Scientific Revolutions* (University of Chicago Press, 1962, 1970).

Kuhn, T. S. *Black-Body Theory and the Quantum Discontinuity 1894–1912* (Clarendon Press, Oxford, 1978).

———— "Revisiting Planck," *Historical Studies in the Physical Sciences,* 14(2) 231–52 (1984).

Lactantius, *The Works of Lactantius,* trans. W. Fletcher, 2 vols. (Clark, Edinburgh, 1871).

Lagrange, J. L. *Mécanique analitique* (Desaint, Paris, 1788).

Landau, L. D. and Lifschitz, E. M. *The Classical Theory of Fields,* trans. M. Hamermesh (Addison-Wesley, Reading, 3d ed., 1971).

Langford, J. J. *Galileo, Science, and the Church* (University of Michigan Press, Ann Arbor, 1971).

Lannutti J. C. and Williams, P. K., eds. *Current Trends in the Theory of Fields* (American Institute of Physics, New York, 1978).

de Laplace, P. S. *Traité de mécanique céleste,* 5 vols. (Paris, Duprat and successors, An VII [1798]–1827); trans. N. Bowditch (Hillard, Gray, Little, and Wilkins, Boston, 1829–39).

———— *Oeuvres,* 7 vols. (Imprimérie Royale, Paris, 1843–47).

Larmor, J. *Aether and Matter* (Cambridge University Press, 1900).

———— ed. *Origins of Clerk Maxwell's Electric Ideas* (Cambridge University Press, 1937)

Lasswitz, K. *Geschichte der Atomistik,* 2 vols. (Voss, Hamburg, 1890).

Lavoisier, A. L. *Elements of Chemistry,* trans. R. Kerr (Creech, London, 3d ed., 1796).

Layard, A. H. *Discoveries Among the Ruins of Nineveh and Babylon* (Harper, New York, 1856).

Leonardo da Vinci, *Notebooks of Leonardo da Vinci,* ed. and trans. E. MacCurdy, 2 vols. (Reynal & Hitchcock, New York, 1938).

Lévi-Strauss, C. *The Raw and the Cooked* (Harper, New York, 1969).

Lewis, C. S. *The Discarded Image* (Cambridge University Press, 1964).

Lindberg, D. C., ed. *Science in the Middle Ages* (University of Chicago Press, 1978).

Lindsay, R. B., ed. *Energy: Historical Development of the Concept* (Dowden, Hutchinson, & Ross, Stroudsburg, 1975).

Lloyd, G.E.R. *Magic Reason and Experience* (Cambridge University Press, 1979).

Lodge, O. *Modern Views of Electricity* (Macmillan, London, 1889).

———— *Ether and Reality* (Doran, New York, 1925).

Loewy, E. *The Rendering of Nature in Early Greek Art,* trans. J. Fothergill (Duckworth, London, 1907).

Longair, M. S., ed. *Confrontation of Cosmological Theories with Observational Data* (Reidel, Dordrecht, 1974).

Loomis, E. S. *The Pythagorean Proposition* (Mohler, Berea, Ohio, 1927, 1940).

Lorentz, H. A., Einstein, A., Minkowski, H., and Weyl, H. *The Principle of Relativity,* ed. A. Sommerfeld, trans. W. Perrett and G. H. Jeffery (Methuen, London, 1923; repr. Dover, New York, 1952).

Lovejoy, A. O. *The Great Chain of Being* (Harvard University Press, Cambridge, 1936).

Lucretius, *The Way Things Are*, trans. R. Humphries (Indiana University Press, Bloomington, 1968).

Mach, E. *Die Mechanik in ihrer Entwicklung historisch-kritisch dargestellt* (Brockhaus, Leipzig, 1883); trans. T. J. McCormack as *The Science of Mechanics* (Open Court, La Salle, 1893 and later eds.)

MacKinnon, E. M. *Scientific Explanation and Atomic Physics* (University of Chicago Press, 1982).

Macrobius, *Commentary on the Dream of Scipio*, trans. W. H. Stahl (Columbia University Press, New York, 1952).

Magie, E. M. *A Source Book in Physics* (McGraw-Hill, New York, 1935).

Maier, M. *Atalanta fugiens* (Galler, Oppenheim, 1617).

Mâle, E. *Religious Art in France in the Thirteenth Century* (Princeton University Press, 1984).

Manuel, F. *Isaac Newton Historian* (Harvard University Press, Cambridge, 1962).

—— *A Portrait of Isaac Newton* (Harvard University Press, Cambridge, 1968).

—— *The Religion of Isaac Newton* (Clarendon Press, Oxford, 1974).

Mather, C. *Memorable Providences Relating to Witchcraft and Possessions* (R. P., Boston, 1689; repr. Parkhurst, London, 1691).

de Maupertuis, P.L.M. *Oeuvres*, 4 vols. (Bruyset, Lyon, 1758; repr. Olms, Hildesheim, 1965).

—— "Les lois du mouvement et du repos, déduites d'un principe de métaphysique," *Mémoires de l'Académie de Berlin* 1746 (1747).

Maxwell, J. C. *Scientific Papers*, ed. W. D. Niven, 2 vols. (Cambridge University Press, 1890).

Mayer, J. R. "On the forces of inorganic nature" *Phil. Mag.* 24: 371–77 (1862). Repr. in Lindsay (1975).

McCloskey, M., Caramazzi, A., and Green, B. "Curvilinear motion in the absence of external forces," *Science* 210: 1139–41 (1980).

McCrea, W. H. and Milne, E. A. "Newtonian universes and the curvature of space," *Q. J. Math* 5: 72–80 (1934).

McEvoy, J. J. *The Philosophy of Robert Grosseteste* (Oxford University Press, New York, 1982).

McMullin, E., ed. *The Concept of Matter in Greek and Medieval Philosophy* (University of Notre Dame Press, South Bend, 1963).

Michelson, A. A. and Morley, E. W. "On the relative motion of the earth and the luminiferous ether," *Am. J. Sci.* 34: 333–45 (1887). Also published in *Phil. Mag.* 24: 449–63 (1887).

Migne, J. P. *Patrologiae cursus completus, Series latina*, 221 vols. (Migne, Paris, 1844–65).

Miller, A. I. *Einstein's Special Theory of Relativity* (Addison-Wesley, Reading, 1981).

—— *Imagery in Scientific Thought* (Birkhäuser, Boston, 1984).

Minkowski, H. "Raum und Zeit," *Physikalische Z.* 20: 104–11 (1908). Trans. in Lorentz et al. (1923).

Misner, C. W., Thorne, K. S., and Wheeler, J. A. *Gravitation* (Freeman, San Francisco, 1973).

More, L. T. *Newton* (Scribner, New York, 1934).

Mott, N. F. "The wave mechanics of α-ray tracks," *Proc. R. Soc. London* A126: 79–84 (1929).

Nasr, S. H. *An Introduction to Islamic Cosmological Doctrines* (Harvard University Press, Cambridge, 1964).

Needham, J. *Science and Civilisation in China*, 6 vols. (Cambridge University Press, 1954–).

Ne'eman, Y. and Kirsh, Y. *The Particle Hunters* (Cambridge University Press, 1986).

Nemerov, H. "Poetry," in *Encyclopaedia Britannica* 14: 599 (1975).

Neugebauer, O. *The Exact Sciences in Antiquity* (Harper, New York, 1962).

—— *A History of Ancient Mathematical Astronomy*, 3 vols. (Springer-Verlag, New York, 1975).

Newton, I. *The Chronology of Ancient Kingdoms Amended* (Tonson, London, 1728).

—— *Opticks*, 4th ed. (Innys, London, 1730; repr. Dover, New York, 1952).

—— *Sir Isaac Newton's Mathematical Principles of Natural Philosophy and his System of the World*, ed. and trans. F. Cajori (University of California Press, Berkeley, 1934).

—— *Isaac Newton's Papers & Letters on Natural Philosophy*, ed. I. B. Cohen (Harvard University Press, Cambridge, 1958).

—— *The Correspondence of Isaac Newton*, 7 vols., ed. H. W. Turnbull et al. (Cambridge University Press, 1959–77).

—— *The Mathematical Papers of Isaac Newton*, 8 vols., ed. D. T. Whiteside (Cambridge University Press, 1967–81).

Newton, R. R. *The Crime of Claudius Ptolemy* (Johns Hopkins University Press, Baltimore, 1977).

—— "Comments on 'Was Ptolemy a fraud?' by Owen Gingerich," *Q.J.R. Astron. Soc.* 21: 388–99 (1980).

Nicholas of Cusa, *see* Cusanus.

Nichomachus of Gerasa. *Introduction to Arithmetic*, trans. M. L. D'Ooge (Macmillan, New York, 1926).

Oakley, F. *Omnipotence, Covenant, & Order* (Cornell University Press, Ithaca, 1984).

Ober, W. B. "Robert Mayer, M.D. and the mechanical equivalent of heat," *N. Y. State J. Medicine* 68: 2447–54 (1968).

O'Connor, D. and Oakley, F. *Creation: The Impact of an Idea* (Scribner, New York, 1969).

O'Donnell, J. J. *Cassiodorus* (University of California Press, Berkeley, 1979).

Oresme, N. *Le Livre du ciel et du monde*, ed. A. D. Menut and A. J. Denomy, trans. A. D. Menut (University of Wisconsin Press, Madison, 1968).

Organ, T. W. *An Index to Aristotle* (Princeton University Press, 1949).

Pais, A. *'Subtle is the Lord. . .'* (Clarendon Press, Oxford, 1982).

—— *Inward Bound* (Clarendon Press, Oxford, 1986).

Panofsky, E. *Meaning in the Visual Arts* (Doubleday, New York, 1957).

———— *Galileo as a Critic of the Arts* (Nijhoff, The Hague, 1954).

———— *Idea: A Concept in Art Theory* (Harper & Row, New York, 1968).

Park, D. *Introduction to Strong Interactions* (Benjamin, New York, 1966).

———— *Introduction to the Quantum Theory* (McGraw-Hill, New York, 2d ed., 1974).

———— *Classical Dynamics and Its Quantum Analogues* (Springer-Verlag, New York, 1979).

———— *The Image of Eternity* (University of Massachusetts Press, Amherst, 1980).

Pauli, W. "Über das Wasserstoffspektrum vom Standpunkt der neuen Quantenmechanik," *Z. Physik* 36: 336–63 (1926). Trans. in van der Waerden (1967).

———— *Theory of Relativity*, trans. G. Field (Pergamon, London, 1958).

———— *Aufsätze und Vorträge über Physik und Erkenntnistheorie* (Vieweg, Braunschweig, 1961).

———— *Scientific Correspondence With Bohr, Einstein, Heisenberg, a.o.*, ed. K. von Meyenn, 2 vols. (Springer-Verlag, Berlin, 1979–).

Pemberton, H. *A View of Sir Isaac Newton's Philosophy* (Palmer, London, 1728).

Petersen, A. *Quantum Physics and the Philosophical Tradition* (MIT Press, Cambridge, 1968).

Planck, M. "Zur Theorie des Gesetzes der Energieverteilung im Normalspektrum," *Verhandl. d. Deutschen Physikal Gesellsch.* 2: 237–47 (1900). Repr. and trans. in Kangro, H. *Planck's Original Papers in Quantum Physics*, trans. D. ter Haar and S. Brush (Wiley, New York, 1972).

———— *Eight Lectures on Theoretical Physics*, trans. A. P. Wills (Columbia University Press, New York, 1915).

———— *Scientific Autobiography*, trans. F. Gaynor (Philosophical Library, New York, 1949).

Plato, *Collected Dialogues*, ed. E. Hamilton and H. Cairns (Princeton University Press, 1961).

———— *Timaeus a Calcidio translatus*, ed. J. H. Waszink (Warburg, London and Brill, Leiden, 1962).

Pliny (C. Plinius Secundus), *Natural History*, trans. H. Rackham and W.H.S. Jones, 10 vols. (Loeb Classical Library, 1938–52).

Plotinus. *Enneads*, trans. A. H. Armstrong (Loeb Classical Library, 1966).

Plutarch. *Moralia*, vol. 12, trans. H. Cherniss and W. C. Helmbold (Loeb Classical Library, 1957); vol. 13, pt. 1, trans. H. Cherniss (1976).

Poincaré, H. "Sur la dynamique de l'électron," *C. R. Acad. Sci. Paris* 140: 1504–508 (1905).

———— *Science and Method*, trans. F. Maitland (Scribner, New York, 1914).

Popper, K. *The Open Society and its Enemies* (Princeton University Press, 1950).

———— *Conjectures and Refutations* (Basic Books, New York, 1962).

———— *The Logic of Scientific Discovery* (Harper & Row, New York, 1965).

Porta, J. B. (G. della Porta), *Natural Magick* (Wright, London, 1658; repr. Basic Books, New York, 1957).

Price, W. C., Chissick, S. S., and Ravensdale, T., eds. *Wave Mechanics: The First Fifty Years* (Wiley, New York, 1973).

Proclus. *Commentary on the First Book of Euclid's Elements*, ed. and trans. G. R. Morrow (Princeton University Press, 1970).

Procopius. *History of the Wars*, vol. 6, trans. H. B. Dewing (Loeb Classical Library, 1935).

Ptolemy (Claudius Ptolemaeus). *L'Ottica di Claudio Tolomeo*, ed. G. Govi (Vigliardi, Torino, 1885).

———— *Opera Astronomica Minora*, ed. J. L. Heiberg (Teubner, Leipzig, 1907).

———— *Tetrabiblos*, ed. and trans. F. E. Robbins (Loeb Classical Library, 1940).

Ptolemy. *The Almagest*; Copernicus, N. *On the Revolutions of the Heavenly Spheres*; Kepler, J. *Epitome of Copernican Astronomy: IV and V; The Harmonies of the World: V*, trans. R. C. Taliferro and C. G. Wallace (Encyclopaedia Britannica, Chicago, 1952).

———— *Ptolemy's Almagest*, trans. G. J. Toomer (Springer, New York, 1984).

Read, J. *Prelude to Chemistry* (Macmillan, New York, 1937).

Redondi, P. *Galileo: Heretic* (Princeton University Press, 1987).

Reeves, H. *Atoms of Silence*, trans. R. A. Lewis and J. S. Lewis (MIT Press, Cambridge, 1984).

Robert Grosseteste, *Die Philosophischen Werken des Robert Grosseteste, Bischofs von Lincoln*, ed. L. Bauer (Aschendorff, Munster i. W., 1912).

———— *On Light, or The Incoming of Forms*, trans. C. G. Wallace (St. Johns Bookstore, Annapolis, 1939).

Roemer, O. "Démonstration touchant le mouvement de la lumière," *J. des Sçavans* 233–36 (1676). Trans. in *Phil. Trans. R. Soc. London* 12: 893–94 (1677).

Rogers, H. J., ed. *Congress of Arts and Sciences, Universal Exposition, St. Louis, 1904*, 8 vols. (Houghton Mifflin, Boston, 1905).

Rohault, J. *Traité de physique*, 2 vols. (Savreux, Paris, 1671).

———— *System of Natural Philosophy*, trans. J. Clarke, 2 vols. (Knapton, London, 2d ed., 1729).

Ronchi, V. *The Nature of Light*, trans. R. Barocas (Heinemann, London, 1970).

Rowan-Robertson, M. *The Cosmological Distance Ladder* (Freeman, New York, 1985).

Rozental', I. L. "Physical laws and the numerical values of fundamental constants," *Soviet Phys. Uspekhi* 23: 296–305 (1980).

Rutherford, E. "The scattering of α and β particles by matter and the structure of the atom," *Phil. Mag.* 21: 669–88 (1911).

Sacrobosco, J. *Sphaera* (Ratdolt, Venice, 1485).

Sambursky, S. *The Physical World of the Greeks* (Routledge & Kegan Paul, London, 1956).

———— *Physics of the Stoics* (Routledge & Kegan Paul, London, 1959).

———— ed. *Physical Thought from the Presocratics to the Quantum Physicists* (Pica, New York, 1974).

de Santillana, G. *The Crime of Galileo* (University of Chicago Press, 1965).

———— *Reflections on Men and Ideas* (MIT Press, Cambridge, 1968).

G. Sarton, "Early observations of sunspots," *Isis* 37: 69–71 (1947).

———— *A History of Science*, 2 vols. (Harvard University Press, Cambridge, 1952).

Scheiner, C. *Rosa ursina sive sol* (Bracciano, 1626–30).

Schilpp, P. A., ed. *Albert Einstein, Philosopher-Scientist* (Library of Living Philosophers, Evanston, 1949).

———— ed. *The Philosophy of Rudolf Carnap* (Open Court, La Salle, 1963).

Schrödinger, E. *Collected Papers on Wave Mechanics* (Blackie, London, 1928).

Shlyakhter, A. L. "Direct tests of the constancy of fundamental nuclear constants," *Nature* 278: 340–41 (1976).

Shumaker, W. *The Occult Sciences in the Renaissance* (University of California Press, Berkeley, 1972).

Sinai, Ya. G. "Dynamical systems with elastic reflections," *Russian Math Surveys* 25: 137–89 (1970).

Smith, C. P. *James Wilson* (University of North Carolina Press, Chapel Hill, 1956).

Sommerfeld, A. *Atomic Structure and Spectral Lines*, trans. from 3d German ed. by H. L. Brose (Dutton, New York, 1923).

Southern, R. W. *Robert Grosseteste* (Clarendon Press, Oxford, 1986).

Stevin, S. *The Principal Works of Simon Stevin*, 5 vols., ed. E. Crone et al., trans. C. Dikshoorn (Swets & Zeitlinger, Amsterdam, 1955–1966).

Strodach, G. K. *The Philosophy of Epicurus* (Northwestern University Press, Evanston, 1963).

Stukeley, W. *Memoirs of Sir Isaac Newton's Life* (Taylor & Francis, London, 1936).

Swenson, L. S., Jr. *The Ethereal Aether* (University of Texas Press, Austin, 1972).

Swerdlow, N. M. and Neugebauer, O. *Mathematical Astronomy in Copernicus's De Revolutionibus*, 2 vols. (Springer, New York, 1984).

Szabó, A. "The transformation of mathematics into deductive science and the beginnings of its foundation on definitions and axioms," *Scripta Math.* 27: 27–49; 113–139 (1964).

———— *The Beginnings of Greek Mathematics*, trans. A. M. Ungar (Reidel, Dordrecht, 1978).

Tacitus, P. C. *Dialogus, Germania, Agricola*, trans. W. Peterson (Loeb Classical Library, 1914).

Taylor, F. S. *The Alchemists* (Heinemann, London, 1951).

Taylor, J. H., Fowler, L. A., and McCulloch, P. M. "Measurement of general relativistic effects in the binary pulsar PSR 1913 + 16," *Nature* 277: 437–40 (1979).

Thackray, A. *Atoms and Powers* (Harvard University Press, Cambridge, 1970).

Thompson, S. P. *Elementary Lessons in Electricity and Magnetism* (Macmillan, London, 1894, 1905).

Thomson W. (Lord Kelvin). *Mathematical and Physical Papers*, 6 vols. (Cambridge University Press, 1882–1911).

———— *Popular Lectures and Addresses*, 3 vols. (Macmillan, London, 1889–94).

———— *Baltimore Lectures on Molecular Dynamics and the Wave Theory of Light* (Clay, London and Johns Hopkins University, Baltimore, 1904).

Thorndike, L. *A History of Magic and Experimental Science*, 8 vols. (Macmillan, New York, 1923–58).

——— *University Records and Life in the Middle Ages* (Columbia University Press, New York, 1944).

Tillyard, E.M.W. *The Elizabethan World Picture* (Chatto & Windus, London, 1945).

Toraldo di Francia, G., ed. *Problems in the Foundations of Physics* (North-Holland, Amsterdam, 1979).

Toulmin, S. and Goodfield, J. *The Architecture of Matter* (Harper & Row, New York, 1962).

Treder, H.-J. "Kepler und die Begrundung der Dynamik," *Die Sterne* 49: 44–48 (1973).

Truesdell, C. *Essays in the History of Mechanics* (Springer, New York, 1968).

Tubbs, A. D. and Wolfe, A. M. "Evidence for large-scale uniformity of physical laws," *Astrophys. J.* 236: L105–108 (1980).

van Melsen, A. G. *From Atomos to Atom* (Duquesne University Press, Pittsburgh, 1952).

Vessot, R.F.C. et al., "Test of relativistic gravitation with a space-borne hydrogen maser," *Phys. Rev. Lett.* 45: 2081–88 (1980).

Vickers, B., ed. *Occult and Scientific Mentalities in the Renaissance* (Cambridge University Press, 1984).

Viennot, L. "Spontaneous reasoning in elementary dynamics," *European J. of Scientific Education* 1: 205–21 (1979).

Vincent of Beauvais. *Speculum quadruplex* (*Speculum majus*), 4 vols. (Bellerie, Douai, 1624; repr. Akademische Druck, Graz, 1964).

Vitruvius. *De architectura libri decem*, ed. F. Krohn (Teubner, Leipzig, 1912).

——— *On Architecture*, trans. F. Granger (Loeb Classical Library, 1931).

Voltaire. *Oeuvres complètes*, vol. 23 (Garnier, Paris, 1879).

van der Waerden, B. L. *Sources of Quantum Mechanics* (North-Holland, Amsterdam, 1967).

Walker, D. P. *Spiritual and Demonic Magic* (Warburg Institute, London, 1958).

Wallace, W. A. *Causality and Scientific Explanation*, 2 vols. (University of Michigan Press, Ann Arbor, 1972–73).

——— *Galileo's Early Notebooks* (University of Notre Dame Press, Notre Dame, 1977).

Weinberg, S. "Photons and gravitons in S-matrix theory: Derivation of charge conservation and equality of gravitational and inertial mass," *Phys. Rev.* B135: 1049–56 (1964).

——— "Photons and gravitons in perturbation theory: Derivation of Maxwell's and Einstein's equations," *Phys. Rev.* B138: 988–1002

——— *Gravitation and Cosmology* (Wiley, New York, 1972).

——— *The First Three Minutes* (Basic Books, New York, 1977).

Weiner, C., ed. *History of Twentieth Century Physics* (Scuola "Enrico Fermi," vol. 57) (Academic Press, New York, 1977).

Weinstock, R. "Dismantling a centuries-old myth: Newton's *Principia* and inverse-square orbits," *Am. J. Phys.* 50: 610–17 (1982).

Weisberg, J. M. and Taylor, J. H. "Observations of post-Newtonian timing effects in the binary pulsar 1913 + 16," *Phys. Rev. Lett.* 52: 1348–50 (1984).

Westfall, R. S. *Never at Rest: A Biography of Isaac Newton* (Cambridge University Press, 1980).

Weyl, H. *Symmetry* (Princeton University Press, 1952).

Wheeler, J. A. and Zurek, W. H., eds. *Quantum Theory and Measurement* (Princeton University Press, 1983).

Whewell, W. *History of the Inductive Sciences*, 2 vols. (Parker, London, 3d ed., 1857).

Whittaker, E. T. *A History of the Theories of Aether and Electricity*, 2 vols. (Nelson, London, 2d ed., 1953).

Wigner, E. P. *Symmetries and Reflections* (Indiana University Press, Bloomington, 1967).

Willey, B. *The Seventeenth Century Background* (Chatto & Windus, London, 1934).

Williams, E. and Park, D. "Photon mass and the galactic magnetic field," *Phys. Rev. Lett.* 26: 1651–52 (1971).

Wittgenstein, L. *Notebooks, 1914–1916*, trans. G.E.M. Anscombe (Blackwell, Oxford, 1961).

Yang, C. N. "Selection rules for the dematerialization of a particle into two photons," *Phys. Rev.* 77: 242–45 (1950).

———— and R. L. Mills, "Conservation of isotopic spin and isotopic gauge invariance," *Phys. Rev.* 96: 191–95 (1954).

Yates, F. *Giordano Bruno and the Hermetic Tradition* (Random House, New York, 1964).

———— *The Occult Philosophy in the Elizabethan Age* (Routledge & Kegan Paul, London, 1979).

Young, T. "On the theory of light and colours," *Phil. Trans. R. Soc. London* 20: 12–48 (1802).

———— *A Course of Lectures on Natural Philosophy and the Mechanical Arts*, 2 vols. (Johnson, London, 1807; repr. Johnson, New York, 1971).

———— *Miscellaneous Works of the Late Thomas Young*, ed. G. Peacock and J. Leitch, 3 vols. (Murray, London, 1855; repr. Johnson, New York, 1972).

Yourgrau, W. and van der Merwe, A., eds. *Perspectives in Quantum Theory* (MIT Press, Cambridge, 1971).

Yukawa, H. "On the interaction of elementary particles, I," *Proc. Physico-Mathematical Soc. Japan* 17: 48–57 (1935). Repr. in Beyer (1949).

Index

Academy, 39, 51; beginning, 40; end, 89

accident, in Aristotle, 44; in Catholic doctrine, 168

action at a distance: affirmed by Boscovich, 201; denied by Newton, 188; of a magnet, 174; in nineteenth century, 271; in quantum mechanics, 341, 423-424

Albertus Magnus, 105, 135

alchemy, 114-121

d'Alembert, Jean LeRond, 248

Alhazen. *See* Ibn al-Haitam

Almagest, 70-73

Anaxagoras of Clazomenae, 55

Anaximander of Miletus, 8, 27

Anaximenes of Miletus, 8

angel, 84, 121

Anselm, Saint, 99

anthropic principles, 383

Apollonius of Perga, 71, 163

appearance of objects in quantum mechanics, 425-426

Aquinas, Saint Thomas: on epicycles, 90; Aristotle Christianized, 105, 135; on magic, 121

Archimedes of Syracuse: on the lever, 101-102; on the size of the universe, 66

Aristarchus of Samos, heliocentric theory, 65, 143

Ariosto, Lodovico, *Orlando Furioso*, 84

Aristotle, 7, 9, 42-53, 172; on causes, 46; on comets, 163; on cosmology, 51, 55, 63-64; gives the earth's diameter, 56; on the elements, 45, 115; on light, 125-126; on mathematics, 101; on motion, 48-50; error concerning local motion, 139, 237; on Plato's Ideas, 237; on the Pythagoreans, 17; on science and art, 51, 109; on sensation, 44; terminology, 43-44; on time, 203; on violent motion, 211

arithmetic, 79

art, 109

astrology, 71, 84, 113, 171-173

atomic theory: in antiquity, 25-31; in Dark Ages, 96-98; in Middle Ages, 104, 110, 132; in 17th century, 194-201; in twentieth century, 310-323

atoms, reality of, 309-310

Augustine of Hippo, Saint, 82, 157; on Genesis, 105; on magic, 121; on number and order, 95, 235, 388; on Plato's Ideas, 40; on time, 224, 230

Autolycus of Pitane, 136

Averroes. *See* Ibn Rushd

Baade, Walter, 373

Babylonian mathematics, 13

Bach, Johann Sebastian, 158

Bachelard, Gaston, 75, 163

Bacon, Francis, 215

Bacon, Roger, 133-135

Barrow, Isaac, 15, 179; on "scientific method," 187

Bede, Saint Venerable, 97-98

Bell, John Stuart, 342-344; theorem, 343

Bentley, Richard, 188, 244

Bernardus Sylvestris, 85

Bernoulli, Daniel: on kinetic theory, 253; on laws of motion, 247

Bernoulli, Jean, 209; on planetary motion, 415-416

"bodies," 37, 43

Boethius, Anicius Manlius Severinus: on Divine love, 50; on music, 21; translates Aristotle, 79

Bohr, Niels, 314; atomic theory, 311-314; complementarity, 327-333

Born, Max, on randomness in quantum mechanics, 322, 341

Boscovich, Roger, 200-201

Boundless, 9

Boyle, Robert, 195-196; law of gases, 253

Bradwardine, Thomas, 136

Brahe, Tycho, 149; cometary parallax, 163-165

de Broglie, Louis, 315-317

Bronowski, Jacob, 13